Active Galactic Nuclei

SCIENCES

Universe, Field Director – Fabienne Casoli

Galaxies, Subject Head – Françoise Combes

Active Galactic Nuclei

Coordinated by
Françoise Combes

iSTE

WILEY

First published 2022 in Great Britain and the United States by ISTE Ltd and John Wiley & Sons, Inc.

ISTE Ltd
27-37 St George's Road
London SW19 4EU
UK

www.iste.co.uk

John Wiley & Sons, Inc.
111 River Street
Hoboken, NJ 07030
USA

www.wiley.com

Library of Congress Control Number: 2022934058

British Library Cataloguing-in-Publication Data
A CIP record for this book is available from the British Library
ISBN 978-1-78945-087-3

ERC code:
PE9 Universe Sciences
PE9_7 The Galaxy
PE9_8 Formation and evolution of galaxies
PE9_9 Clusters of galaxies and large scale structures

Contents

1

Active Galactic Nuclei: Competition Between Accretion and Ejection

Suzy COLLIN-ZAHN
LUTH, Paris Observatory, Meudon, France

Since the discovery of the first "active galactic nuclei" (AGN) in 1943, and of quasars, 20 years later, a lot of progress has been made concerning the understanding of these objects. First we realized that they were powered by a massive black hole, accreting the surrounding gas, causing intense radiation – a model that became a paradigm in the 1980s. At the same time, their apparent diversity was understood as mainly due to the different orientations of their rotational axis. In the years 1990–2000, the presence of "dormant" supermassive black holes, and their coevolution with bulges, was demonstrated in most galaxies. The presence of relativistic jets and winds, as well as the different modes of accretion, were linked to the accretion rate and spin of the black holes.

1.1. History: discovery and initial observations

The discovery of active galaxies can be traced back to 1943, when Carl Seyfert published an article in which he studied six galaxies with a bright point-like nucleus, whose spectrum presented intense emission lines superimposed on a blue continuum (Seyfert 1943). These were the same lines as those of planetary nebulae but much broader. Seyfert attributed this width to speeds of several hundred to several

Active Galactic Nuclei,
coordinated by Françoise COMBES. © ISTE Ltd 2022.

thousand km/s. This interpretation has never been questioned. Figure 1.1 shows the image of one of these galaxies (simply called a "Seyfert galaxy").

Figure 1.1. *Image of the Seyfert galaxy NGC 5548 obtained with the Hubble Space Telescope. We distinguish the diffraction cross proving that the nucleus is unresolved. The galaxy is seen almost from the front (source: NASA). For a color version of this figure, see www.iste.co.uk/combes/nuclei.zip*

The second significant date in history is 1952, the year in which Baade and Minkowski discovered that two intense "radio" sources in the sky were galaxies: M87 and Cygnus A (Baade and Minkowski 1952; Figures 1.2 and 1.3). M87 (also called Virgo A) is characterized by a very fine jet emerging from the heart of the galaxy. Cygnus A is a distant galaxy that seems to consist of two colliding galaxies, and the spectrum of its core shows emission lines (but these are narrower than those of Seyfert galaxies), while most galaxies have absorption lines. These "radio galaxies", as well as those which were subsequently discovered, are much more powerful in the radio domain than other galaxies, typically by a factor of 10^4, and their radio luminosity was immediately attributed to the synchrotron radiation produced by relativistic electrons moving in a magnetic field (see Chapter 4). Geoffrey Burbidge (1959) – we specify his first name to distinguish him from his wife, Margaret, who played a considerable role in the study of quasars and active galaxies – showed that the total energy stored in such a system must be equal to 10^{54} or 10^{55} joules, which corresponds to the complete transformation into energy of 10^7

or 10^8 solar masses. If we correct the inaccurate Hubble constant he used, we find a value almost a hundred times greater.

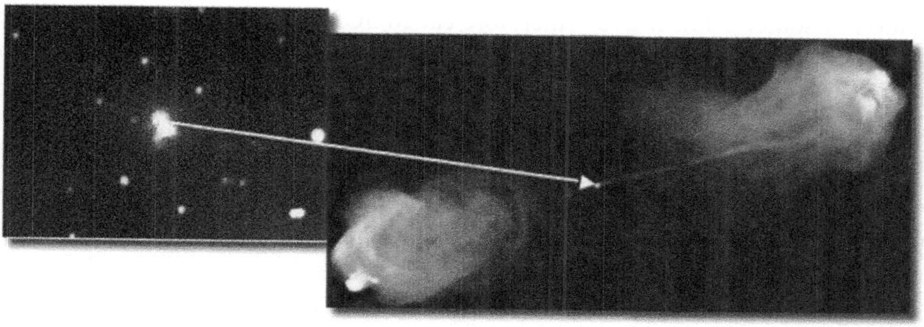

Figure 1.2. *Left, optical image of the galaxy Cygnus A. Right, false-color radio image of the galaxy (which is the small dot in the center), showing large, very bright radio lobes at their ends, and a filament joining the galaxy at the right lobe. The distance projected onto the sky between the edges of the two lobes is 300,000 light years (source: NRA0 for the radio image; Baade and Minkowski 1952). For a color version of this figure, see www.iste.co.uk/combes/nuclei.zip*

Figure 1.3. *Optical image of galaxy M87 taken with the Hubble telescope. We can distinguish the jet coming from the nucleus (source: NASA; Baade and Minkowski 1952). For a color version of this figure, see www.iste.co.uk/combes/nuclei.zip*

The third founding event occurred in 1963. In 1959, Cambridge's third catalog of radio sources, the "3C", had been published, containing 328 radio sources, half of which did not correspond to any known visible object. In 1962, measurements of the position of source number 273 – therefore called 3C273 – during its passage behind the moon, identified it with a very blue point star. Spectra taken with the 5-meter telescope at the Palomar Observatory showed a blue continuum on which bright and broad lines were superimposed, spaced just like the lines of the Balmer series, but all shifted by 16% towards the red compared to their usual wavelengths (Schmidt 1963). Other radio sources were then identified, also exhibiting a spectral shift towards the red (for this, we use the term "redshift", or the letter $z = (\lambda_{obs} - \lambda_{emit})/\lambda_{emit}$, where λ_{obs} and λ_{emit} are the observed and emitted wavelengths, respectively). They were called "quasars", a term contracted from "quasi-stellar objects", indicating that they look like simple stars in the photographs. Almost immediately, Sandage (1965) discovered a much larger population of quasars that do not radiate in the radio domain ("radio quiet").

A very lively discussion then took place on the origin of redshifts which lasted for 20 years: were they gravitational redshifts, or due to a Doppler effect, or even cosmological (i.e. due, say, to the expansion of the Universe), or were they the manifestation of an unknown physical law? The latter explanation was adopted by a group of astronomers led by American astronomer Halton Arp who discovered associations between objects with different redshifts, sky-aligned quasars, etc. (see, for example, Arp 1970). We could indeed ask this question, because the cosmological redshifts led to extraordinary energy (10^{40} joules for 3C273, that is, the energy of 1,000 galaxies) emitted in a tiny region. Indeed, 3C273 was a variable star, already known for a hundred years, whose brightness underwent variations of one magnitude in a week. By virtue of the causality principle, the dimension of the emissive region therefore had to be less than a light week, which seemed impossible. Several scenarios were then proposed (chain exploding supernovae, supermassive stars, galactic eruptions, etc.), but none were convincing. However, two great physicists, the American Salpeter (1964) and the Soviet Zeldovich (1964), immediately thought of black holes (which did not yet bear this name). Most specialists quickly converged on cosmological redshifts, and subsequent observations confirmed this conclusion.

We quickly drew the parallel between Seyfert galaxies and quasars: small emitting regions, blue continuums, same spectral lines (once the redshift of quasars was taken into account). Seyfert galaxies seemed to be "mini-quasars" that are close to us. It was therefore thought that quasars were simply nuclei of galaxies that had undergone a particularly "active" phase in the past, and it was at this time that the term active galactic nuclei was created to designate both quasars and the nuclei of

Seyfert galaxies. This idea was not confirmed until 20 years later, when we could observe the “host galaxies” of quasars.

In fact, it was quickly realized that the optical spectrum of Seyfert galaxies consists of two types of lines, permitted lines and forbidden lines, and that there are two types of Seyfert galaxies: type 2 Seyfert, whose lines – mainly forbidden – are relatively narrow (with a full width at half-maximum (FWHM) of a few hundred km/s at most), and type 1 Seyfert, whose lines are broad (FWHM of several thousand km/s) and are mainly permitted lines (see Figure 1.4). As we will see later, the broad lines and the narrow lines are emitted by different regions: the broad line region (BLR) is dense and compact; the narrow line region (NLR) is diffuse and extended (Dibai and Pronik 1967; Souffrin 1968). Non-radio quasars are related to type 1 Seyfert in their lines, and it was later found that their X-ray and infrared properties, as well as their variability, are also very similar.

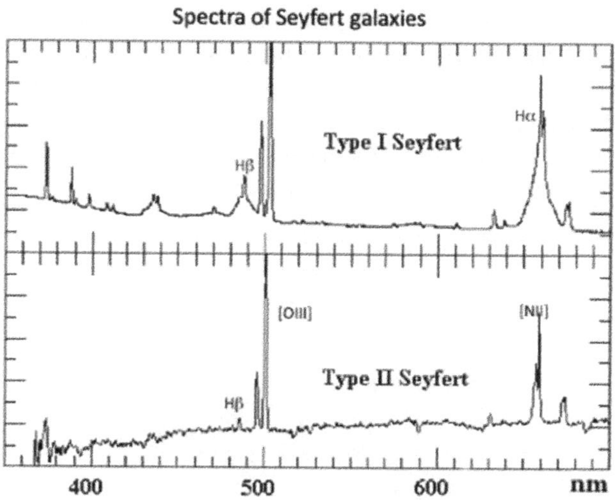

Figure 1.4. *Comparison of the spectra of a Seyfert 1 and a Seyfert 2 galaxy. The Hα and Hβ broad lines do not exist in type 2 Seyfert*

Radio galaxies were soon thought to be linked to quasars, due to the enormous amount of energy stored in radio sources. By analogy with the radiation of radio galaxies, which is clearly of synchrotron origin, all the radiation from quasars and AGNs was then attributed to the synchrotron mechanism, which had serious consequences for the central engine. This idea came from the power law distribution ($F_\nu \sim \nu^{-\alpha}$, where α is the spectral index) and the polarization of visible and ultraviolet radiation. Following this was a period of great confusion, during which many objects were discovered which could be related to quasars by their power or by their spectra. This incongruous group was called "the zoo of active galaxies". We will return to these objects later, whose properties are now understood and fit into a unified classification. It was not until 1978 that the ultraviolet radiation of non-radio AGNs was recognized to have the characteristics of that emitted by an accretion disk (Schields 1978), and it was only then that the paradigm of "*accretion on a supermassive black hole*" was accepted (Rees 1978). It has never been questioned since.

Another important date in the history of AGNs is 1985, when the spectrum of Seyfert 2 galaxy NGC 1068 was discovered to have broad lines in polarized light (Antonucci and Miller 1985). The authors deduced that the type 2 Seyfert also have a BLR, but that this is hidden from our sight by an absorbing medium and that we see the image "reflected" in our direction, as if we were observing it through a periscope (see Figure 1.5). This discovery was the first step in the direction of a unified classification[1]. It was logical, once it was assumed that AGNs were powered by an accretion disk, to deduce that they must have different appearances depending on the direction of the observer relative to the axis of the disk. This definition has been extended to AGNs as a whole, and we now generally refer to type 1 AGNs and type 2 AGNs. However, as we will see, there was still a long way to go before it was figured out how all the incongruous classes of AGNs could fit into a coherent framework based on a small number of parameters.

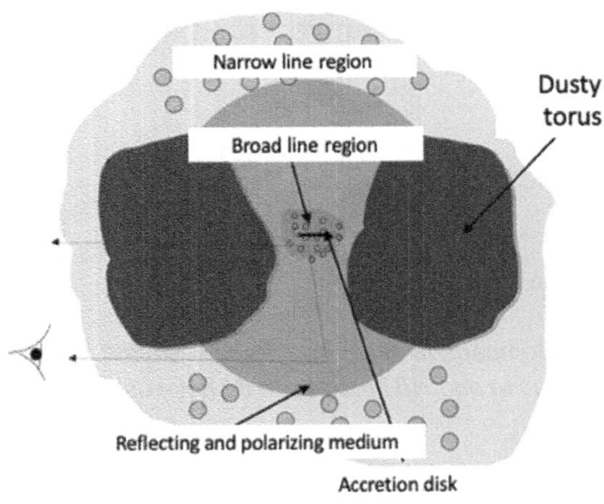

Figure 1.5. *A cross-section of Antonucci and Miller's diagram unifying type 1 and 2 Seyfert. For a color version of this figure, see www.iste.co.uk/combes/nuclei.zip*

1. In fact, we now know that there are "true" Seyfert 2 galaxies that are not Seyfert 1 galaxies "seen from the edge".

1.2. Important observational properties

Let us present a list of properties that we will not understand, for the most part, until the end of the chapter; this is essential if we want to have an overview of AGNs. However, this list corresponds to a simplified idea of reality: in particular, the selection modes of AGNs have a profound influence on their characteristics, and lead to more complex classifications than those given here (see Padovani 2017; Padovani *et al.* 2017).

1.2.1. *AGNs across the electromagnetic spectrum*

The radiation from AGNs ranges from radio waves to high-energy gamma rays. Figure 1.6 gives their spectral energy distribution (SED), expressed in νF_ν per unit frequency interval, a unit that makes it easy to see where the energy is (see the caveat at the end of the chapter). For objects with an intense jet observed in the radio region, this distribution encompasses the entire radio domain and the gamma domain: it is therefore mainly due to the jet. For those without it, it only stretches from infrared to the X-ray region, and dips at 100 keV. In the case of the jet, the spectrum can have interesting shapes, which will be discussed in Chapter 4, and can extend up to TeV. In the second case, the infrared region is dominated by a dust torus, the visible and ultraviolet regions by the accretion disk and the X-ray region by three components (the "soft excess", "corona" and "reflection"): we will show later that all these components are linked to accretion. *The division between radio and non-radio objects is therefore in fact a division between AGNs with the jet and jetless AGNs.* In the rest of the chapter, we will mainly focus on non-jet, or "non-radio" AGNs, radio AGNs being dealt with in Chapter 4.

1.2.1.1. *The radio wave domain*

A parameter generally used to separate radio AGNs (RL) from non-radio AGNs (RQ) is $R=[F_\nu(5GHz)/F_\nu(4400)] = 1.36\ 10^5[\nu F_\nu(5GHz)/\nu F_\nu(4400)]$. The separation chosen, R = 10, corresponds to approximately 10% radio AGNs and 90% non-radio AGNs. As we will see when we look at optical and X-ray emissions, this dichotomy does not make much sense. For example, it cannot be used for type 2 AGNs whose visible and ultraviolet radiations are strongly absorbed.

Unless it is self-absorbed (usually at low frequencies, however, the transition can occur in the high frequency, visible or even X-ray domain), the radio spectrum is represented by a typical $F_\nu \sim \nu^{-\alpha_R}$ power law synchrotron radiation. We call sources with $\alpha_R < 0.5$ or $\alpha_R > 0.5$ flat spectrum radio quasars (FSRQ) or steep spectrum radio quasars (SSRQ), respectively.

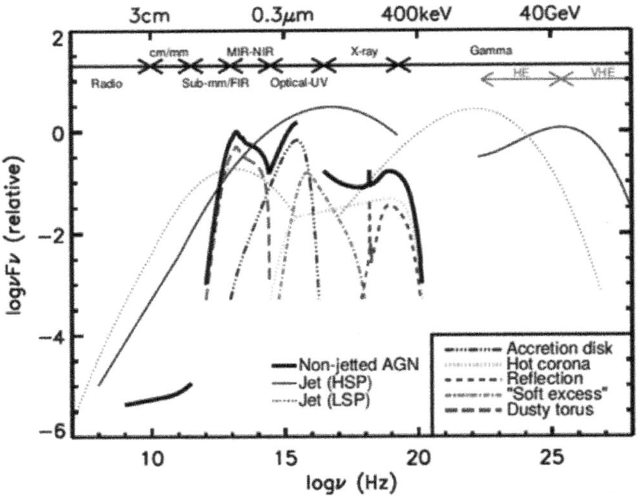

Figure 1.6. *Spectral energy distribution of radio and non-radio AGNs, according to Padovani (2017). Jet (HSP) and Jet (LSP) correspond to two jet spectral distributions. For a color version of this figure, see www.iste.co.uk/combes/nuclei.zip*

All AGNs have a radio source in the form of a jet in their nucleus, but it is very weak in non-radio galaxies, and sometimes unobservable. In radio AGNs, the radio source consists of two intense symmetrical radio jets originating in the nucleus, extending into the host galaxy and possibly even extending far into intergalactic space. Fanaroff and Riley (1974) separated these jets into two categories: FRII radio galaxies and FRI radio galaxies (see Figure 1.2 showing Cygnus A as an example of FRII and Figure 1.7 showing galaxy 3C449 as an example of FRI). FRIIs have shiny edges – called hot spots – while in FRIs the brightness decreases as the distance from the center increases. FRIs are less luminous than FRIIs, and are associated with optical spectra different from those of quasars and Seyfert galaxies: the lines are narrow there, and correspond to more weakly ionized ions. FRIs are generally part of the low ionization nuclear emission region (LINER). The extended jets associated with FRIs are more relativistic than those of FRIIs; on the other hand, on a very small scale (therefore, in the vicinity of the black hole), the FRII jets are relativistic, but are slowed down on their path. Moreover, often the jet is only seen from one side, because of the relativistic amplification; "superluminal" motions are then observed, indicating speeds. Differentiating between FRI and FRII is therefore not a question of differentiation due to the viewing angle, but of their physical properties. They could be connected to the morphology of the host galaxy.

For the spectral index difference, however, the difference is attributed to the angle between the jet and the line of sight. It is believed that FSRQs and SSRQs are FRII radio galaxies whose jet is oriented in a direction close to the line of sight (~<15°) for FSRQ, while this angle is greater for SSRQ (15~< θ ~< 40°). Refer to Chapter 4 for more information.

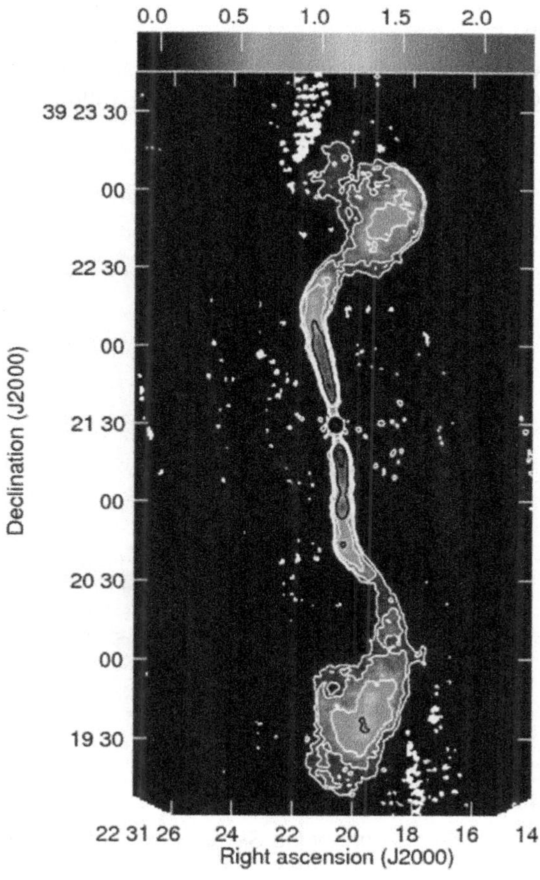

Figure 1.7. *False-color radio map of galaxy 3C449 – type FR I. The intensity is indicated in the scale on the right. It is completely different from Cygnus A, being more intense in the center than at the edges. In the case presented here, the axis of the jet has varied over time. The distance projected onto the sky between the ends of the jets is 200,000 light years. Credit: NRAO. For a color version of this figure, see www.iste.co.uk/combes/nuclei.zip*

1.2.1.2. *Infrared*

While the radio domain has long been covered by ground-based radio telescopes, infrared is primarily observed in space, which explains the relatively late observations. Near-infrared (NIR), from 1 to 10 microns, was observed first on the ground, followed by medium-infrared (MIR) from 10 to 100 microns, observed by the ISO satellite launched in 1990 and Spitzer satellite launched in 2003. Finally, the far-infrared (far-IR) from 100 microns to 1 mm was observed by IRAS, launched in 1983, with a very low spatial resolution, then by Spitzer, and finally by Herschel, launched in 2009. All these observations led to the following vision regarding non-radio AGNs.

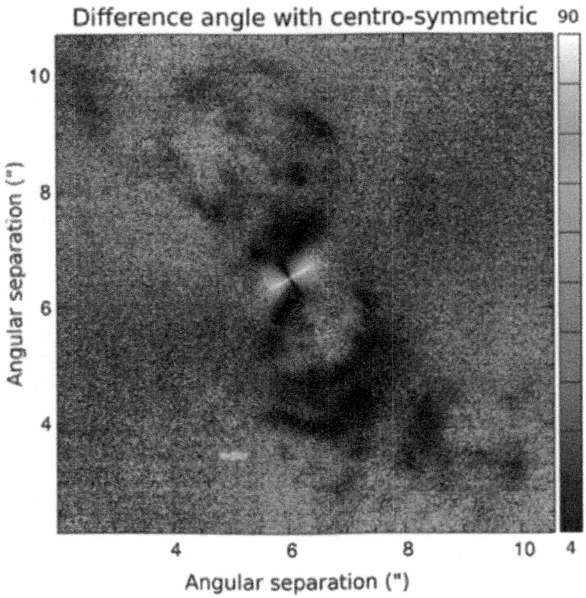

Figure 1.8. *Image showing the core of the Seyfert 2 NGC 1068 galaxy obtained from polarimetric observations made in 2015 with the SPHERE (Spectro-Polarimetric High-contrast Exoplanet REsearch instrument) at the Very Large Telescope (VLT). The image shows, in black, an hourglass-shaped bicone, and, in white, an elongated structure perpendicular to the bicone. Credit: Gratadour, Lesia/Paris Observatory. For a color version of this figure, see www.iste.co.uk/combes/nuclei.zip*

A significant fraction of MIR in non-radio AGNs is due to radiation from dust surrounding the central source and absorbing some – sometimes most – of its optical and ultraviolet radiation, and re-emitting it in the infrared. This dust occupies a molecular "torus", the thickness of which is approximately the same as its distance from the center. This is the prolongation of the accretion disk between the BLR and the NLR and explains the absence of permitted lines in type 2 AGNs, which are seen on the side of this torus. We have direct proof of its existence thanks to the high-resolution images obtained by interferometry (see Figures 1.8 and 1.9). The temperature of the dust heated by the central radiation varies from 3,000 K to 100 K, and as a result it emits between 3 and 100 microns. Figure 1.10 shows the energy distribution of a non-radio AGN: there is a "dip" around 1 micron, at the junction between the optical spectrum produced by the accretion disk and the infrared one due to dust.

On the NIR and MIR continuums, many emission lines are superimposed coming from the BLR (they include hydrogen Paschen and Brakett series) and from the NLR (forbidden lines [NeV], [NeII], [OIV]), as well as coronal lines coming from a hot gas, as well as molecular bands coming from the torus, finally the spectral characteristics of dust grains like those of silicate at 10 microns. Polycyclic aromatic hydrocarbons (PAH) bands are also observed, but it is not clear what fractions are due to the presence of AGN or to star formation in more distant regions.

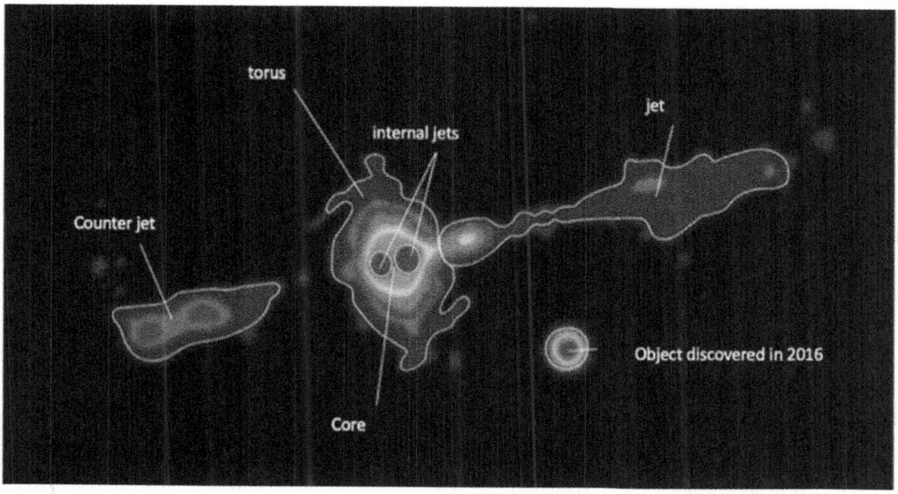

Figure 1.9. *A false-color radio image of the central region of Cygnus A. The torus of around 900 light years is clearly visible as well as a second supermassive black hole discovered as a radio source in 2016 (bottom right). Credit: Carilli et al. NRAO/ AUI/NSF. For a color version of this figure, see www.iste.co.uk/combes/nuclei.zip*

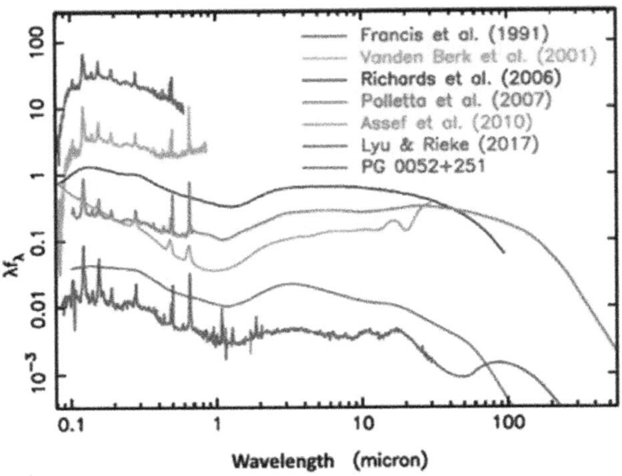

Figure 1.10. *Spectral distributions of non-radio AGNs obtained by several groups and collected by Brown et al. (2019). We can see the 1-micron dip at the junction between the disk emission and the torus emission. For a color version of this figure, see www.iste.co.uk/combes/nuclei.zip*

In addition, very bright far infrared galaxies were discovered in the 1980s, which were called "Ultra Luminous Infrared Galaxies" or ULIRGs, and which were quickly suspected to be quasars that were completely coated in dust, or seen through a very thick torus.

In radio AGNs, the infrared spectrum is a power law and is mainly due to synchrotron radiation.

1.2.1.3. *Visible light and ultraviolet*

Optical images of type 1 AGNs show a point source whose emission often exceeds the integrated emission of the host galaxy. The continuum spectral index is around 0.5, much higher than what one would expect for an accretion disk (α=–1/3; see below this section) which implies that there is an underlying component whose contribution decreases with frequency. On this continuum, many broad permitted lines (FWHM from 1,000 to 10,000 km/s) and narrow forbidden lines (FWHM from 400 to 1,000 km/s) are superimposed. Only narrow lines are present in type 2 AGNs (see Figure 1.4).

The main permitted lines are: in the ultraviolet, the Lyα line at 1,215Å of hydrogen, the lines of heavier elements such as CIV at 1,549Å, the MgII doublet at 2,800Å and the semi-forbidden line CIII at 1,909Å, whose intensity was key in determining the density of the permitted line region[2]. In the optical spectrum, we observe the lines of the Balmer series and many forbidden lines, the most intense of which are: [OIII] 4,959 and 5,007Å, [0II] 3,726 and 3,729Å, [NII] at 6,527 and 6,548Å and [NeV] 3,426 and 3,446Å (see Figure 1.11)[3].

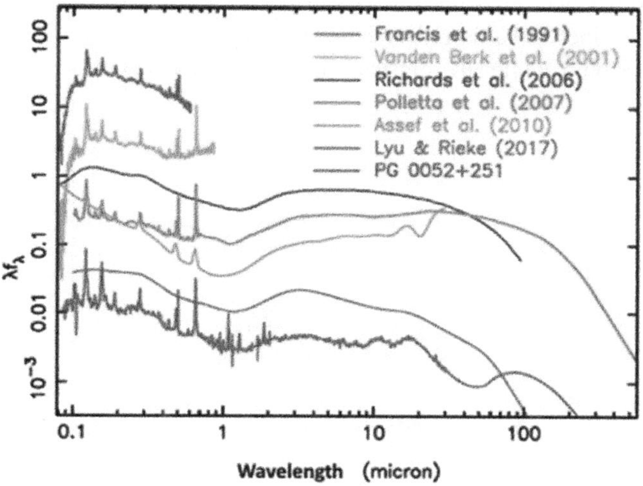

Figure 1.11. *This figure shows the continuum and the main UV and optical lines of a composite quasar spectrum obtained by the addition of 700 SDSS spectra between 1,000 and 8,000 Å. For a color version of this figure, see www.iste.co.uk/combes/nuclei.zip*

The relative intensities of the lines are comparable in types 1 and 2, but the profiles are very variable from one object to another, from Lorentzian profiles with a peak in the center and very wide wings, to profiles that are flat in the center, or even have a peak on either side of the center (see Figures 1.12 and 1.13). We are used to breaking up complex profiles into several Gaussians, allowing us to study the temporal variations of each. Two peak profiles are generally interpreted as being due to binary AGNs (see, for example, Storchi-Bergman *et al.* (2017)), but it is also possible that they are due to the rotation of a disk (Rokaki *et al.* 1993). For type 1 AGNs, it can be seen that the lines of the most ionized elements are broader than those of the less ionized elements, which is easily interpreted in the context of photoionization by the central source (see below). We also note a relationship

2. Note that the ultraviolet lines of high redshift quasars are shifted to the visible range, which allows them to be observed.

3. Note that the flux is carried according to the wavelength.

between the luminosity of the broad lines L(line) and the bolometric luminosity, with L(line) approximately proportional to $\sqrt{L_{bol}}$, which corresponds to an anti-correlation between the equivalent width and the luminosity: discovered in 1977 on the line of CIV1549 and called the "Baldwin effect"; it was subsequently extended to other permitted lines. It is still poorly explained. Finally, we note the motions of the lines towards red or blue: for example, the line CIV1549 is often shifted towards blue, while Hβ is instead shifted towards the red.

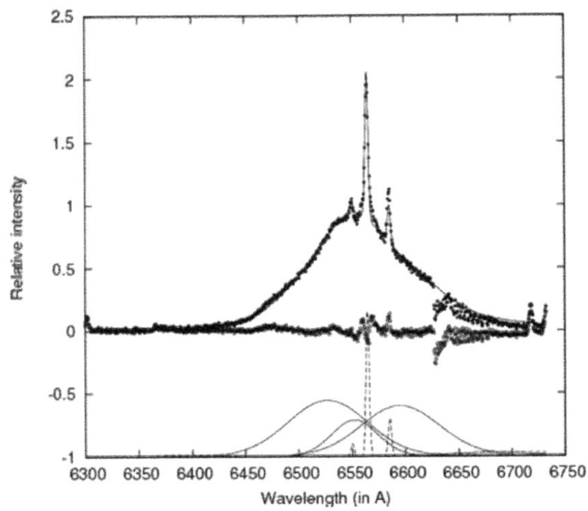

Figure 1.12. *Complex profile of the Hα line in the Markarian 841 galaxy (a Seyfert 1) broken down into several Gaussians. The narrow lines are the two forbidden lines [NII] 6,548 and [NII] 6,584 surrounding the narrow component of the line Hα. Credit: Bon et al. (2009)*

A particular class of AGN 1 is that of narrow line Seyfert 1 (NLS1s). The line intensities are comparable to those of the Seyfert 1 but they are much narrower (typically: FWHM ~ 500 km/s). NLS1s have other peculiarities in the optical and ultraviolet domains, such as very intense FeII lines compared to Balmer lines. These characteristics are found in a principal component analysis (PCA), which was proposed for the first time by Boroson and Green (1992) and showed the existence of correlations between the observed AGN parameters. At one end of the principal component are objects with a high FeII/Hβ and low [OIII]/Hβ ratio, a high X-continuum spectral index and relatively low luminosity. NLS1s fit into this classification. At the other end, we find the opposite properties. A more detailed analysis is proposed by Marziani *et al.* (2018). We will return to the meaning of these correlations later.

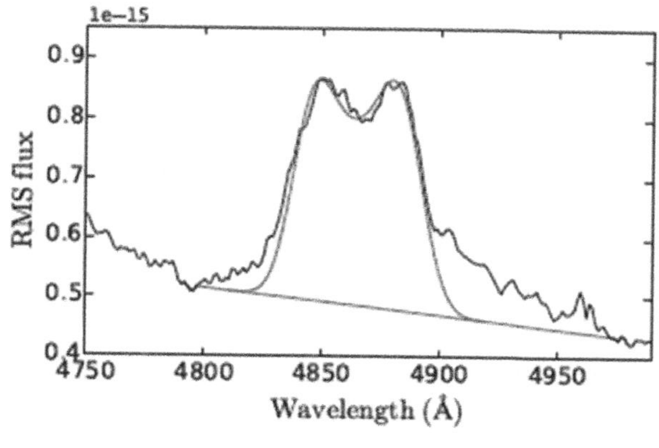

Figure 1.13. *The fitting of an accretion disk model to the profile of Hα in the low luminosity object (LLAGN) NGC 1097. The solid black line is the observed profile; the solid red line is the model (Storchi-Bergmann et al. 2017). For a color version of this figure, see www.iste.co.uk/combes/nuclei.zip*

Finally, we must mention broad absorption lines (BALs), present in the ultraviolet in 10–20% of quasars. They extend into the violet wing of the emitting lines up to velocities of 20,000 km/s with respect to the systemic redshift (see Figure 1.14). Because of their shape, they are called troughs. Historically, BALs whose average velocities are > 2,000 km/s have been separated from NALs whose velocities are <500 km/s and mini-BALs, which are intermediate. They all correspond to winds ejected from the central regions. In some cases, they are also present on the red side of the emission line, implying a complex geometry, possibly infall motions. In some objects, the absorption lines completely cover the BLR, while, in others, they only partially cover it. There are several types of BAL quasars: the majority are HIBALs, in which the BALs are due to strong ionization elements, such as CIV1549, SiIV1397 and NV1240. A small fraction (1%) are LoBALs in which the MgII2799 line is intense. There is not a clear difference between the BALs of radio and non-radio AGNs nor, among radio BALs, between flat and steep spectra (see Rochais *et al.* 2014), implying that the viewing angle is not the predominant parameter. In large surveys, like that of the SDSS (Trump *et al.* 2006), which contains about 40,000 AGNs with BALs, their properties are quantified by indices such as "balnicity index" (BI), "absorption index" (AI) and other indices characterizing the equivalent widths of the lines, their blue shifts or their linewidths. It is not possible to give more details about these properties here.

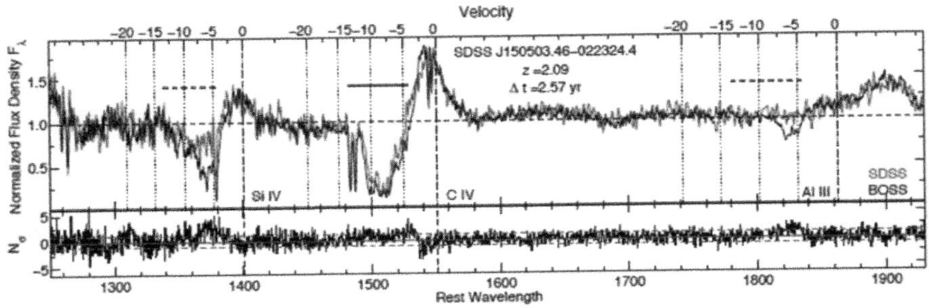

Figure 1.14. *The spectra of an AGN BAL obtained at two epochs separated by 2.5 years (black and red). We can distinguish the variations in absorption in the violet region of the lines of CIV, SiIV and Al III. The x axis shows the rest wavelengths at the bottom and the blue shifts at the top with respect to the emission lines, expressed in velocities (1,000 km/s). Credit: N. Filiz Ak et al. (2018). For a color version of this figure, see www.iste.co.uk/combes/nuclei.zip*

In quasars we also observe many narrow absorption lines that are shifted towards the blue with respect to the systemic redshift: they have nothing to do with the quasar itself, but are due to clouds or galaxies located on the line of sight.

1.2.1.4. *X-ray domain*

It is traditionally considered that the X-ray domain extends from 0.1 keV up to 512 KeV, where the gamma domain begins. With regard to non-radio AGNs, it makes more sense to stop it at around 100 keV, because the spectrum collapses at these energies, while in radio AGNs, we have seen that it continues up to much higher energies, and is of non-thermal origin. X-rays below 500 eV are absorbed by interstellar matter in our galaxy and the host galaxy. This absorption must be corrected. It generally corresponds to a column density $N_H \sim 10^{21}$ cm^{-2} in type 1 AGN, but can reach 10^{23-24} cm^{-2} in type 2 AGN due to the absorption by the torus. The primary emission is then completely absorbed around 2 or 3 keV.

It can be seen that the spectral index α_{ox}=[log(L_ν(2keV)-log(L_ν(2500 Å)]/2,605)] (representing the ratio between UV and X-ray luminosity) varies between 1 for objects of low luminosity and 2 for those of high luminosity, that is, the brighter objects radiate less in the hard X-ray region. However, X-ray luminosity is still a small fraction of the bolometric luminosity. For a long time, we have probably overestimated the energy contained in the UV and soft X-ray domains, because one was influenced by the spectral distribution theoretically predicted for the accretion disk. A continuum proposed by Mathews and Ferland (1987) constituting what was called the "Big Blue Bump" was considered to be standard for several decades and served as the basis for calculating photoionized models. It competed with the continuum proposed by Zheng *et al.* (1997), which was less rich in UV, and was also used (see Figure 1.15).

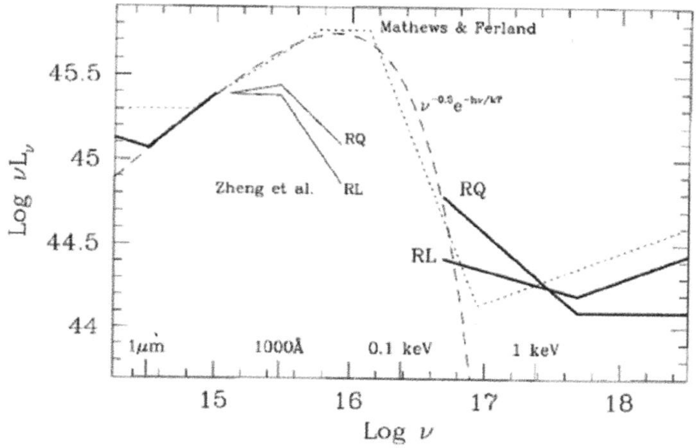

Figure 1.15. *Two possible energy spectral distributions for AGNs, according to Laor et al. (1997)*

X-ray observations underwent a real revolution in 1999 when the two satellites Chandra and XMM-Newton were launched, which provided data with a spectral resolution reaching 1,000 at 1 keV. It was thus discovered that the X-ray spectra of type 2 AGNs show many emission lines in the soft X-ray domain: they come from a hot region located beyond the torus. They are probably emitted and/or reflected by the warm region which also reflects the broad lines (see Figure 1.16). Other intense lines are observed between 2 and 10 keV, in particular, at 6.4 keV, the Kα line of Fe in different ionization states. They are associated with a "bump" around 30 keV[4].

The AGN-1 spectra show an intense continuum in the soft X-ray domain, and, in about 50% of the cases, numerous absorption lines of different elements such as strongly ionized carbon and oxygen (OVII and OVIII). These are the counterparts of the BALs observed in the UV domain. They are shifted to short wavelengths, and, like BALs, correspond to gas ejected from AGNs. This region is called the warm absorber (WA). The Kα line of iron is weaker in AGN-1 than in AGN-2.

4. We refer to hump, probably to distinguish between the bump in the soft X-ray domain.

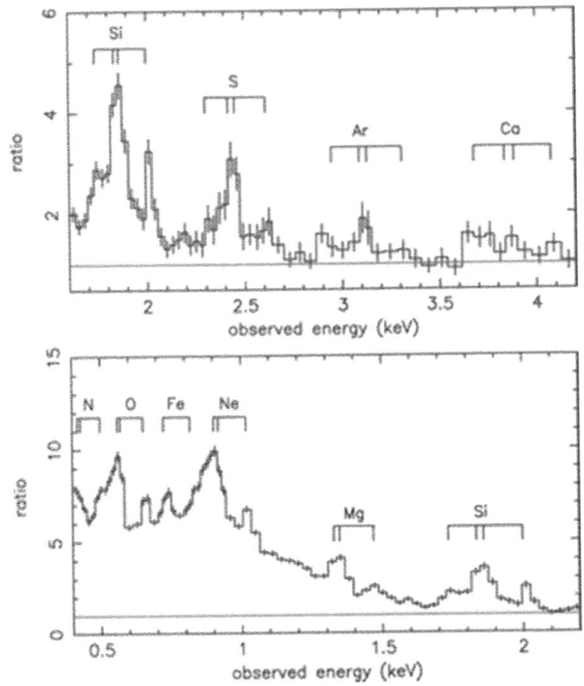

Figure 1.16. *X-ray spectrum of the Seyfert 2 galaxy NGC 1068 obtained via satellite XMM-Newton, which reveals fluorescence lines of silicon, sulfur, calcium and argon. Credit: ESA and University of Leicester. For a color version of this figure, see www.iste.co.uk/combes/nuclei.zip*

1.2.1.5. *Gamma domain*

Figure 1.6 shows the spectral distribution of some radio AGNs. In those where the axis of the jet is close to the line of sight, there are two "humps", each extending over several decades in frequency. We will not go into these observations in detail as they are discussed in Chapter 4.

1.2.2. *Variability (non-radio AGN)*

AGN-1s are variable in all frequency bands, with the fastest variations occurring at the highest energies. In AGN-2s which are not too absorbed, the variations in hard X-rays can also be very rapid. When we have light curves of the visible continuum and broad lines, we see that the variations in the continuum always precede those of the lines. We also note that the light curves of the visible continuum, after a very short time, follow those of the ultraviolet, proof that they are all produced by variations in more energetic central radiation (see Figure 1.17).

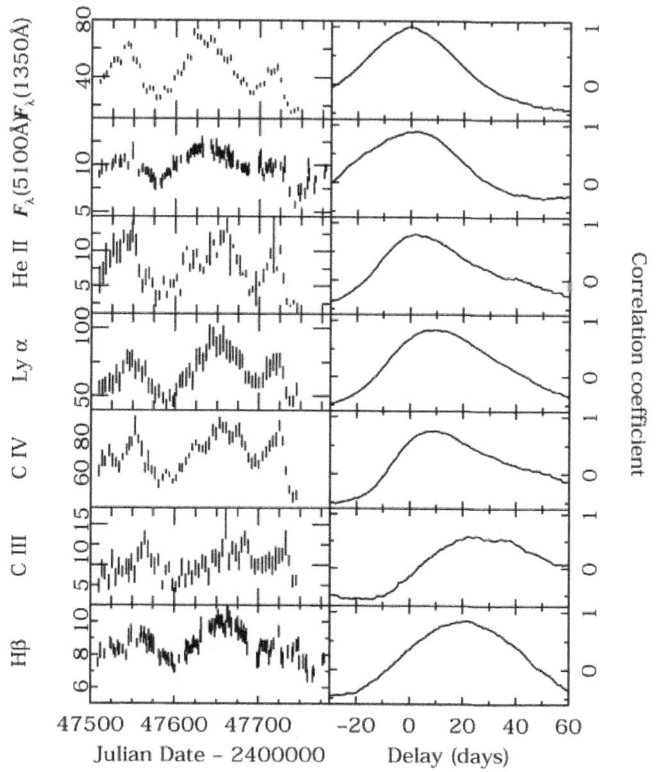

Figure 1.17. *Light curves of the Seyfert 1 NGC 5548 on the left and corresponding CCFs on the right (see reverberation mapping below), obtained in 1988–1989. From the top, the light curves of UV continuum at 1350 Å, optical continuum at 5,100 Å, He II 1640, Lyα 1215, CIV 1549, CIII] 1909 and Hβ 4861. Continuum fluxes are in units of 10^{-15} ergs s^{-1} cm^{-2} $Å^{-1}$, and line fluxes are in units of 10^{-13} ergs s^{-1} cm^{-2}. CCFs are calculated with respect to the continuum UV at the top (source: Peterson (2014))*

1.3. Interpretation I: thermal processes

1.3.1. *General information*

Observation and interpretation of AGN spectra has led to the publication of thousands of articles and the holding of dozens of conferences, and they continue to grow with the advent of large surveys. Although only due to "thermal" processes, the radiation of non-radio AGNs is completely different from stellar radiation. This is what makes it possible to distinguish an AGN from a region of star formation (a "starburst"), which can have the same compactness.

A characteristic of AGNs is that the media from which the radiation originates are relatively dilute (10^4 cm^{-3} for the forbidden line region, 10^6 cm^{-3} for the molecular torus, $10^{10 \text{ to } 12}$ cm^{-3} for the permitted line region, $10^{12 \text{ to } 15}$ cm^{-3} for the accretion disk).

The processes involved are:

1) atomic and molecular processes: emission and absorption of radiation by particles: free–bound and free–free continua, or bound–bound which gives rise to spectral lines;

2) emission or absorption by dust, and Thomson and Compton scattering by electrons.

These processes are similar to those that occur in nebulae or in stellar atmospheres, with some peculiarities that must be taken into account, which explains why sophisticated computer codes had to be developed specifically for AGNs. Consider the following different regions:

– NLR: this looks like a planetary nebula, but the photoionizing radiation is different from that of a hot star: it contains soft and hard X-ray photons which cause a very different stratification of the medium: several degrees of ionization coexist in the same region, unlike nebulae where we can define, for certain ions, a "Strömgren zone". However, the codes that are used for planetary nebulae can be used for NLR, with the addition of some atomic processes. The intensities of the spectral lines are different. Proof of this is the famous "BPT" diagram. Since they know that the ionizing continuum is very different from that of hot stars in AGNs, Baldwin *et al.* (1981) measured the line intensity ratios in AGNs and in nebulae. Figure 1.18 shows that indeed it is possible to construct line ratio diagrams where the HII regions and AGNs occupy very distinct zones, without needing to correct the intensities of a reddening which is often poorly known, because the lines used have similar wavelengths. Almost 30 years after the introduction of the BPT diagram, this is how we still separate AGNs from star-forming galaxies, or Seyfert galaxies from LINERs, and this is the case for the tens of thousands of objects observed in large surveys!

– BLR: besides the fact that it is also photoionized by a continuum containing X photons, it is much denser than planetary nebulae, which leads to the existence of different processes (molecular processes, Auger ionizations, dielectronic recombination, triple recombination, ionizations by secondary electrons, etc.) and, above all, it is much thicker optically, which requires a calculation of the transfer for the lines, avoided in nebulae.

– The atmosphere of the accretion disk differs considerably from a stellar atmosphere. The underlying continuum extends into the X-ray domain, and photons experience Compton scattering as they travel outward. The surface is irradiated with hard X-rays (a very hot corona; see below as well as Chapter 3). For the most central regions of the disk, the relativistic effects must be taken into account.

– The BAL region resembles the atmosphere of a P Cygni star, but the motions are faster, the geometry and dynamics more complex, and the magnetic field probably plays an important role.

– The hot gas emitting the X-rays in AGN has a temperature and a density close to those of the solar corona, but it is photoionized and not collisionally ionized.

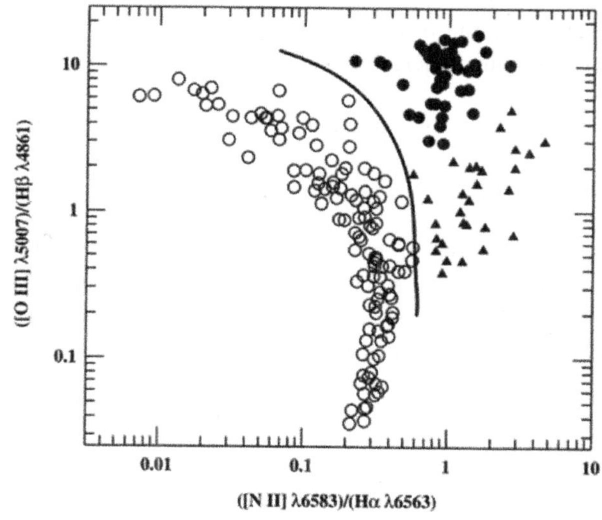

Figure 1.18. *The BPT diagram for emission lines in galaxies. Open circles represent HII regions, closed circles are AGNs, triangles represent LINERs. Adapted by B.M. Peterson in "An Introduction to Active Galactic Nuclei" (Peterson 1997)*

We will focus on BLR and NLR, which is the most discussed topic of non-radio AGNs.

1.3.2. *Formation of BLR and NLR*[5] *spectra*

Note that what is described in this section can be found in part in the book by Osterbrock and Ferland (2006), "Astrophysics of Gaseous Nebulae and Active Galactic Nuclei". There we can find, among others, the equations which are not detailed here concerning the thermal and ionization equilibria.

We will discuss the BLR and NLR models later, but, for now, we will take the simple image of a set of animated clouds of high random velocities, occupying about 10% of the line of sight of the central source. These clouds therefore do not obscure each other, but each is optically thick in the Lyman continuum, so, together, they absorb 10% of the source. To avoid errors, we can treat each cloud in plane-parallel geometry. We will also assume stationarity, which, most of the time, is justified because atomic processes are much faster than dynamic processes.

The medium is not very dense; therefore, it is far from thermodynamic equilibrium. It is therefore necessary to calculate the populations of atoms and ions by considering, at each point of the gas, all the processes involved in the statistical equilibrium of free or bound energy levels.

Three types of equilibria come into play.

a) Ionization equilibrium

The fractional ionization of an ion is given by the equality between the photoionizations coming from the central source and the Auger ionizations from all levels (in general, only the fundamental level counts), and the radiative and dielectronic (and possibly collisional) recombinations on all levels. We write an equilibrium equation for each ion i of a given element E, which forms a series of equations that we close by adding for each element E the equation $\Sigma_{Ni,E} = N_E$.

An important quantity that comes into this balance is the *ionization parameter*. It corresponds to the ratio between the density of photons and the density of the gas. There are several versions. The first that was used is dimensionless and corresponds to the ratio between the density of photons and those of hydrogen particles:

$$U = \int_{E1}^{E2} \frac{(L_E/E)dE}{4\pi r^2 c N_H}$$

5. We assume that the readers have some understanding of spectroscopy and are familiar with the basics of transfer. If not, they will need to consult specialized books.

where c is the speed of light, r is the distance from the source, E is the energy of photons, L_E is the luminosity per energy interval, and E1 and E2 are the given limits, for example, the ionization threshold of hydrogen, 13.6 eV, for E1. Under these conditions, we find, for both BLR and NLR, that U ~ 0.01–0.1.

Subsequently, when looking at the state of the hotter gas radiating in the X-ray domain (the atmosphere of the irradiated accretion disk or the warm absorber), the following expression was used instead:

$$\xi = \int_{E1}^{E2} \frac{L_E dE}{r^2 N_H}$$

ξ is approximately 10^3 ergs s cm^{-1} in the regions considered.

b) Thermal equilibrium

The heating is due to the photoionizations by the incident radiation and the diffuse radiation (if the medium is optically thick at the frequency considered), and the cooling is due to the radiative recombinations and the de-excitations of the bound levels. For hot and thick media like the atmosphere of the accretion disk, we must also take into account the cooling by free electrons and Compton scattering, which transfers part of the energy of the photons to the electrons.

c) Statistical equilibrium of the bound levels

It is written by equaling the number of all radiative and collisional transitions starting from one level to the number of all transitions arriving on this level, per unit of time and volume (taking into account the continuous terms).

If we assume the atom is reduced to two energy levels, the statistical equilibrium is written as:

$$N_m \left(B_{mn} \int J_\nu \phi_\nu d\nu + A_{mn} + N_p C_{mn}\right) = N_n \left(B_{nm} \int J_\nu \phi_\nu d\nu + N_p C_{nm}\right)$$

where B_{mn} and A_{mn} are Einstein's coefficients for induced and spontaneous de-excitation, B_{nm} is the coefficient for induced excitation, N_p is the numerical density of disturbing particles (usually electrons, unless the gas contains many neutral atoms, which occurs in certain regions), N_pC_{mn} and N_pC_{nm} are the rates of de-excitations and collisional excitations, J_ν is the intensity integrated on the angles at the frequency ν, ϕ_ν is the emission profile of the spectral line, which we have here assumed to be identical to the absorption profile (coherent emission). The coefficients C_{mn} (respectively, C_{nm}) are given by the expressions C_{mn} (respectively, C_{nm}) = $<\sigma_{mn}\ v>$ (respectively, $<\sigma_{nm}\ v>$) $s^{-1}\ cm^3$, where $<\sigma_{mn}\ v>$ (respectively, $<\sigma_{nm}\ v>$) are the cross-sections integrated on the particle velocity distribution function (here, Maxwellian), from the energy E_{nm} for C_{nm}. C_{mn} and C_{nm} depend only on temperature T.

At thermodynamic equilibrium, the radiative and collisional processes are balanced separately, which makes it possible to calculate the relations between C_{mn} and $C_{nm,}$ on the one hand, and B_{mn}, B_{nm} and $A_{mn,}$ on the other, using the relations of Saha, Boltzmann, Maxwell and Planck. We then find, for the source function S, after a few algebraic manipulations (still assuming the emission and absorption profiles to be identical, which means that the source function is constant in a spectral line):

$$S(\text{line}) = \varepsilon_\nu/\kappa_\nu = \frac{N_m A_{mn}}{N_n B_{nm}} \frac{1}{1 - \frac{N_m B_{mn}}{N_n B_{nm}}}$$

where ε_ν and κ_ν are, respectively, the emissivity and the absorption coefficient per unit of frequency, volume and solid angle, or:

$$S\,(\text{line}) = \frac{\int J_\nu \phi_\nu\, d\nu + \eta B_\nu(T)}{1+\eta} \quad \text{with } \eta = \left(1 - \exp(-\frac{E_{nm}}{kT})\right)\frac{N_p C_{nm}}{A_{mn}}$$

where B_ν is the Planck function.

This equation highlights the double structure of the source function, consisting of a diffusion term $(\int J_\nu \phi_\nu d\nu)$ and an absorption term ηB_ν. The first term corresponds to the radiative absorptions of the photon E_{nm}, which are followed, by emission, by spontaneous de-excitation without the change of frequency. The second term is a source term due to collisions. **The term of diffusion will prove to be fundamental for the BLR**. For more details, see Box 1.1 on radiation transfer.

When several other levels (including one continuous level) are coupled with the two levels m and n, it is necessary to take them into account in the equilibrium equation of the two levels. We can then reduce the source function to a form similar to the previous one:

$$S\,(\text{line}) = \frac{\int J_\nu \phi_\nu \, d\nu + \eta B_\nu(T) + \eta'}{1 + \eta + \eta''}$$

in which η' and η'' contain all the terms relating to the other levels, which makes it possible to solve the problem by successive iterations.

1.3.3. *Permitted and forbidden lines*

1.3.3.1. *Permitted lines: two processes govern their formation*

a) Recombination

This is the process by which lines of ions which have a low abundance, but whose upper ion is dominant (see Figure 1.19) are formed. The most significant are the lines of HI, HeI and HeII.

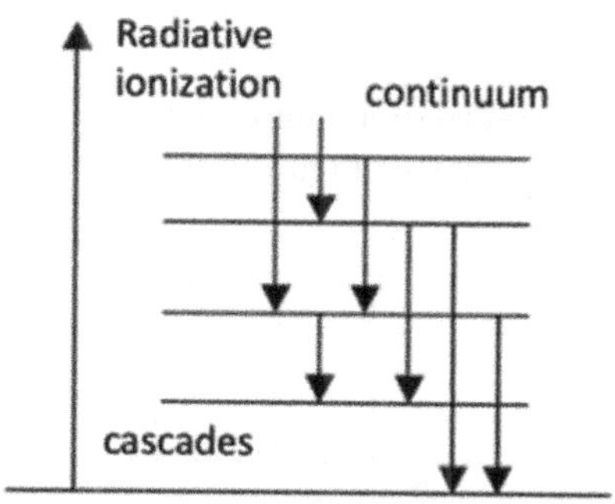

Figure 1.19. *Permitted lines formed by recombination followed by radiative cascades. Photoionization usually occurs from the ground level*

The statistical equilibrium of the levels is written as:

$$N_m \sum_{n=1}^{n=m-1} A_{mn} = N_x + N_p\,\alpha_m(T) + \sum_{k=m+1}^{k=\infty} N_k\,A_{km}$$

where $\alpha_m(T)$ is the recombination coefficient, not dependent on temperature[6] and N_x+ is the numerical density of ions in the higher state. The other excitation and de-excitation processes are negligible in the absence of diffuse radiation. We can therefore calculate the populations of all levels of the atom by inverting this "matrix of cascades", and express the emissivity of the various transitions in the form:

$$\varepsilon_{mn} = (1/4\pi)N_e N_{H^+}\ \alpha_m(T,N_e)\ E_{nm}$$

where α_m is the "effective" recombination coefficient on level m. For example, the line emissivity Hβ is:

$$\varepsilon(H\beta) = (1/4\pi)N_p\ N_{H^+}\ 1.38710^{-25}\ T_4^{-0.983}\ 10^{-0.0424/T_4}\ \text{erg cm}^{-3}\ \text{s}^{-1}\ \text{ster}^{-1}$$

with $T_4 = T/10^4$K. These emissivities depend little on T, so the ratio $\varepsilon(H\alpha)/\varepsilon(H\beta)$ is almost constant and close to 3. We can use this ratio to determine the reddening because the two lines are quite far apart in the spectrum. However, it is often found to lead to a greater reddening in BLR than that found by other methods, which has been a problem for a long time. A probable explanation is the approximation to the escape probabilities (see Box 1.1 on radiation transfer).

b) Collisional excitation

The permitted lines of ions corresponding to the dominant degrees of ionization are formed by collisional excitation (see Figure 1.20).

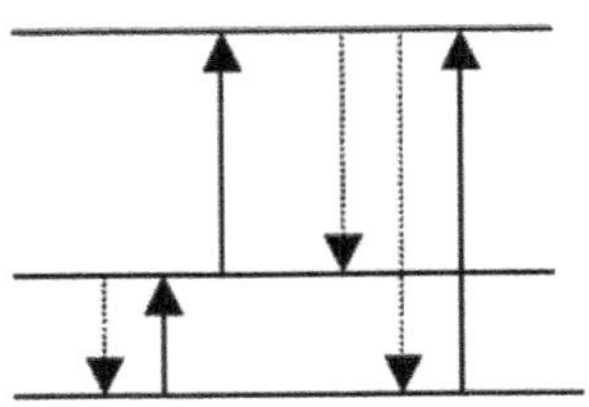

Figure 1.20. *Permitted lines formed by collisions. Solid lines: excitation by collision. Dotted lines: radiative de-excitations*

6. For the hydrogenoids, it is necessary to separate the different levels of L, which leads to, among other things, taking into account the de-excitation of the 2s level with two photons, which produces a continuum.

The statistical equilibrium of the levels is then written as:

$$N_m \sum_{n=1}^{n=m-1} A_{mn} = \sum_{k=1}^{k=m-1} N_k N_e C_{km}$$

In fact, it can be shown that, for these ions and under the BLR and NLR density conditions, the induced radiative terms are negligible in the absence of diffuse radiation. By considering the case of the atom at two levels, the statistical equilibrium is then written as:

$$N_m A_{mn} = N_n N_p C_{nm} \text{ and } \varepsilon_{nm} = (1/4\pi) N_n N_p C_{nm} E_{nm}.$$

1.3.3.2. *The forbidden lines*

The forbidden lines correspond to transitions for which the probabilities of spontaneous excitation and radiative excitation and induced at certain levels are very low. Again, assuming an atom at two levels, the equilibrium of the higher level is thus established $N_n N_p C_{nm} = N_m (A_{mn} + N_p C_{nm})$, which gives the emissivity:

$$\varepsilon_{nm} = \frac{1}{4\pi} N_n E_{mn} A_{mn} \frac{N_p C_{nm}}{N_p C_{mn} + A_{mn}}$$

It can be seen that, below a certain density $N_{crit} = A_{mn}/C_{mn}$ known as the *critical density* of the transition, the de-excitations by collisions are negligible, and the emissivity is then $\varepsilon_{nm} = (1/4\pi) N_n N_p C_{nm} E_{nm}$.

We can now understand why the forbidden lines become as intense as the permitted lines (or more intense) at low density. Indeed, if we compare the emissivity of a forbidden line with that of a permitted line also formed by collisional excitation, the ratio between the forbidden and permitted emissivities is equal to C_{nm}(forbidden)/C_{nm}(permitted). This is approximately equal to 1, or even greater. If the permitted line is formed by recombination, like Hβ, we find a ratio which is also close to 1 for ions with an abundance of 10^{-4} with respect to hydrogen, like oxygen or carbon.

The critical density varies from 10^4 to 10^6 cm^{-3} depending on the lines considered. For semi-forbidden lines such as CIII] 1909, it is around 10^8 cm^{-3}. As this line is relatively narrow, it was deduced that the density in the BLR is greater than 10^8 cm^{-3}. In fact, other considerations suggest that it is approximately 10^{10} cm^{-3}.

Using different lines and different ions, we can determine the density and temperature of the gas, as well as the abundance of the elements. For example, consider the forbidden lines of ions of electron configuration p^2, p^3, p^4. This is the case, among others, for ions Ne^{++}, Ne^{+++}, Ne^{+4}, O^+, O^{++}, N^+, S^+. The fundamental levels of these ions consist of three sub-levels 1, 2 and 3 (see Figure 1.21). For the ratios between the emissivities of the lines 3-1 and 3-2 we find the following expression:

$$\varepsilon_{31}/\varepsilon_{21} = \frac{E_{13}}{E_{12}} \frac{A_{31}}{A_{32}} \frac{C_{13}A_{21} + N_e C_{21} C_{13} + N_e C_{12} C_{23}}{C_{12}(A_{31} + A_{32} + N_e C_{31} + N_e C_{32})}$$

where N_e is the electron density. An identical expression is found for the ratio $\varepsilon_{32}/\varepsilon_{21}$. These ratios are rapidly varying functions of temperature (terms in exp (-E/kT) in C_{12}, C_{23} and C_{13}), and independent of the density below a critical value. The typical temperature of the NLR is thus found to be close to 10,000 degrees. In ions with configuration p^3 such as O^+ and S^+, levels 2 and 3 are each divided into two sub-levels. The ratio between the two lines coming from the second S^+ level, when they are resolved, makes it possible to determine the density with good precision, in the domain 10^3–10^5 cm^{-3}.

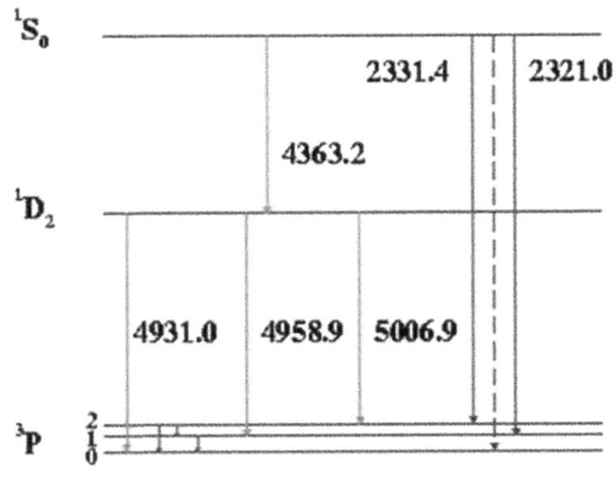

Figure 1.21. *Fundamental levels of the ion OIII. The ratio between the strongest lines 2-1, 5007+4959 and line 4363 strongly depends on temperature. The doublet 2331/2321, when resolved, allows the density to be obtained. For a color version of this figure, see www.iste.co.uk/combes/nuclei.zip*

1.3.3.3. *Results for BLR and NLR*

For density in the NLR and the BLR, if we take the following values: 10^4 and 10^{10} cm^{-3}, respectively, and if we remember that the two regions have roughly the same ionization parameter (proportional to $L_{ion}/(r^2N_H)$ with the same L_{ion}, we find that the BLR is, at a distance, 1,000 times smaller than the NLR. The latter is resolved in nearby AGNs, and its size is a few hundred parsecs. The size of the BLR is therefore from a few light months to a few light years, which is confirmed by the variability studies which will be discussed in section 1.3.4. Moreover, by identifying the values calculated and observed for the luminosities of the lines, we find that the filling factor of the emitting volume is around 1 for the NLR, and much smaller for the BLR ($\sim 10^{-6}$). We find that the mass of the NLR is around 10^6 solar masses, but that of the BLR is only 1–10 solar masses.

For the BLR, we cannot directly deduce the temperature from the intensity ratios of the permitted lines, and only the models provide it. It is also around 10,000 degrees K, like that of the NLR. An important parameter which results from this is the pressure $\sim nkT \sim 1$ in CGS.

One problem regarding BLR is called "the energy budget problem". Indeed, if we add the emission of all the permitted lines, we find that it is greater than the energy absorbed in the Lyman continuum. One solution would be that we do not see the same continuum that the BLR "sees", or it is anisotropic or poorly estimated. Another would be the presence of mechanical heating, due to shock waves, for example. This issue has not yet been solved.

1.3.4. *Variability of the allowed lines and of the continuous, reverberation mapping*

Andrillat and Souffrin (1968) predicted that the BLR lines would vary over a few years, in response to changes in the ionizing ultraviolet continuum (assuming that it varied roughly like the brightness). They even thought they had observed such a variation in the spectrum of NGC 3615 compared to the 1943 article by Seyfert, but it is likely that the variation was an artifact due to the displacement of the spectrograph slit during very long exposures. However, in 1982, Blandford and McKee predicted that changes in the central ionizing source should generate changes in the intensities and profiles of the emission lines, and that these should be delayed relative to changes in the continuum, due to the travel time. They therefore proposed to record the continuum and the lines as a function of time, and they showed that we could deduce the structure of the BLR.

From that moment on, intensive surveillance campaigns, some lasting 20 years, were organized, using different telescopes, some of which were even entirely dedicated to these observations. This method was called "reverberation mapping" (RM). The basic idea is that the lines respond to variations in the ionizing continuum with a delay depending on the distance between them, on the structure of the BLR and on the position of the observer with respect to the source and the BLR. Fortunately, the ionizing continuum is strongly correlated with the visible continuum, as both are due in part to the illumination of the accretion disk by the central X-ray source, as suggested by Collin-Souffrin (1991). It is also necessary to assume that the structure of the BLR does not change during the duration of the observation, which is legitimate, given the propagation time of the dynamic disturbances, and, on the contrary, that the response of each point to the variations of the continuum (corresponding to the ionization and recombination time) is much faster than the typical time of the variations, which is also valid for the densities of the BLR.

We therefore expect to be able to measure, for each point of the BLR, animated by a velocity V projected on the line of sight (therefore for each point of a spectral line) the delay between the light curve of the continuum $L_c(t)$, and that of the corresponding point of the spectral line $L_r(V,t)$. In fact, we do not have enough precise information to establish this delay, except for a few very bright AGNs, and so we are forced to limit ourselves to the integrated luminosity of the line, $L_r(t)$. The equation which links it to that of the continuum is therefore:

$$L_r(t) = \int_{-\infty}^{+\infty} \Psi(\tau) L_c(t-\tau)\, d\tau$$

where Ψ is the *transfer function*, the response to an instantaneous impulse. Figure 1.22 shows the different moments when a distant observer perceives a variation for a spherical envelope containing emitting clouds. To determine the delays between a line r and the continuum c, we use the cross-correlation function (CCF): $CCF(\tau) = \int_{-\infty}^{+\infty} L_r(t)\, L_c(t - \tau)\, dt$.

Thus, in Figure 1.16, we see the light curves of NGC 5548, with the corresponding CCFs between each line and the continuum. Figure 1.23 shows an example of a velocity-delay map, making it possible to deduce that the most distant part of the BLR moves towards the source. Some close AGNs have been observed in a very detailed way (in particular, NGC 5548, shown in Figure 1.1), which made it possible to verify that the lines of strong ionization form closer to the center than those of weak ionization, and that they are larger, as expected, but without clear conclusions about the structure of the BLR. Analysis of the profiles of these objects over time has shown, however, that there is often a mixture of rotation and random movements or ejection (Shapovalova *et al.* (2019), and previous articles).

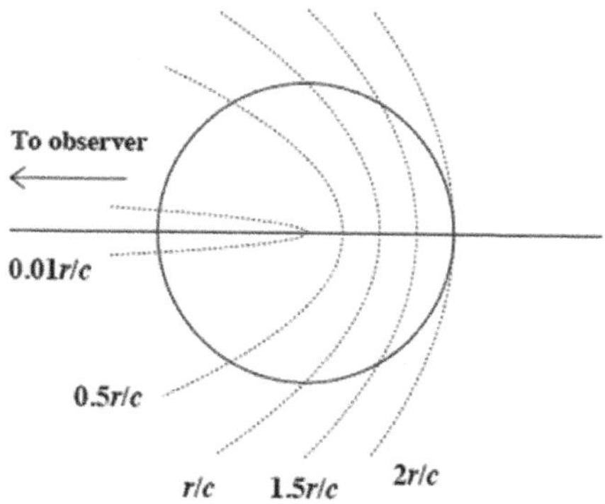

Figure 1.22. *The circle represents the section of a spherical shell containing emissive clouds. The continuous source is in the center of the shell. After a variation in the continuum, the observer, placed far to the left, sees the response of the clouds, which are on an iso-delay surface. Here, five iso-delay surfaces are shown, each being shown with the delay time relative to the source (expressed in shell radius). The furthest point on the shell responds with a delay equal to 2r/c (source: Peterson 2001)*

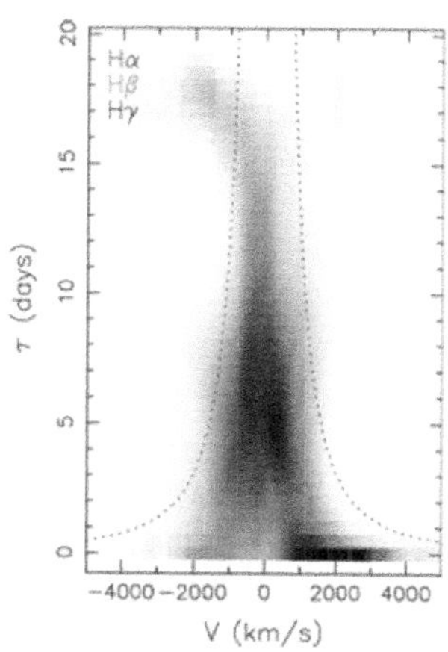

Figure 1.23. *Map showing the delay time as a function of speed for three Balmer lines in Arp 151. We see that the Hα clouds which have the greatest delays are moving towards the source. Credit: Bentz et al. (2010). For a color version of this figure, see www.iste.co.uk/combes/nuclei.zip*

Although the observation campaigns did not reveal the detailed structure of the BLR as expected (except for a few AGNs which were studied very precisely), they were able to give the size of the BLR for about 50 objects (see Peterson (2013)). Particularly, they showed the existence of an empirical correlation between the radius of the BLR and the luminosity (see Figure 1.24): we find that R (BLR) is approximately proportional to $L^{1/2}$. Netzer (2013) mentions more specific relationships:

$$R_{BLR}(H\beta) = 0.12L^{0,6\pm0,1}pc \quad \text{and} \quad R_{BLR}(CIV1549) = 0.04L^{0,6\pm0,1}pc,$$

where L_{46} is the bolometric luminosity expressed in units of 10^{46} ergs, the median value between close AGNs and quasars.

This relation, which spans five orders of magnitude in luminosity, is poorly explained. It is certainly due to the fact that the BLR can only exist around a particular value of the ionization parameter $L_{ion}/(nr^2)$. With other similar relationships, this is an issue that is still debated.

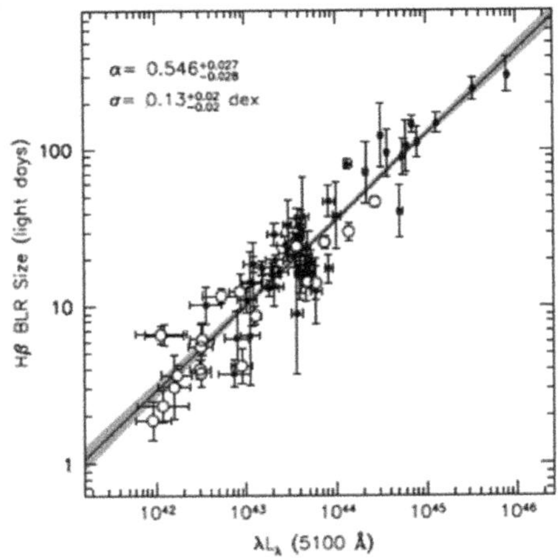

Figure 1.24. *Relationship between the size of the BLR and the luminosity measured by the response of the Hβ line, as a function of the luminosity of the AGN. The open circles correspond to AGNs added between 2009 and 2013. Credit: Bentz et al. (2013)*

An unexpected result of RM was to directly provide the masses of about 50 black holes in weaker quasars and AGNs, and to indirectly determine the mass of black holes of tens of thousands of others for which we have only one spectroscopic observation. Indeed, it is possible to estimate the mass of the central black hole (it is assumed that it dominates the BLR) using the relation:

$$M_{BH} = f\,R_{BLR}\,\Delta V^2/G,$$

where ΔV is the Doppler line width, and f is a unity factor which depends on the structure, kinematics and inclination of the BLR. By using the relation between L and R_{BLR}, we find that:

$$M_{BH} = f\, L^{1/2} \Delta V^2, \text{ or } M_{BH} = f\, L^{1/2} (FWHM)^2,$$

which no longer requires knowledge of R_{BLR}. The question is to find the best value for f. Now, we can get an idea, because there is another way to determine M_{BH} when you can estimate the velocity dispersion σ_* of stars in a region dominated by the mass of the black hole in the center of the host galaxy. This region, with radius $R_{BH} = GM_{BH}/\sigma_*^2$, is resolved in nearby galaxies. This could be measured in about 30 AGNs where M_{BH}/f was determined by the RM. Ultimately, we obtain, for example, for Hβ:

$$M_{BH}/M_o \sim 10^8\, [L(5100)/10^{46} ergs/s]^{0.65}\, [FWHM(H\beta)/10^3 km/s]^2$$

where L(5100) is the luminosity at 5,100 A. The masses of black holes determined using a single measurement are considered to be subject to a factor of three uncertainties. The masses determined in this way are between 10^6 and 10^9 M_o. Where NLS1s have low masses, we find the nuclei of elliptical galaxies such as M87 to be on the large side, with $M_{BH} = 6\ 10^9\ M_o$ (Figure 1.25).

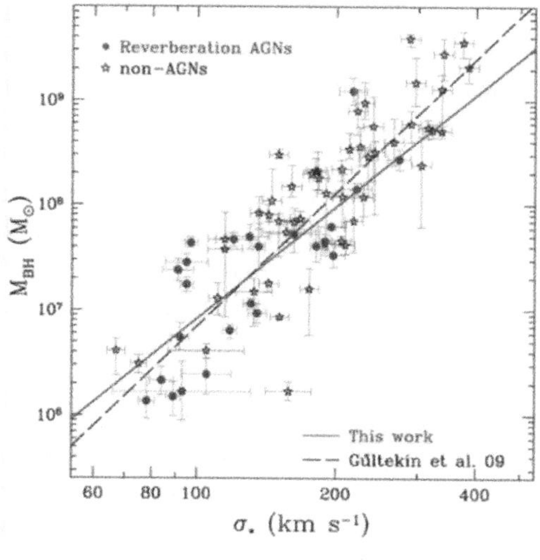

Figure 1.25. *Relation between M_{BH} and σ_* for AGNs, where M_{BH} was determined by RM (in blue), compared with inactive galaxies where M_{BH} was determined by the velocity dispersion. For the masses determined by RM, we have assumed f = 0.72. The solid line gives the best correlation for AGNs, and the dashed line gives the best correlation for inactive galaxies in the sample (source: Woo (2010)). For a color version of this figure, see www.iste.co.uk/combes/nuclei.zip*

Moreover, the galaxy bulge mass, M_{bul}, is given by $M_{bul} \sim G/(R_{BH}\,\sigma_b^2)$. It was thus discovered that there was a strong correlation between M_{BH} and M_{bul}. This fundamental result proves that central black holes and galaxy bulges follow a parallel evolution. This will be discussed in Chapter 5.

1.3.5. *Hot gas*

There are three aspects of AGNs that seem to represent the same type of medium or at least a medium with a common origin originating above the torus or accretion disk: BALs, the warm absorber, and the hot medium in AGN-2s. All three correspond to winds that can reach several thousand km/s.

1.3.5.1. *BALs*

Many studies of BALs relate to particular AGNs or to important surveys (obtained, in particular, with the SDSS). The most difficult task is to separate the column density from the density to deduce the inhomogeneity of the gas, and to deduce the rate of ejected mass. A few lines provide important information. The phosphorus doublet PV1118-1128 is present in many objects with a strong intensity up to saturation, and indicates column densities of around 10^{22} cm^{-2} and large ionization parameters U ~ 1. The different optical thicknesses of saturated lines show partial coverage on scales of 0.01 pc, while the excited states of Silicon SiIV 1073 and CIII 1176 indicate typical densities of 3 10^5 cm^{-3} and small distances from the center, of around a fraction of a parsec.

Ultimately, we can conclude that the BALs correspond to a photoionized gas located between 1 and 10,000 pc of the black hole, with densities of around 10^4–10^6 cm^{-3} and column densities of 10^{21}–10^{23} cm^{-2}. The profiles vary on scales from a few days to a few years. These variations are often attributed to clouds crossing the line of sight, but they can also be due to variations in the ionizing flux.

Non-BAL quasars are always stronger in X-rays than BAL quasars, which indicates that the latter are absorbed. On the other hand, the intensity of BALs increases with increasing luminosity. More precisely, BALs always correspond to high values of the L/M_{BH} ratio, that is, very bright quasars, or to not very massive NLS1s (Grupe and Nousek 2015). It should also be noted that BAL quasars have the same spectral characteristics as NLS1s, namely, a high FeII/Hβ ratio and a low [OIII]/Hβ ratio.

1.3.5.2. *Warm absorber (WA)*

The X-rays of WA come from a hot gas with a higher ionization parameter than that of BALs. In some objects, the velocities of the UV and X-ray emitters are identical, leading to the conclusion that the two media coexist in the same place. We can indeed imagine an inhomogeneous flow, with dense clouds giving BALs, surrounded by a diluted medium corresponding to the WA, the thermal instability which we will discuss later, which could explain such a coexistence. Sometimes, however, the X and UV emitters correspond to different velocities, which means that they are formed in different places. It is difficult to go into more detail on the basis of observations alone.

1.4. Interpretation II: theoretical models

1.4.1. *Black hole*

Details on the environment close to a black hole can be found in Chapter 3. We give here only a few basic parameters.

The spin of the black hole is given by the angular momentum $J = aGM_{BH}^2/c$, where a is a parameter between 0 and 1. For $a = 0$, we are dealing with a Schwarzschild black hole without rotation, and for $0 < a \leq 1$, a Kerr black hole. The spin can be determined by the profile of the Fe Kα. When the line is emitted near the inner radius of a Kerr disk, it exhibits a large red wing, due to gravitational reddening (see Figure 1.27).

The gravitational radius of the black hole R_G, equal to half of the "Schwarzschild radius" R_S, which is the radius of a non-rotating black hole below which photons cannot escape: $R_G = GM/c^2 = 1.5\ 10^{13}\ M_8$ cm, where G is the gravitational constant and M_8 is the black hole mass expressed in $10^8\ M_O$.

The Eddington luminosity, L_{Edd}, corresponds to the equality between the radiation pressure which repels the accreted gas, and the gravitational force per unit area which attracts it. For a completely ionized gas whose molecular weight μ corresponds to the solar composition, we obtain: $L_{Edd} = 4\pi\ GM_{BH}\ \mu\ m_p/\sigma_T \sim 1.5\ 10^{46}\ M_8$ ergs/s, where m_p is the proton mass and σ_T is the Thomson cross-section. It is the maximum luminosity of an object drawing its energy from the accretion (in fact, the luminosity can be higher than L_{Edd} when the accretion is variable, and this is probably partly what explains why some quasars have luminosities equal to 2–3 L_{Edd}). On the other hand, it can be much lower if the accreted gas is not completely ionized, because the radiation pressure is then greatly increased compared to the

Thomson or Compton absorption alone, or if the accreted gas does not have the time to radiate below a radius much greater than R_G.

The accretion rate is $M' = L/(\eta c^2)$, and the "critical accretion rate", $M'_{crit} = L_{Edd}/(\eta c^2)$, where η is the conversion efficiency of mass into energy. This efficiency varies from 0.056 for a black hole without rotation to 0.30 for a black hole with maximum rotation, but it can be much lower if the accretion rate is very high or, on the contrary, very low. Unlike L_{Edd}, M'_{crit} is not a limit, because the efficiency of mass-energy conversion is of the order of 0.1 only when $10^{-3} < L/L_{Edd} < 0.3$. For $L/L_{Edd} > 0.3$ and $L/L_{Edd} < 10^{-3}$, the efficiency can be much lower (see Chapter 3). The accretion rate corresponding to the Eddington luminosity is $M'_{Edd} = 2\ M_8\ [\eta/0.1]^{-1}\ M_O$/year.

The mass is from $10^6\ M_o$ to $10^{10} M_o$. There are several ways to determine it, besides RM and measuring the velocity dispersion in the bulge. Several methods, such as measuring the rotational velocity of gas or stars in the central region, are well suited to quiescent black holes because it is not affected by the strong luminosity of the nucleus. Certain AGNs show molecular spectral lines in the radio domain amplified by a maser effect due to the central source. In one of them, NGC 4258 (M106), where we could observe both the proper motions and the radial velocity of the maser sources, we saw that they were distributed on a warped disk and that their velocities strictly followed a Keplerian law. The mass of the black hole was deduced with great precision ($3.8\ 10^7\ M_o$).

In addition to supermassive black holes, AGNs contain the following components:

– a jet in the center, small for $10^{-3} < L/L_{Edd} < 1$ (non-radio AGN), widely extended for $L/L_{Edd} << 10^{-3}$ (radio AGN);

– an accretion disk, which may or may not be geometrically thin depending on the value of L/L_{Edd}; the disk is geometrically thick and optically thin if $L/L_{Edd} << 10^{-3}$ and it has a tunnel in its center which allows relativistic particles to escape (it is a radio AGN). It is geometrically thin and optically thick if $10^{-3} < L/L_{Edd} < 0.1$. It is geometrically thick and optically thick if $0.1 < L/L_{Edd} < 1$. Finally, it is almost spherical if $L/L_{Edd} > 1$ (see Chapter 3);

– the BLR, made up of relatively dense clouds moving with velocity that is close to Keplerian, at a distance of about $10^{3\text{–}4} R_G$ from the black hole;

– a molecular torus containing dust, in the extension of the accretion disk, at a distance of $10^{4\text{–}5} R_G$ from the black hole;

– NLR forming a more or less spherical and continuous diluted medium, at a distance of $10^{5-6}R_G$ from the black hole;

– hot wind with an unknown distance and density, occupying about 50% of the line of sight of the black hole for $10^{-3} < L/L_{Edd} < 0.3$.

We will see that we must add a very hot corona located above the disk to this list of components, for $10^{-3} < L/L_{Edd} < 0.3$.

To facilitate the understanding of this complex structure, Figure 1.26 shows a one-dimensional diagram for the case $10^{-3} < L/L_{Edd} < 0.3$. The distances, independent of the mass of the black hole, are expressed in R_G, and in parsec for a 10^8 M_o black hole (note that the coordinates are logarithmic).

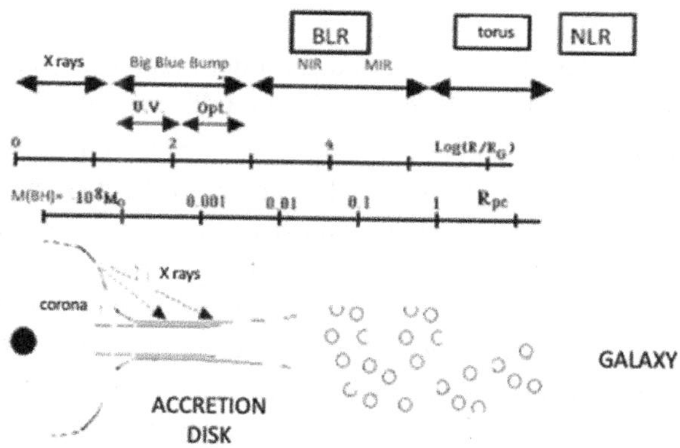

Figure 1.26. *One-dimensional diagram showing the distances to the black hole of the different components in a typical AGN. The radii expressed in parsec correspond to $M_{BH} = 10^8$ M_o*

1.4.2. *The accretion disk*

This subject will be discussed in detail in Chapter 3. Since we are interested in the observational aspects here, it is necessary to say a few words about the spectral distribution expected for the disk, because it poses a problem.

Consider the case $10^{-3} < L/L_{Edd} < 0.3$ of quasars and Seyfert galaxies. Accretion occurs in principle via a "standard" disk (see Shakura and Sunayev (1973)) which is optically thick and geometrically thin. By disregarding the relativistic effects (which is valid for $r > 10\ R_G$), we easily show that the energy radiated per unit of time at a distance r from the center is:

$$\frac{dL_r}{dr} = \frac{3}{2}\frac{GMM'}{r^2}\left(1-\sqrt{\frac{r_{in}}{r}}\right)$$

where r_{in} is the internal radius of the disk. Provided that the optical thickness of the disk is sufficient to radiate like a black body ($\tau >> 1$), we can write the following: $dL_\nu = 4\pi^2 B_\nu(T)\ rdr$, where we multiply the emission by two to take into account the two sides of the disk, which gives, for the total monochromatic luminosity emitted by the disk,

$$L_\nu = \frac{8\pi^2 h\nu^3}{c^2}\int\frac{rdr}{\exp(h\nu/kT)-1},$$

the integral to be taken between r_{in} and the outer edge r_{out}. Applying the Stefan–Boltzmann law to radius r, $\int \pi B_\nu\, d\nu = \sigma T^4$, we find that T varies as $r^{-3/4}$ except near the inner edge (where the term in parentheses is close to zero and relativistic effects must be taken into account). Therefore the "color" of the disk reddens as you move away from the center. As T becomes small at the outer edge, its contribution is small. We find that between the two frequencies $\nu_{out} = \nu(r_{out})$ and $\nu_{in} = \nu(r_{in})$, where $\nu_{out} << \nu_{in}$, the emitted spectrum varies as $\nu^{1/3}$. It collapses exponentially for $\nu > \nu_{in}$ (Wien's law) and varies as ν^2 for $\nu < \nu_{out}$ (Rayleigh–Jeans law). Between these two frequencies, L_ν is proportional to $(MM')^{2/3}\nu^{1/3}$, and can be expressed as follows: $M_8 M'_o = 2.6\ (L_{5100.45}/\cos i)$, where M'_o is the accretion rate per year in solar mass, $L_{5100.45}$ the luminosity at 5,100 Angströms in units of 10^{45} ergs/s, and i the disk tilt angle on the line of sight. These variables are typical for a Seyfert galaxy.

The inner edge of the disk where the flow ceases to be rotational and becomes radial (consequently, the gas falls into the black hole without having the time to radiate) is equal to $6R_G$ for a non-spinning black hole, and $1.2R_G$ for a Kerr black hole with maximum spin (this is why the efficiency of a Kerr black hole is higher than that of a Schwarzschild black hole). This radius is called innermost stable circular orbit (ISCO). When the gas approaches the ISCO around a spinning black hole, the Kα X-ray line of iron takes a particular shape, with a very broad red wing due to gravitational reddening, and two peaks, the blue one being amplified by the relativistic motion (see Figure 1.27).

The outer edge of the disk corresponds to the radius where the vertical component of the disk's gravity outweighs that of the black hole. It then breaks up into clouds as a result of gravitational instability. This radius is approximately $10^{3-4}R_G$.

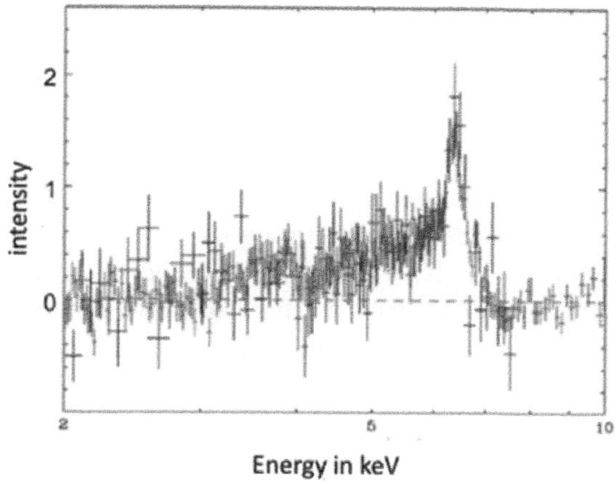

Figure 1.27. *Example of a Kα line of iron with a broadened red wing and an amplified blue peak, which corresponds to gas in rotation very close to the black hole, and confirms that it is a Kerr black hole. For a color version of this figure, see www.iste.co.uk/combes/nuclei.zip*

The spectral distribution described above does not involve general relativity nor comptonization, which modify it at high energies, but it is approximately correct in the visible and infrared domains. Numerous calculations have been carried out, the most precise of which are those of Hubeny *et al.* (2000), which take into account the relativistic effects and the transfer in the atmosphere of the disk. Figure 1.28 shows the contribution of different rings of the disk around a Kerr black hole, and we see that the most distant ($\geq$100 R_G) have a negligible contribution. This is the case of an AGN with a high L/L_{Edd} (0.3) value, which explains that the $\nu^{1/3}$ part of the spectrum only exists in the ultraviolet. The visible and infrared spectra increase with frequency (we are close to the Rayleigh–Jeans law), and it collapses between 1 and 10 keV. It is clear that the spectral distribution due to the disk alone is incompatible with what is observed.

Several observations can guide us towards an explanation. We have seen that the optical light curve follows that of UV. The UV curve itself does not strictly follow that of the X-ray, but the explanation is surely that the latter varies faster due to its small dimensions. It is therefore very likely that it is the X-radiation which causes the variations of the optical (see Rokaki and Magnan (1992), Rokaki *et al.* (1993) and

Figure 1.29). A second proof comes from the X-ray spectrum, where the 30 keV bump and the Kα iron line (as well as other lines) provide proof of the reflection of an X-ray source by a cold medium. The idea, proposed by Haardt and Maraschi (1991), is that a very hot thin gas (a "corona") exists above the disk, created by the turbulent magnetic field. This corona transforms the UV photons emitted by the disk into hard X-rays by inverse comptonization, and emits these hard X-ray photons which will hit the disk, and are reflected by it (Figure 1.30). This explains why the UV band is emitted before the optical band, and the infrared band after the optical band: the proportion reflected in each of the bands reflects their distances from the center, which increase with the wavelength. Figure 1.31 shows why we can then report the observations. The NIR and optical are due, in large part, to the illumination and reflection of the disk. As the disk's own luminosity decreases r^{-3} and, moreover, its thickness increases with the radius (see Chapter 3), the proportion of reflected light increases with wavelength. The torus thus begins to dominate above 3 microns, while the disk dominates in the far-UV. The corona takes over from the disk above 1 keV, with the X-rays and the 30 keV bump due to the reflection on the disk and on the torus.

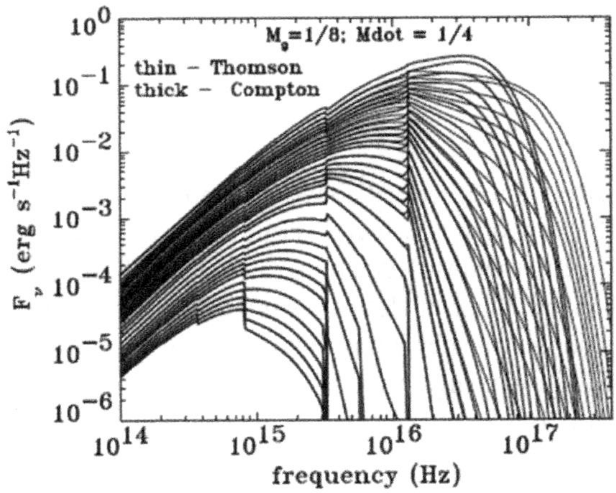

Figure 1.28. *Flux emerging from each ring of the disk for a Kerr black hole with M_8 = 1.2 and L/L_{Edd} = 0.3. From top to bottom, the curves correspond to the rings R/Rg = 1.5, 2, 2.5, 3, 3.5, 4, 5, 6, 7, 8, 9, 10, 12, 14, 16, 18, 20, 25, 30, 40, 50, 60, 70, 80, 90, 100, 120, 140, 160, 180 and 200. Thin lines represent the model without comptonization, and thick lines the model with comptonization. Credit: Hubeny et al. (2000)*

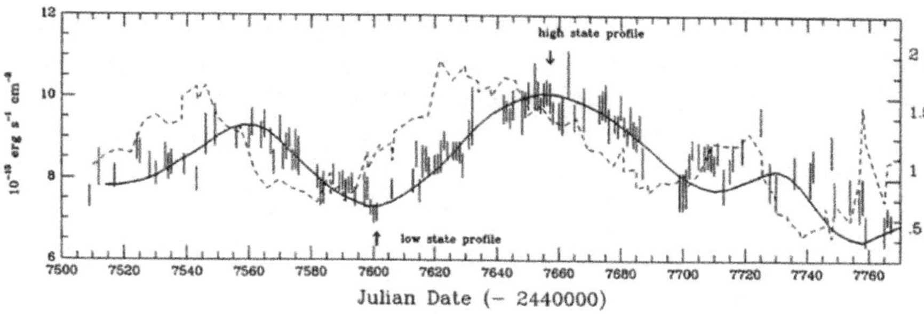

Figure 1.29. *Calculated light curve of Hβ (solid line) and compared to observations (bars) for an X-ray curve (dotted line) in phase with UV, with the same amplitude and variance. The right scale gives the values of X-ray flux (Rokaki et al. 1993)*

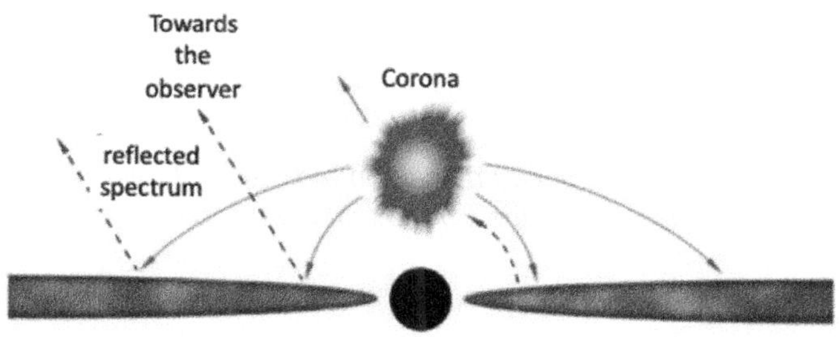

Figure 1.30. *The corona seen directly and by reflection on the disk. Note that the corona is probably in contact with the disk. For a color version of this figure, see www.iste.co.uk/combes/nuclei.zip*

The disk is therefore irradiated with X-rays, which completely changes the structure of its atmosphere and even its entire vertical structure, if the energy provided by irradiation is greater than that of the viscosity. Irradiation is taken into account by the Titan code (Dumont *et al.* 2000), but in an approximate way with regard to comptonization. Ross and Fabian (2005) developed a code that takes into account the presence of an underlying black body representing the radiation from the disk, and comptonization by a transfer equation (Kompanetz). They calculated a grid of models that could be used at different points on an irradiated disk to obtain indications concerning the parameters of the observed objects. Figures 1.32–1.34 show the calculated spectra for some of these models, where a bump appears around 30 keV and the Kα line of iron. The intensity of the bump increases with the ionization parameter, and the iron line loses its strong ionization components around

7 keV when the X-ray spectral index is high and the hard X-ray radiation disappears. These spectra are, for some, at least, in agreement with the spectral index observed around 10 eV (note that they are given in EF(E) and not in F(E)!).

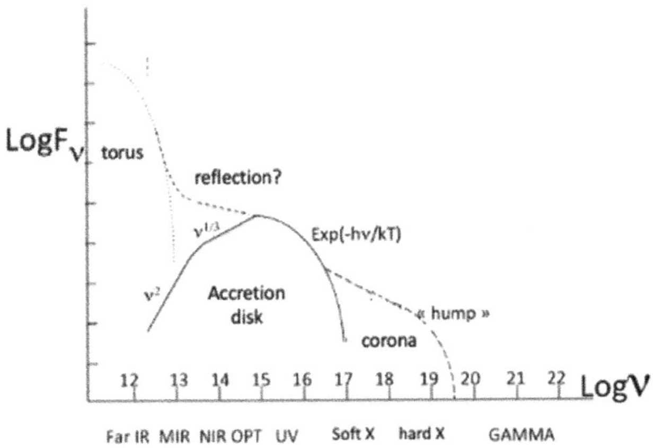

Figure 1.31. *Spectrum of a non-radio AGN-1, taking into account the corona and the reflection*

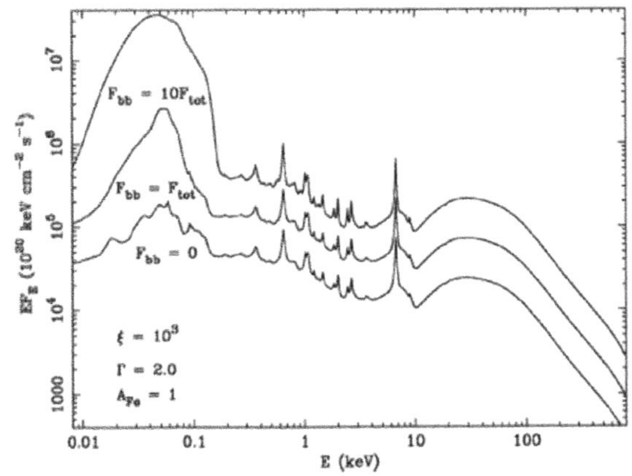

Figure 1.32. *Emerging and reflected spectra, with a black body of temperature 10 eV injected at the bottom of the atmosphere, for 3 flux values. The ionization parameter (external) ξ is equal to 10^3 erg cm s^{-1}, the corresponding spectral index in photon is equal to 2 (therefore the frequency index defined at the beginning of the chapter is equal to 1), and the iron abundance is normal. The spectra have been shifted by a factor of 3 for a clearer presentation (Ross and Fabian 2005)*

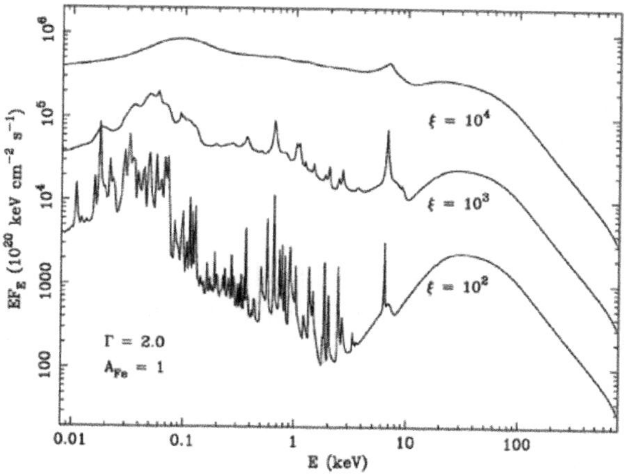

Figure 1.33. *Reflected spectrum, for three ionization parameter values ξ. The spectral index in photon is equal to 2, and the abundance of iron is normal (Ross and Fabian 2005)*

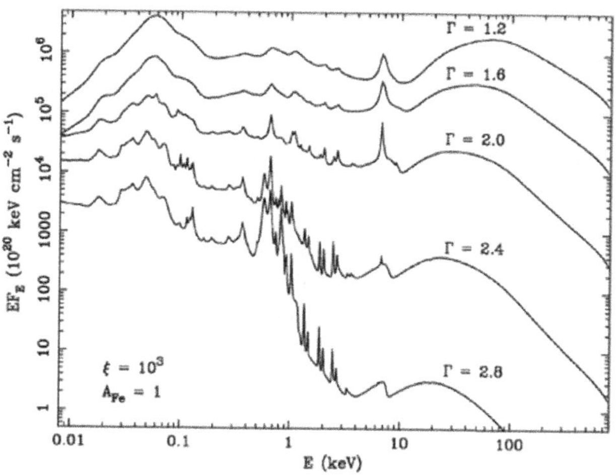

Figure 1.34. *Reflected spectrum, for five spectral index values in photons, the ionization parameter ξ is equal to 10^3 erg cm s^{-1} and the abundance of iron is normal. The spectra have been shifted by a factor of 5 for a clearer presentation (Ross and Fabian 2005)*

1.4.3. *BLR and NLR*

Much information has already been given about these components, but very little has been given on their motions and physical nature. A fundamental problem is that of their structure, clouds or continuous flow. At present, it is the clouds that are focused on for the BLR, and the flow for the NLR.

The issue with clouds is that of confinement. It can be gravitational or magnetic. Magnetic confinement requires a magnetic field of the order of one Gauss. For now, there is no indication of this. Regarding gravitational confinement, it was thought for a long time that the BLR could consist of atmospheres of giant stars inflated by ultraviolet radiation from the central source, known as "bloated stars", forming somehow self-gravitating clouds. This assumption was eliminated because it corresponded to a large number of clouds (10^{6-8}), while the lines are relatively smooth and do not allow many components. Dumont and Collin-Souffrin (1990 and previous articles) proposed that the BLR consists of the irradiated surface of the accretion disk, so that the profile of the lines is due to rotation; but this assumption is only valid for a radius smaller than that at which the gravitational instability of the disk is established. On the other hand, we will see later that wind can be produced on the surface of the disk, and, in this case, the motion will be a mixture of rotation and ejection. It is probable that the BLR participates in such a wind.

Confinement is resolved by assuming that the BLR clouds are located within a very hot and dilute medium of the same pressure (Krolik *et al.* 1981). A thermal instability proposed by Fields (1965) thus explains the presence of dense and cold interstellar clouds in a hot environment.

As the ionization parameter U is proportional to $L_{ion}/(cr^2n)$, that is, at P_{rad}/n, where P_{rad} is the radiation pressure, we find that the equilibrium between the gas pressures nT of a hot medium and a cold medium, photoionized and located at the same distance from an ionizing source, is carried out if $U_{hot}/T_{hot} = U_{cold}/T_{cold} = P_{rad}/P_{gas}$. We can calculate T as a function of U/T assuming equilibrium between the losses and the gains of the medium. We then obtain a curve which possibly presents points for which U/T is the same for several values of T: this means that there are several media in pressure equilibrium at different temperatures (and therefore also densities). Such curves (called stability curves) are given in Figure 1.35, for different ionizing continuums. On the right of the curves, the heating outweighs the losses, on the left it is the opposite. We see that in the parts where the curve has a negative slope, if we increase T, the medium will tend to heat even more, and if we decrease T, it will cool more. These parts of the curve therefore correspond to unstable phases which cannot exist. Where the curve has a positive slope, the medium is stable. Two (or three) stable phases can coexist in pressure equilibrium[7]. In the present case of heating by an ionizing continuum, the only constraint is that it can raise the temperature of the diluted medium above 10^6 degrees. Such a model involves permanent exchanges between hot and cold media, triggered by changes in ionizing continuum.

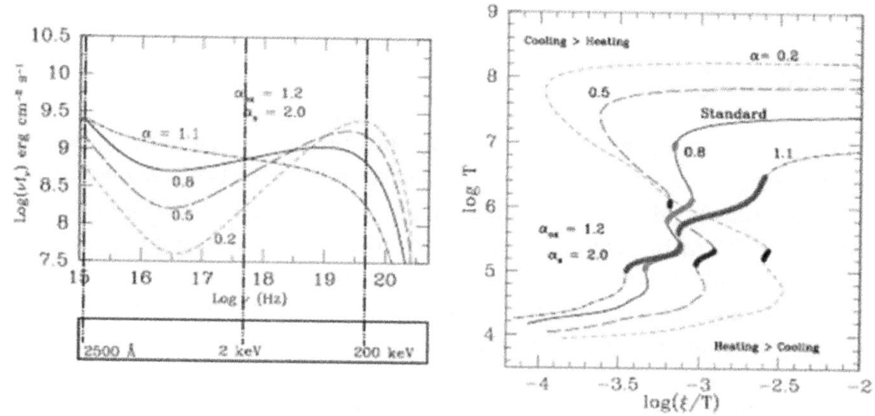

Figure 1.35. *Different spectral energy distributions (SED) and corresponding stability curves, for different slopes of the X-ray continuum and different slopes between the UV and X-ray domain. The left figure gives the SEDs, with α_{ox} values between 2,500 A and 2 keV. The right figure gives the corresponding stability curves. The colors black on green, blue on red, gray on black and cyan on magenta allow the stable phases to be easily visualized (Chakravorty et al. 2009). For a color version of this figure, see www.iste.co.uk/combes/nuclei.zip*

Other possibilities have been suggested, such as that of a wind containing dust and rising above the disk because of the radiation pressure, then falling back ballistically after the sublimation of the dust, as a kind of "failed wind" (Czerny and Hryniewicz 2011; Czerny *et al.* 2017). Magnetic confinement has also been considered, and acceleration models have been proposed for the BLR, BALs and the WA. Finally, we should mention the locally optimally emitting cloud (LOC) hypothesis of Baldwin *et al.* (1995), proposing that clouds of different densities exist everywhere, but that only those which favor the emission of lines are observable.

7. This curve can also be formed using the ratio ξ/T instead of U/T.

The situation is different for the NLR. The forbidden lines [OIII] 4959–5007 often show an asymmetry towards the blue, indicating an outflow. This depends on the L/L_{Edd} ratio which, as we will see later, is the fundamental parameter characterizing an AGN. With the development of multi-image spectroscopy and the increase in spatial resolution, NLR begins to be mapped in nearby AGNs, and it appears to participate in the outflow.

1.4.4. *The dusty torus*

Recent interferometric observations show that the torus has a radius of less than 1 or 10 parsecs, smaller than what was estimated in the past (Burtscher *et al.* 2013). Since it completely obscures the central regions, it is deduced that its column density is between 10^{22} and 10^{25} cm^{-2}. If its density is intermediate between the BLR and the NLR, from 10^4 to 10^7 cm^{-3}, this means that its thickness can take intermediate values between 10^{15} and 10^{21} cm. For small values, it must consist of small clouds, and for large ones, it could be a continuous medium.

The question has been discussed a lot, and infrared emission can help answer it. Indeed, in the case of a continuous environment, the temperature of the dust should be higher inside than outside. Therefore, the flux in the MIR band should be higher at high energy in AGN-1s, which are seen face-on, than in AGN-2s, which are seen edge-on. This does not appear to be the case. High silicate absorption, which is not observed, would also be expected in the MIR emission of AGN-2. On the contrary, in the case of a medium made up of clouds, the surface of each cloud would receive roughly the same amount of energy from the central source, which would explain that we observe a comparable MIR spectral distribution in AGN-1 and AGN-2. On the other hand, the models accounting for the ratio of Fe Kα and Si Kα fluorescence lines due to the torus in AGN-2 also seem to prove that it is formed of dense clouds (Liu *et al.* 2016).

The opening angle of the torus can be deduced from the fraction of the luminosity emitted in the infrared compared to the bolometric luminosity. Several authors agree on a fraction between 30% and 50%. The intense Fe Kα line in many AGN-2s comes from the reflection of the central source on the torus. A Baldwin effect (decrease in the equivalent width of the line when the luminosity increases) would be observed, showing the decrease in the coverage factor of the torus when the luminosity increases. The size of the torus could be estimated in a few cases by a method similar to RM, showing that it was of the order of a few times the size of the BLR, which corresponds to the minimum size necessary to explain the disappearance of the broad lines in AGN-2. Furthermore, the internal radius of the torus is determined by the distance from the center at which the dust is sublimated, that is, 1,500 degrees for graphite grains. This temperature corresponds to a sharp emission at 2 microns, responsible for the rise in the infrared of the spectral distribution of non-radio AGNs. For an ultraviolet luminosity of 10^{46} ergs/s, this radius is one parsec. This distance is not far from the limit of the accretion disk. A fairly natural idea is therefore that the torus is at the origin of the flow which originates from the inner regions of the host galaxy and which supplies the disk with gas. It can be noted that there does not seem to be any difference in the properties of the torus in radio and non-radio AGNs, which corroborates the idea that it does not depend on the properties of the more internal regions determining the presence of the jet, but on external regions linked to the host galaxy.

Finally, a possibility that was mentioned in the past but which is no longer considered is that of a warped disk which would hide part of the central region in certain directions. Such a disk is observed in the galaxy NGC 4258, the AGN where the mass of the black hole has been determined with the greatest precision, but it is very thin and the probability of seeing the central regions through it is weak, and moreover it would not emit enough infrared.

1.4.5. *Hot environment: wind*

The velocities observed in both BALs and WA indicate that it is a wind. This can be created by thermal pressure, by radiation pressure or by magnetic forces. To successfully escape the environment of the black hole, it must exceed the escape velocity $v_{esc} = (2GM_{BH}/r)^{1/2}$. The closer it is created to the black hole, the higher the escape velocity, so the more efficient the gas acceleration mechanism will have to be.

Consider first the thermal pressure. A disk in hydrostatic equilibrium can lose its mass if its atmosphere is irradiated, which is indeed the case for quasars and Seyfert galaxies. X-rays heat the gas via Compton diffusions to a temperature of 10^7 K (called the Compton temperature). A wind occurs if the thermal velocity corresponding to this temperature exceeds the escape velocity. As v_{esc} is a function of the radius, the mechanism will be effective only at large radii. We therefore expect low velocity winds causing absorptions in the X-ray continuum. The mechanism is therefore not sufficient to explain all the observations.

Radiation pressure acceleration can act in a cooler gas and cause ultraviolet absorptions. Radiation pressure can indeed overcome gravity even for $L << L_{Edd}$

provided that the opacity of the gas is greater than that of Thomson or Compton scattering. This is the case if $T < 10^5$ K, because the gas is not completely ionized and it contains ions with strong resonance transitions or ionization thresholds capable of absorbing the radiation and the corresponding kinetic energy. The opacity can thus be multiplied by a factor of 10^3 compared to a fully ionized gas (this is also called the "multiplier factor"). We know that such transitions exist since they are observed in BALs and WA. Murray *et al.* (1995) therefore proposed that wind is launched from the disk by the radiation pressure, at a radius where the disk radiates into the UV domain. The problem is that the wind density is low and the central X-ray source will easily bring it up to Compton temperature, thus removing all absorbers. To avoid this, Murray *et al.* suggested the existence of an interposing medium between the wind and the central source, which acts as a shield for X-rays. More elaborate simulations then built on the same principle (see Proga *et al.* (2000) and Proga (2007)) showed that the shield is still necessary for acceleration to occur and for it to create absorption lines in the ultraviolet. Another problem with radiative acceleration is that the wind direction is tilted with respect to the plane of the disk, limiting BALs to an angle where the torus should prevent viewing or simply stop the wind. In fact, it seems that radiation pressure acceleration may not be suitable for all types of AGN, and that we should probably look to magnetic acceleration in other cases.

Magneto-rotational instability (Balbus and Hawley 1991; see Chapter 3) is probably the mechanism by which the turbulence necessary for accretion is produced in the disk. Presumably, the magnetic field also plays an important role in the ejection of gas from the disk. The advantage of these models is that the magnetic field can provide confinement, so the medium can be made up of small dense clouds, as in the BLR. A characteristic of magneto-centrifugal acceleration is that it creates a wind co-rotating with the disk. An important parameter is the ratio between the rotational velocity and the expansion velocity. The angular momentum is conserved when there is no co-rotation, while the rotational velocity is conserved if the gas is co-rotating, so the latter has a higher terminal velocity because of the centrifugal force. We can therefore hope to distinguish between the two types of winds.

Ultimately, if the wind comes from the disk, it seems that at least two forms of acceleration, radiation pressure and magnetic acceleration, must play a role, and we are not yet close to the solution. Another possibility is that the wind is coming from the torus, as the opacity of the dust can provide an acceleration mechanism. The wind would then be ejected by evaporation from the torus heated by the radiation from the central source. In this case, we would expect a correlation between reddening and wind speed, but at the current moment, there is no evidence of this.

The WA also appears by the numerous emission lines observed in the X-ray domain in the Seyfert 2 galaxy due to the absence of an underlying continuum (Figure 1.16 at the beginning of the chapter and Figure 1.36). They do not allow us to call it a wind, nor can it be deduced from the lines of the BLR. They have been used to determine the physical conditions of the emitting medium, in particular through the intensity ratios of the helium triplets (Kimkhabala *et al.* 2002). However, in warm and relatively thick environments like this, these ratios can be an order of magnitude wrong when using the escape probabilities to calculate them (see Box 1.1 on radiation transfer and Figure 1.37).

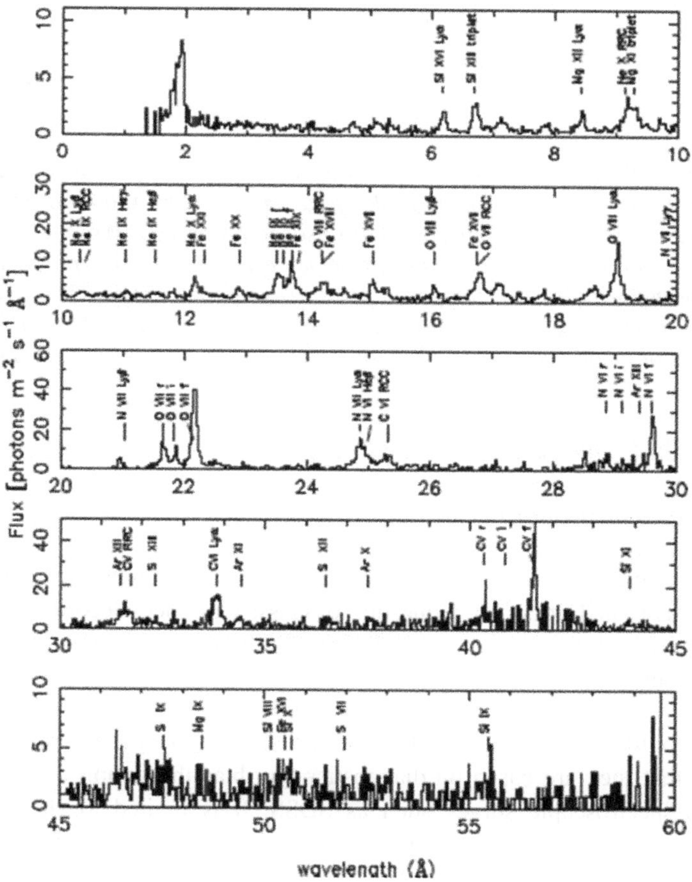

Figure 1.36. *X-ray spectrum of Seyfert 2 galaxy NGC 1068. We can distinguish strong emission lines, observable thanks to the absence of an underlying continuum. Credit: Brinkman et al. (2002)*

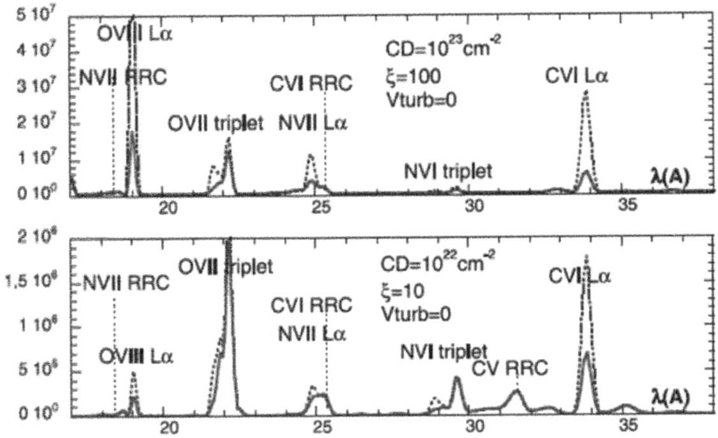

Figure 1.37. *Reflected spectrum for two models corresponding to the hot medium in the Seyfert 2, one in red, calculated taking into account the radiative transfer, the other in black, with the approximation of the escape probability. We see that the line ratios of the CVI and OVII helium triplets calculated with the escape probabilities can be false by a factor of 10 compared to the exact calculation. The ordinate is proportional to* νF_ν. *The spectra are shown with a spectral resolution of 30 (Collin et al. 2004). For a color version of this figure, see www.iste.co.uk/combes/nuclei.zip*

1.5. Towards a unified vision of AGNs

1.5.1. *General information*

In the previous sections, a series of objects and a large number of observed data have been mentioned. It is now a question of classifying them according to a small number of physical parameters. Table 1.1 recalls the different types of AGNs that we have encountered.

A major hurdle when unifying AGNs, especially with the generalization of large surveys, is that they are based on diagnostics that bias the samples. Consider, for example, a spectroscopic survey such as SDSS. It is limited to small redshifts because, for $z > 0.4$, the Hα line passes in an unobservable domain, even if the objects are luminous. Furthermore, AGNs with weak lines (such as LINERS) are hardly detected. An interesting comparison was made regarding the classification as AGN-1 or AGN-2 on the basis of their X-ray or optical properties (Merloni *et al.* 2014; see Figure 1.38). It shows that 30% of AGNs are classified differently by the two methods. This is why we should not neglect the theoretical approach, which complements the often-biased empirical approach, in the hope that the two converge towards a common solution.

Type	Acronym Meaning	Properties
Quasar	QUASi-stellAR objet	Radio or non-radio
RQ AGN	Radio Quiet AGN	AGN silent in the radio domain
Sey1	Seyfert 1	FWHM $\geq$ 1,000 km/s
Sey2	Seyfert 2	FWHM < 1,000 km/s
ULIRG	Ultra Luminous Infrared Galaxy	Quasars coated in dust
QSO2	Quasi-Stellar Object 2	Powerful Sey2
Type 1 AGN	Sey1 and quasars	FWHM $\geq$ 1,000 km/s
Type 2 AGN	Sey2 and QSO2	FWHM < 1 000 km/s
BAL	Broad Absorption Line quasar	Broad absorption lines
Non-jetted AGN	Jetless AGN	Without strongly relativistic jet
LINER	Low Ionization Nuclear Emission Region	Low ionization lines
NLS1	Narrow Line Seyfert 1	Seyfert 1 lines, but straight
RL AGN	Radio Loud AGN	Radio quasar or powerful radio galaxy
Jetted AGN	AGN with jet	Strongly relativistic jet
BLRG	Broad-line radio galaxy	Sey 1 radio
FR I	Fanaroff-Riley type I	Edge-darkened radio source
FR II	Fanaroff-Riley type II	Edge-brightened radio source
FSRQ	Flat Spectrum Radio Quasar	Radio AGN, $_{R} < 0.5$
SSRQ	Steep Spectrum Radio Quasar	Radio AGN, $_{R} > 0.5$
BL Lac	BL Lacertae	Quasar with weak lines
Blazar	BL Lac and Quasar	BL Lac and FSRQ, lines present

Table 1.1. *Different types of AGNs*

When Urry and Padovani published their immediately famous "Unified Scheme" in 1995 (see Figure 1.39), experts believed that a definite step was taken in understanding AGNs. It made it possible to understand not only the separation between AGN-1 and AGN-2, but also that of radio AGNs, between normal radio quasars and blazars, where the relativistic jet points in a direction close to the line of sight.

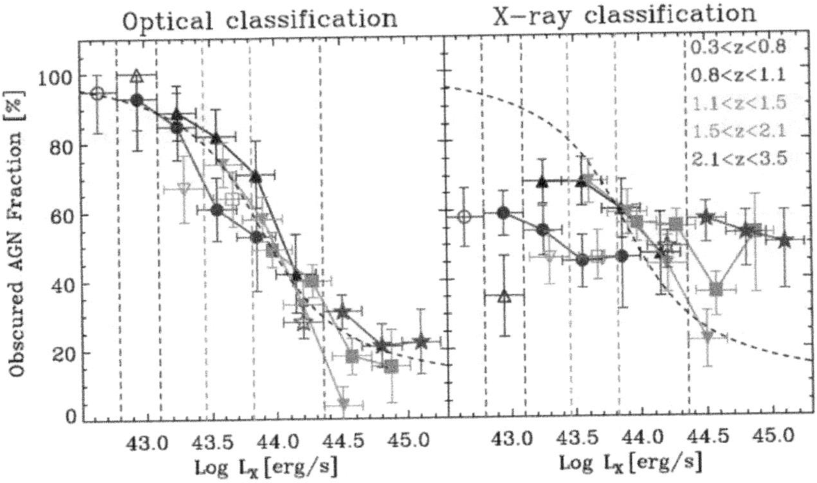

Figure 1.38. *The figure on the left gives the fraction of AGN-2 determined by optical classification, as a function of the X-luminosity, for different redshifts. The vertical dashed lines give the luminosities above which the samples are complete for each redshift. The dashed line shows the best fit for all redshifts. The figure on the right gives the fraction of AGN-2 determined by an X classification. We see that the two classifications give different results (source: Merloni et al. 2014)). For a color version of this figure, see www.iste.co.uk/combes/nuclei.zip*

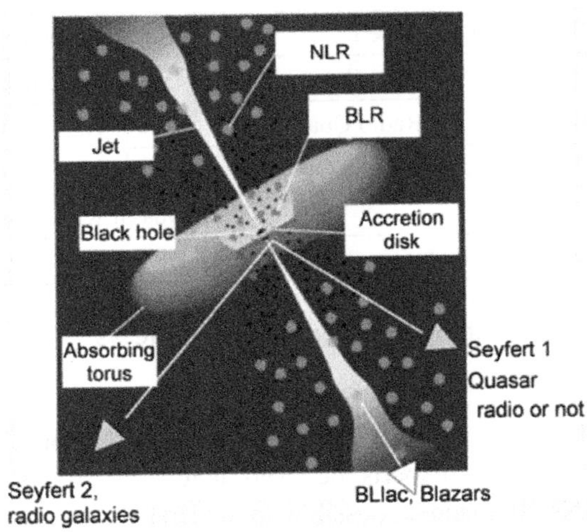

Figure 1.39. *The unified scheme based on the direction of observation proposed in 1995 by Urry and Padovani. For a color version of this figure, see www.iste.co.uk/combes/nuclei.zip*

Unfortunately, that was nowhere near enough, since recent observations of large samples of AGNs and detailed information obtained from some sources have led to the conclusion that this pattern requires major modifications. While highlighting the influence of the viewing angle, it neglects several other important factors: the Eddington ratio, the mass of the black hole, its rotation, the accretion rate and the nature of the host galaxy and its environment.

There are two ways of approaching the problem: the theoretical approach and the empirical approach. Both have their flaws and their advantages. The first allows a physical model to account for the observations, but this model can be incorrect, at least in part. The second makes it possible to find real links between observations, but does not necessarily succeed in explaining them.

1.5.2. *Theoretical approach*

This is the approach proposed by Giustini and Proga (2019). It is summarized in Table 1.2, which gives the main theoretical characteristics of AGNs as a function of the Eddington ratio L/L_{Edd} (or dot(m)). We can see its influence on distinguishing between radio/non-radio, or on the kinetic L_{kin} or radiative L_{rad} energy emitted, as well as on the intensity and geometry of the winds in non-radio AGNs.

It is based on the accretion disk models as a function of mass and accretion rate (described in Chapter 3), as well as on current theories concerning ejected winds discussed previously, and on the observations of BALs or WA which were not completely detailed. The easiest way is to show their model. Even if it is not entirely right, we might think that it is not completely wrong, and it is difficult to go beyond it now (see Figure 1.40).

$\dot{m}$ range (1)	Accretion/ejection flow (2)	Feedback (3)	Examples (4)
very low $\dot{m} \approx 10^{-8}$ ($\ll 10^{-6}$)	non-radiative hot accretion flow relativistic polar jet	L_{kin}	Quiescent/inactive, Sgr A*
low $\dot{m} \approx 10^{-4}$ ($10^{-6} \lesssim \dot{m} \lesssim 10^{-3}$)	outer cold disk at ~1000s R_g, inner hot flow relativistic polar jet	$L_{kin} \gg L_{rad}$	LLAGN M81*, M87
moderate $\dot{m} \approx 10^{-2}$ ($10^{-3} \lesssim \dot{m} \lesssim 10^{-1}$)	outer cold disk at ~10s R_g, extended hot corona weak/moderate LD wind depending on small/large M_{BH}	$L_{kin} \ll L_{rad}$	Seyfert/mini-BAL QSO NGC 5548/PG 1126-041
high $\dot{m} \gtrsim 0.25$ ($0.1 \lesssim \dot{m} \lesssim 1$)	cold accretion disk down to ISCO, compact hot corona moderate/strong LD wind depending on small/large M_{BH}	$L_{kin} < L_{rad}$	NLS1/BAL QSO I Zw 1/PDS 456
very high $\dot{m} \gg 1$ ($1 \lesssim \dot{m} \lesssim 100$)	outer thin disk, inner slim disk, very compact hot corona strong outflows, both polar and equatorial	$L_{kin} \lesssim L_{rad}$	Super-Eddington RX J0439.6-531

Table 1.2. *Main theoretical characteristics of AGNs as a function of the Eddington ratio L/L_{Edd} (source: Giustini and Proga (2019))*

When L/L_{Edd} is very low, the radiative efficiency η is very low, with $M'/M'_{Edd} >> L/L_{Edd}$. The accretion disk is absent, and the black hole is surrounded by a very hot gas emitting weak X-rays. Relativistic electron jets are accelerated in the tunnel in the center, but the kinetic energy released is low. The galaxies are not active.

For $10^{-5} < L/L_{Edd} < 10^{-3}$, the accretion disk is reduced to an outer part which radiates very little, and the black hole is surrounded by a very hot gas emitting in the X-ray domain, with a low component in the UV–visible domain. A polar relativistic jet is produced and carries kinetic energy by creating FRII and FRI radio sources. We then have a low luminosity AGN (LLAGN), in the first case, or a LINER exhibiting low ionization lines, possibly due to shocks, in the second case.

For $10^{-3} < L/L_{Edd} < 0.1$, the efficiency is around 0.1, and we have $M'/M'_{Edd} = L/L_{Edd}$. We are dealing with a quasar or a Seyfert galaxy, fully described in this chapter. The maximum emission occurs in the UV–visible domain for $L/L_{Edd} \sim 10^{-3}$ and shifts towards soft X-rays when L/L_{Edd} increases. The central corona radiates into the hard X-rays and irradiates the disk. The ratio of UV to hard X-rays increases as luminosity increases. We start to see BALs and a WA which are larger as the mass of the black hole increases (they are proportional to the rate of accretion).

For $0.1 < L/L_{Edd} \leq 1$, the efficiency η starts to fall below 0.1. This is the case of NLS1s for small masses, and luminous quasars for large masses, with powerful winds. The spectral distribution includes strong UV radiation compared to X-ray radiation.

For $L/L_{Edd} > 1$, this is the "super-Eddington" case for which the radiative efficiency η is very low and $M'/M'_{Edd} >> 1$. Strong winds are emitted in all directions. UV radiation is intense compared to X-ray radiation.

Of course, we should not forget the influence of orientation. It plays a big role in how the winds and jets appear to us. Giustini and Proga have taken care to compare their model in detail with the observations, and propose AGN-types for each case.

Let us cite yet another example where the theoretical approach agrees with observations: the influence of the black hole rotation on the radio/non-radio distinction. The Blandford–Znajek effect produces relativistic particle jets by converting the black hole rotation via the magnetic field into relativistic particle energy (see Chapter 3). It will be shown, in Chapter 5, that the formation of pairs of supermassive black holes and their mergers forming a rapidly rotating black hole could explain why relativistic jets are preferentially found in elliptical galaxies, often resulting from the major merger of spiral galaxies.

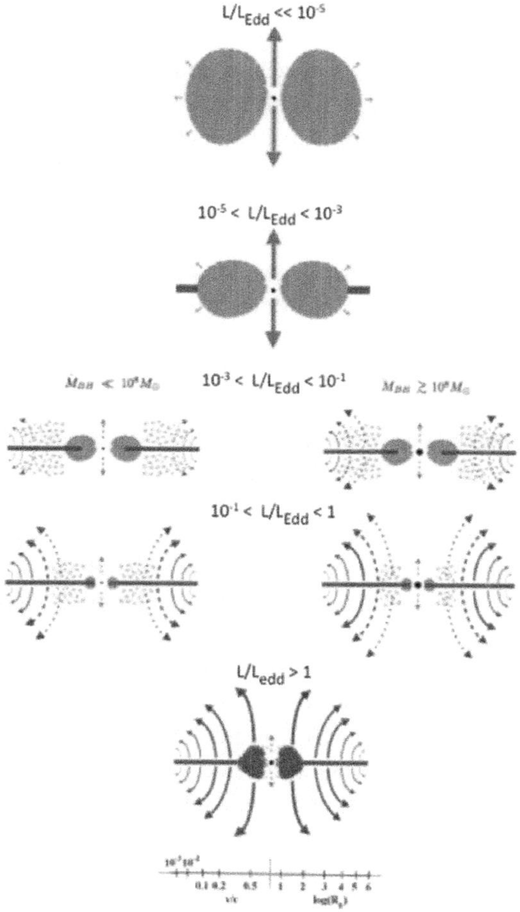

Figure 1.40. *Logarithmic view of the interior parsec of the accretion/ejection flow for the five L/L_{Edd} regimes. The warm, optically thin medium is colored red, and the optically thick medium in green. The wind coming from the disk and the failed wind are in blue. For the optically thick and geometrically thin disk, two extremes are represented: on the left, the case of a very small mass, on the right, that of a very large mass. Arrows represent streamlines, with thickness and length reflecting their intensity and terminal velocity. Solid streamlines represent persistent winds, and dashed and dotted lines represent transient or sporadic winds, while small meandering clouds represent completely failed wind streamlines. The outermost streamlines correspond to thermally or magnetically driven winds from a great distance, from 10^4 R_G to 10^5 R_G, and reach terminal velocities of 100–1,000 km/s. Inner streamlines are winds blown magnetically or by radiation pressure from a small distance, less than 10^3 R_G, and reach velocities of 10,000–100,000 km/s (Giustini and Proga 2019). For a color version of this figure, see www.iste.co.uk/combes/nuclei.zip*

1.5.3. *Empirical approach*

An example of an empirical approach is that of Sulentic *et al.* (2001 and subsequent articles): they looked for the main parameter hiding behind the eigenvector E1 deduced from the principal component analysis (see section 2.1.3.). They thus identified the following characteristics defining vector E1: (1) the velocity dispersion in the BLR, (2) the intensity of the weak ionization lines in the BLR, (3) the intensity of the excess soft X-rays ("Big Blue Bump") and (4) the amplitude of systematic motions in high ionization gas. This led them to separate AGN-1s into two classes: Population A, with FWHM (broad Hβ) ≤ 4,000 km/s and average values of the parameter E1: R(FeII) = W(FeII λ4570)/W(broad Hβ) ≈ 0.7, a large spectral index of the soft X-ray spectrum ≈ 2.8, and a shift of 800 km/s towards the blue of CIV1549; and Population B, with FWHM (broad Hβ) ≥ 4,000 km/s and mean values of the parameter E1: R(FeII)=0.4, a small spectral index of the soft X-ray spectrum ≈ 2.3, and no shift in the line CIV1549.

However, the interpretation of these results is not obvious. If we can understand why the soft X-ray spectral index is related to the shift of the CIV1549 line – for example, more X-ray radiation is absorbed when the wind is directed towards the observer – we do not know why the FeII lines are more intense in this case. In some objects (the most extreme NLR1), R(FeII) is much greater than 1 (see Figure 1.41), which is difficult to explain in the context of photoionization, and leads us to suggest an additional mechanism of BLR heating, which could be due to an intense star formation, especially as the NLS1 accretes at very high rates for low-mass black holes (Collin and Joly 2004). A strong overabundance of iron does not solve the problem, because it leads to the cooling of the medium which partially cancels out the effect of the overabundance. In addition, the iron lines in the X-ray domain do not seem to show a strong overabundance (although the abundance is difficult to estimate given the lack of knowledge of parameters such as the proportion of reflected radiation).

Via a semi-empirical approach using models calculated by Cloudy, Panda *et al.* (2019) have shown that the main parameter is the L/L_{Edd} ratio, provided that the orientation, shape and density of the BLR are taken into account, and that the sources with R(FeII) > 1 correspond to high densities, to an overabundance of iron and to an L/L_{Edd} ratio of 1. This is great progress, but the cause of these properties remains unidentified, and the iron problem, in particular, is not solved for extreme NLS1s.

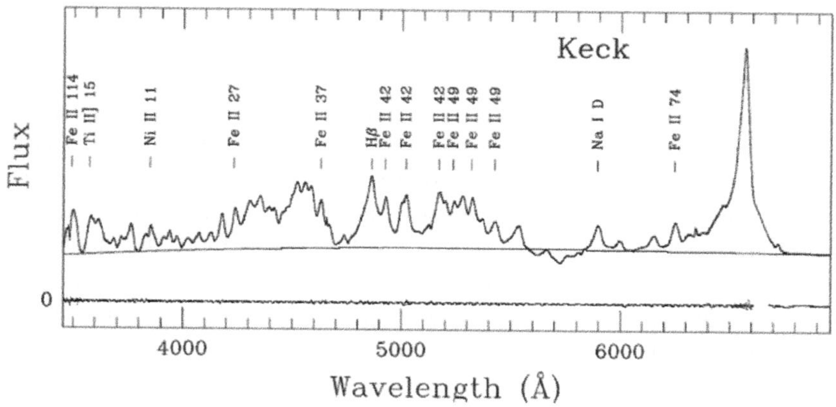

Figure 1.41. *NLS1 IRAS 07598+6508 spectrum corrected for redshift, with its adjustment and residues. The main lines have been identified: we see that the Hβ line is very weak compared to all the iron lines (Veron-Cetty et al. 2006)*

1.5.4. *Conclusion*

Ultimately, if we want to summarize this chapter in a few sentences, we can say that a fundamental mechanism in non-radio AGNs is the presence of winds carrying non-negligible kinetic energy in the galaxy, despite relatively low velocities (at least compared to those of the relativistic jets of radio AGNs), due to the high rate of ejected mass. All the environments that we observe, through BALs, NALs, X-rays, NLRs, contribute to this wind, with the possible exception of a part of the BLR. These winds undoubtedly have a significant impact on galaxies, as Chapter 5 will show.

The fact remains that AGNs are incredibly complex objects and there are still many unsolved matters regarding their physics. Many of them are likely to be solved in the coming years. The interferometry techniques now in place on large telescopes will allow their immediate environment to be studied; the multiplication of large surveys and the commissioning of extremely large optical (ELT) and radio (SKA, CTA) telescopes will make it possible to observe AGNs at high redshifts, to follow their cosmic evolution and determine their impact on the evolution of galaxies; the development of hydrodynamic and magnetohydrodynamic simulations will make it possible to better identify the accretion and ejection flows, as well as the accretion mechanisms, etc.

This is probably the most delicate point regarding emitting regions in AGNs. A number of codes had been constructed in the past to calculate the structure and emission of planetary nebulae and HII regions, that is, dilute and optically thin media (except in the Lyman continuum) ionized by the radiation of a hot star. At the end of the 1970s, observations of compact X-ray sources, implying the presence of a thick and hot medium, prompted the development of transfer codes using an approximation, the escape probability, which allows diffusion and absorption to be taken into account. It has been widely used for BLR in lines and sometimes also in the continuum (Ross 1979; Kwan and Krolik 1981; Kallman and McCray 1982). The most popular and comprehensive of these codes is Cloudy, designed by Ferland and regularly updated (see Ferland 1996). These codes are not only used for relatively cold clouds like BLR, but also for hot photoionized media (accretion disk, BALs and WA). We are going to outline this method.

The transfer equation takes the form

$$\frac{dI_\nu}{d\tau_\nu^{tot}} = I_\nu - S_\nu \text{ where } d\tau_\nu^{tot} = -\kappa_\nu^{tot}\, dl,$$

dl being the length of the element traversed in the direction of radiation, κ_ν^{tot} the absorption coefficient taking into account all sources of opacity, including scattering. The challenge of the transfer is that it is an integrodifferential equation, because we have seen that the specific intensity is present in the source function. Let us consider the case of a homogeneous layer with a flat and infinite surface, the angle between the normal to the surface and the light rays being θ. The formal solution of the transfer equation becomes:

$$\left[I_\nu \exp\left(-\frac{\tau_\nu}{\mu}\right)\right]_{\tau_{\nu 1}}^{\tau_{\nu 2}} = -\int_{\tau_{\nu 1}}^{\tau_{\nu 2}} S_\nu \exp\left(-\frac{t_\nu}{\mu}\right)\frac{dt_\nu}{\mu},$$

where we set $\mu = \cos\theta$, $\tau_{\nu 1}$ and $\tau_{\nu 2}$ as the optical depths at the entry and exit of the layer. We are going to assume that there is no matter between the source and the studied medium (this is the case of the BLR and the NLR); therefore, $\tau_{\nu 1} = 0$, and no radiation entering at the line frequency. By integrating this equation on the angles, we obtain for the outgoing flow $F_\nu = 2\pi S_\nu[0.5 - E_3(T_\nu)]$, where E_3 is the integro-exponential function of order 3, and T_ν the optical thickness of the layer. We have two extreme cases:

1) The medium is optically thin at the frequency ν.

This gives: $F_\nu = 2\pi S_\nu T_\nu = 2\pi\varepsilon_\nu H$; hence, $L_\nu = 4\pi\varepsilon_\nu \times V$, where L_ν is the luminosity of the layer (assuming the surface is finite), H is the thickness of the layer and V is the emissive volume.

2) The medium is optically thick at the frequency ν; therefore, $F_\nu = \pi S_\nu$ and $L_\nu = \pi S_\nu \times A$, where A is the emitting surface.

This can be summed up as follows: in the case of a thin layer, we "see" all the emissivity, and the power is proportional to the volume, while in the case of a thick layer, we "see" the source function, and the power is proportional to the area.

The "thin" case is that of the NLR. Only the Lyman continuum is optically thick, which brings us back to the classic case B of nebulae. For a transition between two levels m and n, we will therefore simply have, once the statistical equilibrium equation of level m is solved: $L_{line} = 4\pi N_m A_{mn} E_{mn} V$. In the case of the BLR, many permitted lines are optically thick, meaning that each point in a layer receives and absorbs photons from the lower and upper layers. In addition, the wings of the lines are generally optically thin, while the center is optically thick. It is therefore necessary to calculate the source function at any point of the layer and at any point of the spectral line! To get around this hurdle, "escape probabilities" are used. Consider the variable:

$$\rho_{mn} = \frac{N_m (B_{mn}\int J_\nu \phi_\nu d\nu + A_{mn}) - N_n (B_{nm}\int J_\nu \phi_\nu d\nu)}{N_m A_{mn}},$$

called "net radiative rate". It measures the balance of radiative transitions [m → n – n → m] compared to spontaneous radiative transitions. We can identify the integrated emissivity on the spectral line at $N_m \rho_{mn} A_{mn} E_{mn}$, so we simply write that $L_{line} = 4\pi N_m \rho_{mn} A_{mn} E_{mn} \times V$, which comes down to the optically thin case.

Using the source function (assumed constant in the line, assumption of the identity of the emission and absorption profiles), we can write the following: $\rho_{mn} = 1 - [\int J_\nu \phi_\nu d\nu / S(line)]$ which shows that ρ_{mn} is always less than 1 and that when the line becomes very thick (i.e. S(line) and $\int J_\nu \phi_\nu d\nu$ tend towards B), the escape probability tends towards zero.

The definition of ρ_{mn} implies that it is an average variable, assuming a homogeneous layer. However, the layer is inevitably inhomogeneous when it is thick, because of the absorption of radiation in the light path. An improvement in the approximation involves defining a local probability, using the results of the different equilibria calculated at each point. We can even (this is the case with Cloudy, for example) go back and forth between the bottom and the top of the layer, which better takes into account the diffuse flow.

What are the disadvantages of such an approximation? A first drawback is that the approximation for escape probabilities is not well defined, and that various codes use different approximations. For example, Netzer (his 2013 book and previous articles) suggests using, as a "reasonable" hypothesis, $\rho_{mn} = 1/(1+b\tau_{nm})$, where b is a parameter of around 1. The main drawback is that a line emitted at one place in the layer is not always reabsorbed there but in another place, and that this reabsorption will lead to additional heating and possibly ionization at the place where it occurs. Regarding the line flux ratios, Collin-Souffrin *et al.* (1981) showed that the ratio Hα/ Hβ is overestimated by a factor of up to 10 in a calculation using the escape probabilities. Dumont developed a code (Titan) which completely takes into account the transfer (Dumont and Collin 2001). Naturally, this calculation must be carried out to the detriment of atomic data, because the calculation time is considerably lengthened: thus, hydrogen and hydrogen ions are limited to six levels, helium to six levels and lithium to five levels, the higher levels being treated by approximations, to compare them with the dozens of levels present in Cloudy, which is intensively used. This can lead to errors in the intensities of the calculated lines and the conclusions that can be drawn from them, not only for the optical lines, but also in the X-ray domain, as seen in Figure 1.37. Hubeny *et al.* (2000 and previous articles) developed a code for the calculation of the vertical structure of the accretion disk where the transfer of the continuum is taken into account.

Box 1.1. *Radiation transfer*

NOTE.– Astronomers have a bad habit, not only of using different units in each wavelength domain (MHz and GHz in the radio domain, Angströms in the visible or X-ray domain, keV in the X-ray domain, and, in general, CGS units instead of MKS units), but also they do not typically represent the spectra in comparable forms. Thus, we can find F_λ as a function of λ, F_ν as a function of ν or λ or in the X-ray domain the number of photons per second as a function of ν (which brings us back to F_ν/ν), etc. It is therefore necessary to be careful when comparing the energy distributions published in the literature, because their appearance is very different depending on the units used. We see many examples of this in this chapter!

1.6. References

Andrillat, Y. and Souffrin, S. (1968). *Astrophys. Lett.*, 1, 111.

Antonucci, R.R. and Miller, J.S. (1985). *ApJ*, 297, 621.

Arp, H. (1970). *Nature*, 225, 1033.

Baade, W. and Minkowski, R. (1954). *ApJ*, 119, 215.

Balbus, S.A. and Hawley, J.F. (1991). *ApJ Suppt.*, 376, 214.

Baldwin, J.A., Phillips, M.M., Terlevich, R. (1981). *PASP*, 93, 5.

Baldwin, J.A., Ferland, G., Korista, K. *et al.* (1995). *ApJ Lett.*, 455, 119.

Bentz, M.C. Walsh, J.L., Barth, A.J. *et al.* (2010). *ApJ*, 716, 993.

Bentz, M.C., Denney, K.D., Grier, C.J. *et al.* (2013). *ApJ*, 767, 149.

Blandford, R.D. and McKee, C.F. (1982). *ApJ*, 255, 419.

Bon, E., Gavrilovic, N., La Mura, G. *et al.* (2009). *New Astron. Rev.*, 53, 121.

Boroson, T.A. and Green R.F. (1992). *ApJ Suppt.*, 80, 109.

Brinkman, A.C., Kaastra, J.S., van der Meer, R.L.J. *et al.* (2002). *A&A*, 396, 761.

Brown, M.J.I., Duncan, K.J., Landt, H. *et al.* (2015). *Proceedings IAU Symposium No. 341.*

Burbidge, G. (1959). *ApJ*, 129, 849.

Burtscher, L., Meisenheimer, K., Tristram, K.R.W. *et al.* (2013). *A&A*, 558, 149.

Chakravorty, S., Kembhavi, A.K., Elvis, M. *et al.* (2009). *MNRAS*, 393, 83.

Collin, S. and Joly, M. (2000). *New Astron. Rev.*, 44, 531.

Collin, S., Dumont, A.M., Godet, O. (2004). *A&A*, 419, 877.

Collin-Souffrin, S. (1991). *A&A*, 249, 344.

Collin-Souffrin, S., Delache, P., Frisch, H. *et al.* (1981). *A&A*, 104, 264.

Czerny, B. and Hryniewicz, K. (2011). *A&A*, 525, 8.

Czerny, B., Li, Y.-R., Hryniewicz, K. *et al.* (2017). *ApJ*, 846, 154.

Dibai, E. and Pronik, V. (1967). *Astronomicheskii Zhurnal*, 44, 952.

Dumont, A.M. and Collin, S. (2001). *ASP Conference Series*, 247, 231.

Dumont, A.M. and Collin-Souffrin, S. (1990). *A&A*, 229, 313.

Dumont, A.M., Abrassart, A., Collin, S. (2000). *A&A*, 357, 82.

Fanaroff, B.L and Riley, J.M. (1974). *MNRAS*, 167, 31.

Ferland, G.J. (1996). University of Kentucky Internal Report.

Field, G.B. (1985). *ApJ*, 142, 531.

Filiz Ak, N., Brandt, W.N., Hall, P.B. (2018). VizieR on-line data catalog. *ApJ*, 791, 88.

Giustini, M. and Proga, D. (2019). *A&A*, 630, 94. arXiv:1904.07341.

Grupe, D. and Nousek, J.A. (2015). *AJ*, 149, 85.

Haardt, F. and Maraschi, L. (1991). *ApJ Lett.*, 380, 51.

Hubeny, Y., Agol, E., Blaes, O. *et al.* (2000). *ApJ*, 533, 710.

Kallman, T.R. and McCray, R. (1982). *ApJ Suppt.*, 50, 263.

Kinkhabwala, A., Sako, M., Behar, E. *et al.* (2002). *ApJ*, 575, 732.

Krolik, J.H., McKee, C.F., Tarter, C.B. (1981). *ApJ*, 249, 422.

Kwan, J. and Krolik, J.H. (1981). *ApJ*, 250(Part 1), 478.

Laor, A., Fiore, F., Zlivis, M. *et al.* (1997). *ApJ*, 477, 93.

Liu, J., Yuan Liu, Y., Xiaobo, L. *et al.* (2016). *MNRAS Lett.*, 459, 100.

Marziani, P., Dultzin, D., Sulentic, J. *et al.* (2018). *Front. Astron. Space Sci.*, 5, 6.

Mathews, W.G. and Ferland, G.J. (1987). *ApJ*, 323, 456.

Merloni, A., Bongiorno, A., Brusa, M. *et al.* (2014). *MNRAS*, 437, 3550.

Murray, N. and Chiang, J. (1995). *ApJ Lett.*, 454, 105.

Padovani, P. (2017). *Nature Astronomy*, 1, 194.

Padovani, P., Alexander, M., Assef R.J. *et al.* (2017). *A&A Rev.*, November, 25.

Panda, S., Marziani, P., Czerny, B. (2019). *Proceedings of the International Astronomical Union*, 7–11 October, Addis Ababa, Ethiopia, 356, 77. arXiv:1905.01729.

Peterson, B. (2001). *Advanced Lectures on the Starburst-AGN Connection*, Peterson, B.M., Pogge, R.W., Polidan, R.S. (eds). Astronomical Society of the Pacific, San Francisco, CA, USA.

Peterson, B. (2014). *Space Sci. Rev.*, 183, 253.

Proga, D. (2017). The central engine of active galactic nuclei. *ASP*, 267.

Proga, D., Stone, J.M., Kallman, T.R. (2000). *ApJ*, 543, 686.

Rees, M. (1978). *The Observatory*, 98, 210.

Rochais, T.B., DiPompeo, M.A., Myers, A.D. *et al.* (2014). *MNRAS*, 444, 2498.

Rokaki, E. and Magnan, C. (1992). *A&A*, 261, 41.

Rokaki, E., Collin-Souffrin, S., Magnan, C. (1993). *A&A*, 272, 8.

Ross, RR. (1979). *ApJ*, 233, 334.

Ross, R.R. and Fabian, A.C. (2005). *MNRAS*, 358, 211.

Salpeter, E. (1964). *ApJ*, 140, 796.

Sandage, A. (1965). *ApJ*, 97, 141, 1560.

Schields, G. (1978). *Nature*, 272, 706.

Schmidt, M. (1963). *Nature*, 197, 1040.

Seyfert, C. (1943). *ApJ*, 97, 28.

Shapovalova, A.I., Popovic, L.C., Afanasiev, V.L. *et al.* (2019). *MNRAS*, 485, 4790.

Souffrin, S. (1968). *Annales d'astrophysique*, 31, 569.

Storchi-Bergmann, T., Schimoia, J.S., Peterson, B.M. *et al.* (2017). *ApJ*, 835, 236.

Sulentic, J.W., Calvani, M., Marziani, P. (2001). *The Messenger*, 104, 25.

Trump, J.R., Hall, P.B., Reichard, T.A. (2006). *ApJ Suppt.*, 165, 1.

Urry, C.M. and Padovani, P. (1995). *PASP*, 107, 803.

Véron-Cetty, M.P., Joly, M., Véron, P. (2006). *A&A*, 451, 851.

Woo, J.-H., Treu, T., Barth, A.J. *et al.* (2010). *ApJ*, 716, 269.

Zeldovich, Y.B. (1964). *Soviet Physics Doklady*, 9, 195.

Zheng, W., Kriss, G.A., Telfer, R.C. (1997). *ApJ*, 475, 469.

Recommended books

Krolik, J.H. (2003). *Active Galactic Nuclei – From the Central Black Hole to the Galactic Environment.* Princeton University Press, NJ, USA.

Netzer, H. (2013). *The Physics and Evolution of Active Galactic Nuclei.* Cambridge University Press, Cambridge, UK.

Osterbrock, D. and Ferland, G. (2006). *Astrophysics of Gaseous Nebulae and Active Galactic Nuclei*, 2nd edition. University Science Books, Melville, NY, USA.

Peterson, B.M. (1997). *An Introduction to Active Galactic Nuclei.* Cambridge University Press, Cambridge, UK.

2

The Galactic Black Hole

Mark Morris

Department of Physics and Astronomy, UCLA, Los Angeles, USA

The black hole at the center of our Milky Way Galaxy – the Galactic Black Hole (GBH) – is a rather modest representative of its class. With a mass of 4×10^6 $M_{\odot}$, it is well over a thousand times less massive than the most extreme supermassive black holes known to be powering the most luminous quasars. Furthermore, the Galactic black hole has a remarkably dim accretion flow, and its luminous energy output is overwhelmed by the dense cluster of bright stars and red giants that surround it, except at radio wavelengths. However, the proximity of the GBH compensates for its restrained activity; as it is over 100 times closer than the next-nearest supermassive black hole in a galactic nucleus, it offers us an unparalleled opportunity to observe its behavior in detail. Consequently, far more observational attention has been paid to the GBH and its entourage of stars and gas than to any other single object outside the solar system.

2.1. Historical emergence of the Galactic black hole

Overwhelming extinction towards the Galactic center (~30 visual magnitudes) dictates that the GBH can only be studied using radio, infrared or X-ray energies above ~2 keV. Consequently, the location of the Galactic center was not really known until the advances in infrared astronomy in the 1960s made it possible to see the

For a color version of the figures in this chapter, see www.iste.co.uk/combes/nuclei.zip.

Active Galactic Nuclei,
coordinated by Françoise Combes.

Galaxy's central concentration of stars (Becklin and Neugebauer 1968). Radio astronomers had already suspected that the dominant radio source complex in that same direction, Sagittarius A, which had been known since the 1950s, coincided with the Galactic nucleus (Downes and Maxwell 1966).

Radio interferometer observations of Sgr A with improved spatial resolution showed that this 10-pc scale source is composed of both thermal and non-thermal substructures, but it was only in 1974 that sufficiently long interferometer baselines made it possible to discover a bright, non-thermal point source lying amidst the extended emission (Balick and Brown 1974)[1]. This point source – later named Sgr A* by Brown (1982) to emphasize its significance within the Sgr A complex – was found to coincide with the center of the Galaxy's stellar cluster within a few arcseconds (the IRS16 complex; Becklin and Neugebauer 1975); indeed, we now know that it defines the dynamical center of the Galaxy. Sgr A* is unique in the Galaxy because its radio brightness is too large to be attributable to a pulsar or any other known type of stellar object, and its spectral energy distribution was found to be rising with frequency, unlike most non-thermal radio sources. The discovery of Sgr A* therefore aroused considerable interest, an interest compounded by the fact that its existence had been somewhat anticipated by Lynden-Bell and Rees (1971), who followed the then-trending hypothesis, originally advanced by Salpeter (1964) and independently by Zel'dovich (1964), that quasars might be powered by accretion onto supermassive black holes residing in galactic nuclei[2]. Lynden-Bell and Rees posited that the prevailing "black hole formation theory of quasars should be extended downwards to black holes of considerably smaller masses". Among their suggested observable manifestations of such a "small" supermassive black hole was that it could appear as a very compact, bright radio source, and that it would be distinctly variable at infrared wavelengths. So, the stage was well set for Sgr A* to be considered a strong candidate as a black hole, and indeed, this point of view was presented in a review by Oort (1977), but the decisive fulfillment of that hypothesis would require a demonstration that the object be endowed with a substantial mass confined to a sufficiently small volume. It was quickly determined from VLBI observations that Sgr A* is very compact (< 200 AU; Lo *et al.* 1975), but not that it had a large mass.

1. See Goss *et al.* (2003) for a detailed account of the discovery of Sgr A*.

2. More precisely, the theory invoked by Lynden-Bell and Rees (1971) was that black holes are formed when quasars die, and that "each dead quasar disappears down its Schwarzschild throat and is surrounded by stars".

2.2. The mass of the Galactic black hole

The first inklings of a large mass at the Galactic center came from measurements of the velocity dispersion of gas clouds occupying the central parsec. Measurements of the velocities and velocity dispersions of the 12.8 μm fine-structure line of ionized neon were reported in a series of papers by a group at UC Berkeley (Wollman *et al.* 1976; Lacy *et al.* 1979, 1980), who favored a model with a point mass of several million solar masses surrounded by a comparable mass of stars within 1 parsec. However, they noted that a model in which all of the mass was in the form of stars distributed throughout the central parsec was only slightly less probable. Subsequently, Serabyn and Lacy (1985) observed the same line with higher spatial resolution and were able to identify orbiting streams of gas – the "Western Arc" and the "Northern Arm" – and fit their velocity distributions with orbits responding to a point mass of $\gtrsim 3.5 \times 10^6$ $M_\odot$ plus a smaller amount of distributed stellar mass[3]. Serabyn and Lacy concluded that the mass distribution is too centrally concentrated to be consistent with a spherical isothermal central cluster (i.e. a cluster having density $\propto r^{-2}$, which is roughly consistent with the observed 2 μm light distribution on larger scales; Becklin and Neugebauer 1968). They therefore attributed their inferred central point mass to a black hole located, within ~2" accuracy, at the position of the IRS16 complex and Sgr A*.

Stellar velocities were also used to assess the mass distribution, in order to check whether the velocity measurements of gas might be affected by non-gravitational effects, such as jets, magnetic fields and supernova remnant dynamics. McGinn *et al.* (1989), Sellgren *et al.* (1990) and Haller *et al.* (1996) measured the velocity dispersion of stars using the shape of the 2.3 μm CO bandheads in the integrated starlight entering apertures of various sizes. Their preferred model invoked a central point mass of 2–2.5 × 10^6 $M_\odot$, somewhat smaller than what had been deduced from the measurements of gas.

However, a much clearer demonstration that a point mass was present came when near-diffraction-limited speckle-imaging observations were undertaken over a period of several years by two groups: Eckart and Genzel (1997, 1998), Genzel *et al.* (1997, 2000) and Ghez *et al.* (1998). The proper motions of ~100 stars located within several arcseconds of Sgr A* were measured, and the velocity dispersion as a function of radius, r, was found to closely follow an $r^{-1/2}$ law, which was interpreted to imply the presence of a dark mass of at least 2.6×10^6 $M_\odot$ confined within 0.1 arcsec of Sgr A*. The proper motion velocities agreed statistically with spectroscopically determined velocities for a Galactic center distance of 8 kpc, so

3. Roberts *et al.* (1996) got a very similar result for the concentrated mass using high-resolution observations of a radio recombination line from gas orbiting even closer to Sgr A*.

the three-dimensional velocities of the stars in this cluster, which came to be known as "S" stars surrounding Sgr A*, appeared to be isotropically distributed.

The next step in refining the stellar motions was to measure the nonlinearity of the proper motions: accelerations. Ghez *et al.* (2000) reported the accelerations of three stars, which provided three projected force vectors that triangulated the location of the point mass to lie within 0.05 ± 0.04 arcseconds of Sgr A*. Later, the time baseline of the astrometric measurements became sufficient to fit 3D orbits (Schödel *et al.* 2002; Ghez *et al.* 2005). Finally, the spectroscopy of the orbiting stars was utilized, and the radial velocity information obtained from spectra led to refined S star orbits (Eisenhauer *et al.* 2005; Paumard *et al.* 2006; Ghez *et al.* 2008). The radial velocity measurements of orbiting stars introduced another important parameter into the orbital fits: the distance, R_o, to the Galactic center. This distance can be determined geometrically by relating the astrometric measurements of proper motion – an angular velocity on the plane of the sky – to the radial velocity measurements obtained from spectroscopy. As a result, the orbital fits yield both the GBH mass and R_o, although these two parameters are inevitably strongly correlated, as has been shown in all of the treatments to date (Eisenhauer *et al.* 2003, 2005; Ghez *et al.* 2008; Gillessen *et al.* 2009, 2017; Boehle *et al.* 2016; Do *et al.* 2019a; GRAVITY Collaboration 2018, 2019). Except for Boehle *et al.* (2016), the analyses have all been focused on a single star, S0-2 (or S2), chosen because it is the brightest S star and has a relatively short orbital period of 16 years. The most recently determined values for the GBH mass and R_o converge on $4.1 \pm 0.1 \times 10^6$ $M_\odot$ and 8.1 ± 0.1 kpc, respectively, with the uncertainties expressed here representing the range needed to accommodate the values of the two reporting groups. The formal statistical uncertainties are far less than this, and indeed the Gravity Collaboration (2019) claims an overall uncertainty in R_o of only 0.3%, but unaccounted systematic uncertainties by one or both groups are probably responsible for the divergence between their values. In any case, the accuracy should continue to improve with more orbital data and improved control of systematics, and the determination of R_o should soon be extraordinarily precise.

The GBH mass determined from stellar orbits exceeds that determined earlier from statistical analyses and Jeans modeling of stellar proper motions by ~50% (although the mass determined statistically from radial velocity measurements alone – $3.45 \pm 1.5 \times 10^6$ $M_\odot$ – was in slightly better agreement; Figer *et al.* 2003). This discrepancy is attributable to the fact that the radial density distribution of the relatively luminous stars that were being used for the analyses with respect to galactocentric radius is flatter than was initially assumed in the statistical treatments (Do *et al.* 2009b, 2013a, 2013b; Buchholz *et al.* 2009; Fritz *et al.* 2016), causing the innermost volume where higher-velocity stars would be present to be relatively underrepresented in the early analyses. Indeed, the luminous late-type stars in the entourage of the GBH – mostly giants – are distributed in a core-like structure that flattens within 8–12" (0.3–0.5 pc; Figer *et al.* 2003), rather than in a cusp structure peaking sharply on the GBH that is theoretically predicted for a dynamically relaxed distribution (Bahcall and Wolf 1976). However, Gallego-Cano *et al.* (2018) have argued that faint stars do appear to form a cusp, so some as-yet-unidentified process might have acted to alter the distribution or the intrinsic luminosity of the giants.

Can we be certain that the implied compact mass is indeed a black hole? Other exotic alternatives have been suggested: a cluster of compact stellar remnants (Maoz *et al.* 1998), a degenerate fermion ball composed of massive dark particles such as sterile neutrinos (Tsiklauri and Viollier 1998), or a boson ball (Torres *et al.* 2000), but observations of stellar orbits having periapse passages that come within 50–125 AU of the center of mass constrain the volume occupied by the mass to such an extent that such possibilities became clearly inviable (Schödel *et al.* 2002; Ghez *et al.* 2005). All would quickly collapse to a general-relativistic black hole. Much tighter constraints on the size of the Sgr A* radio source from VLBI radio observations (e.g. Shen *et al.* 2005) made the implied minimum density of the central mass even larger. Finally, now that an actual supermassive black hole has made an appearance in the galaxy M87, thanks to the Event Horizon Telescope Collaboration (2019), there is no impediment to assigning black hole status to Sgr A*.

2.3. General relativity and the Galactic black hole

Precision measurements of stellar orbits close to the GBH provide an opportunity to explore general relativity (GR) in a poorly tested regime. To date, the GR measurements relating to the GBH have all been focused on the brightest S star, S0-2, which has a compact, eccentric orbit (semi-major axis ~1000 AU, e = 0.89). The first post-Newtonian test was the measurement of the gravitational redshift, reported by two groups following the 2018 periapse passage of S0-2 at a distance of 120 AU from the GBH (GRAVITY Collaboration 2018a; Do *et al.* 2019). At that point, the star was moving at 2.7% the speed of light. Figure 2.1 shows the projected orbit and the excursions of radial velocity as a function of time from the measurements used by the GRAVITY Collaboration (2018a), including the precision measurements around periapse by the GRAVITY interferometer. The conclusion of both groups was that the data deviate very significantly from a Keplerian orbit, and by an amount that is consistent with the predictions of special and general relativity. Figure 2.2 shows the deviations from a Keplerian orbit measured with the Keck Telescopes by

Do *et al.* (2019). The deviations are parameterized by the quantity ϒ, such that the radial velocity deviation as a function of time can be expressed as:

$$\text{Radial velocity deviation as a function of time} = \Upsilon\left[\frac{V(t)^2}{2c} + \frac{GM_{GBH}}{cR(t)}\right], \qquad [2.1]$$

where the first term is the transverse Doppler shift predicted by special relativity, and the second term is the gravitational redshift predicted by GR. $V(t)$ is the 3D velocity of S0-2 at time t, $R(t)$ is the 3D distance between Sgr A* and S0-2, and M_{GBH} is the GBH mass. The prediction of classical special and general relativity is that the parameter ϒ will be exactly 1, and the orbital fitting from both groups is consistent with that value within their 1σ uncertainties.

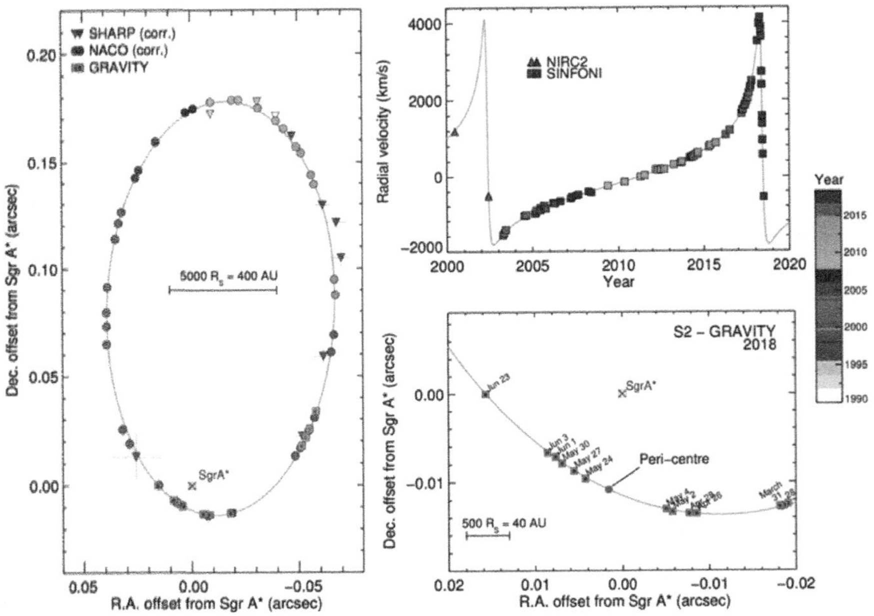

Figure 2.1. *Orbit of S0-2 around Sgr A*. Left: orbit projection on the sky. Top right: radial velocity measurements as a function of time. Bottom right: precision positional offset measurements between S0-2 and Sgr A* by the GRAVITY interferometer on the VLT during the passage near periapse (source: GRAVITY Collaboration (2018a))*

The next most challenging GR test using the GBH is to find the Schwarzschild precession of the orbits of nearby stars. The GR prediction is that the periapse of a stellar orbit will undergo a prograde precession. However, that effect will compete with a *retrograde* precession that would be caused by the presence of any significant extended mass distribution (e.g. Rubilar and Eckart 2001). The GRAVITY Collaboration (2020a) recently reported a retrograde shift in the major axis of the orbit of S0-2 of 12 arcminutes per orbital period, which is consistent, within the measurement uncertainties, with the pure GR prediction. That study also concluded that the total amount of extended mass within the orbit of S0-2 must be at least 1,000 times less than the mass of the GBH, which places a tight constraint on the abundance of stars and stellar remnants in that volume. Improvements in that limit are sure to be had with the continued observational refinement of the orbit of S0-2, and the addition of more orbits that sample larger and smaller volumes than S0-2. In principle, with accurate measurements of the Schwarzschild precessions of several stars, the radial distribution of extended mass can be determined, and any possible effect that mass may have on the GR test will be eliminated.

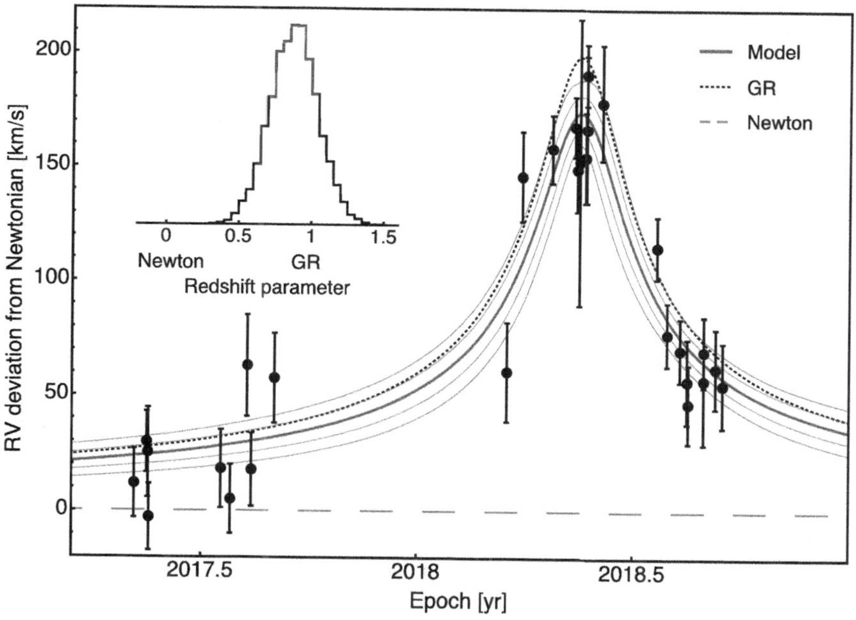

Figure 2.2. *Measured deviation of the orbit of S0-2 from Newtonian predictions. The green line shows the best-fitting orbit model corresponding to equation [2.1], and the green- and gray-shaded regions correspond to the 68% and 95% confidence intervals of the model. The inset shows the posterior probability distribution for Υ, which is within 1σ of the predicted relativistic value of 1 (source: Do et al. (2019))*

On the more distant horizon, stellar orbits might eventually be used to measure higher-order effects, such as the Lense–Thirring precession that accompanies the spin of the black hole. However, because this effect falls off rapidly with distance, it would likely require a sample of much more tightly bound stars than we have at present.

2.4. The accretion flow: radiative signatures of the Galactic black hole

The accretion flow onto the GBH emits across the spectrum, though its total luminosity is quite modest, estimated at a mere few times 10^{35} $L_\odot$ (Serabyn *et al.* 1997), which is only ~ 10^{-9}–10^{-10} of its Eddington luminosity. Clearly, the accretion rate is quite small, but in addition, the accretion flow is likely to be radiatively inefficient, as discussed below.

Accessible windows onto the spectrum include the use of radio through submillimeter, the near- and mid-infrared, and X-rays. The spectrum at radio wavelengths longer than 20 cm is not well determined because of the confusing bremsstrahlung emission from the surrounding gas, coupled with strong interstellar scattering that broadens the source, while the infrared from 10 to 350 μm is so far overwhelmed by the dust emission from the surrounding Sgr A West HII region. Gamma-ray emission has been reported towards the location of the GBH, but with a relatively low spatial resolution that makes it difficult to unambiguously identify the source as the GBH; there are alternatives, including a nearby pulsar wind nebula (Wang *et al.* 2006; Acero *et al.* 2010; see Figure 2.6).

The spectrum of Sgr A* has a peak of ~5×10^{35} ergs s^{-1} in νL_ν at around 350 μm (Serabyn *et al.* 1997), known as the "submillimeter bump", where the partially polarized emission (e.g. Marrone *et al.* 2007) is attributable to synchrotron radiation (see Figure 2.3). Early radio results at relatively long radio wavelengths found that the emission from Sgr A* is variable (Brown and Lo 1982), but that variability was appropriately attributed to refractive interstellar scintillation caused by inhomogeneities in the intervening interstellar medium (Zhao *et al.* 1989). Because of the λ^2 scaling of refractive scintillation, the intrinsic variability of Sgr A* emerges at higher frequencies; the intrinsic radio emission through the submillimeter bump is variable at the 30–40% level on time scales of minutes to hours (e.g. Zhao *et al.* 2003; Macquart and Bower 2006). The statistical characterization of the millimeter-wave emission was found to be describable in terms of a red-noise power spectrum, in which the power spectral density, P, at temporal frequency f is expressible as $P(f) \propto f^{-1}$ (Mauerhan *et al.* 2005).

2.4.1. *Infrared emission from Sgr A**

The infrared counterpart to Sgr A* was discovered independently by Genzel *et al.* (2003) and Ghez *et al.* (2004). Both used new systems that employed adaptive optics on large telescopes for the first time. In both reports, it was clear that the infrared emission is highly variable. Like the millimeter-wave emission, the infrared follows a red-noise power spectrum, although with $P(f) \propto f^{-2}$ (Do *et al.* 2009; Witzel *et al.* 2018), much steeper than for millimeter waves.

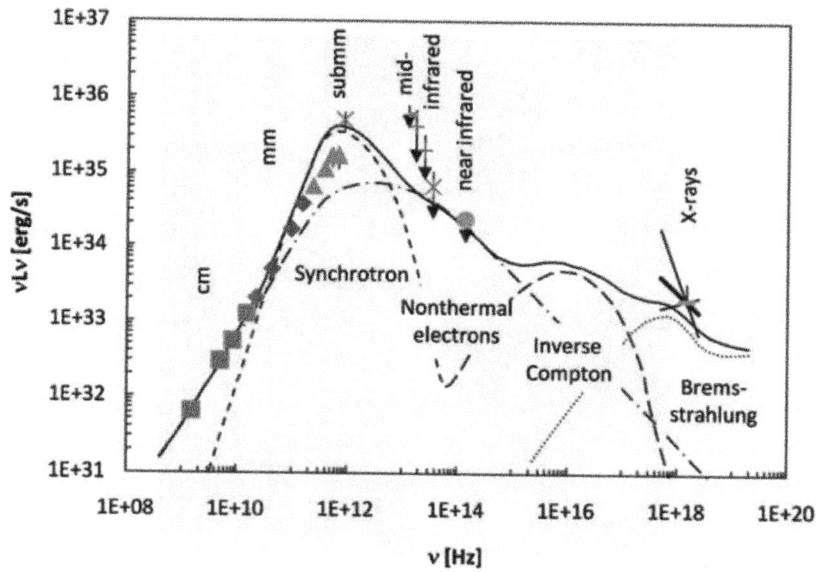

Figure 2.3. *Spectral energy distribution of Sgr A* in its steady state. The quiescent X-ray flux level shown here is attributed to thermal Bremsstrahlung from distances out to the Bondi radius, but the X-ray flares, arising from within several Schwarzschild radii (see section 4.2), have luminosities rising up to several hundred times that level. A model of the emission from Yuan et al. (2003) is superimposed, showing the various processes that comprise the overall spectral energy distribution. Emission in the mid- and far-infrared is difficult to measure because of the bright thermal dust emission surrounding Sgr A* (Schödel 2011), but has recently been reported by von Fellenberg et al. (2018) (source: Genzel et al. (2010))*

Sgr A* stands out in the near-infrared (NIR) for being very red compared to the stars around it (see Figure 2.4). Hornstein *et al.* (2007) found that its intensity rises with increasing wavelength through the NIR, and they concluded that the spectral index appears constant. However, a later treatment of much more data led Witzel *et al.*

(2018) to conclude that the spectral index depends on the source intensity (see also Ponti *et al.* (2017) and the references therein). The source becomes bluer in brighter states, although the spectrum is always rising with wavelength through the infrared band. In spite of the increasing flux density with wavelength, Sgr A* has not yet been detected reliably at wavelengths greater than 5 μm, largely because in the mid-infrared, thermal dust emission in the complex Galactic center region overwhelms the emission from Sgr A* (Schödel *et al.* 2011). However, detecting the variable emission from Sgr A* may soon be possible with the James West Space Telescope because of its expected pointing and gain stability.

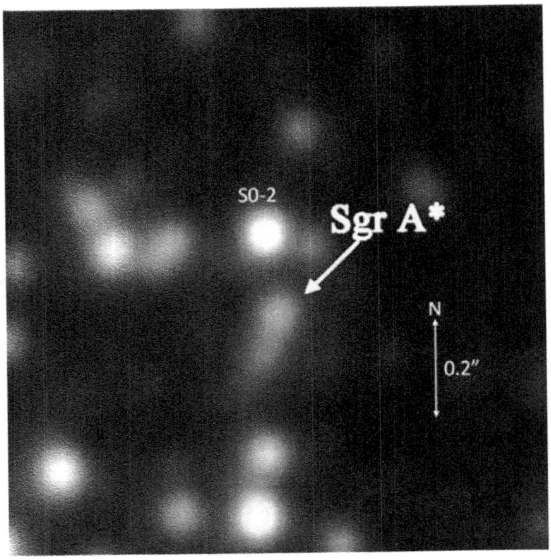

Figure 2.4. *Three-color image of the central arcsecond, as observed in July 2005 at a moment when Sgr A* was particularly bright. Blue: H-band (1.63 μm), green: K' band (2.12 μm), red: L' band (3.78 μm). The relatively bright star to the north of Sgr A* is S0-2, whose orbit has been used to measure the mass of the GBH, and to explore the effects of general relativity. The extended red emission to the southwest of Sgr A* is a dust emission feature associated with the likely stellar source, G1 (Witzel et al. 2017) (source: Hornstein et al. (2007))*

The statistical characteristics of the infrared variability of Sgr A* have been investigated on several occasions using data from large telescopes equipped with adaptive optics (Dodds-Eden *et al.* 2011; Witzel *et al.* 2012), and by the Hubble Space Telescope, but the most complete characterization to date used eight days of observations at a wavelength of 4.5 μm from the Spitzer Space Telescope between 2013 and 2017 (see Figure 2.5; from Witzel *et al.* (2018)), along with eight years of K_s-band (2.18 μm) data from the VLT and 10 years of K'-band (2.12 μm) data from the Keck telescope. As previously discovered (Do *et al.* 2009), Sgr A* was shown to be continuously variable, but typically exhibits several prominent maxima per day. Those maxima are often termed "flares", but that designation does not capture the continuous and random nature of the variable emission process, which shows variation across a broad range of temporal frequencies. Early reports of a quasi-periodic infrared signal from Sgr A* (Genzel *et al.* 2003) have not been confirmed; Do *et al.* (2009), and later, Witzel *et al.* (2018) found no excess power in their periodogram analyses of their extensive databases at frequencies that could correspond, for example, to the orbital frequency of the accreting material near the innermost stable circular orbit (ISCO) around the GBH. This is all the more curious since we might expect to see periodicities from rapidly orbiting gas blobs near the ISCO. Indeed, the GRAVITY Collaboration (2018b) reported orbital motions of blobs near the ISCO having periods of 33–65 minutes, which is just outside the ISCO for a Schwarzschild (zero-spin) black hole of 4 million solar masses. Perhaps such blobs do not survive long enough to leave a periodic signal.

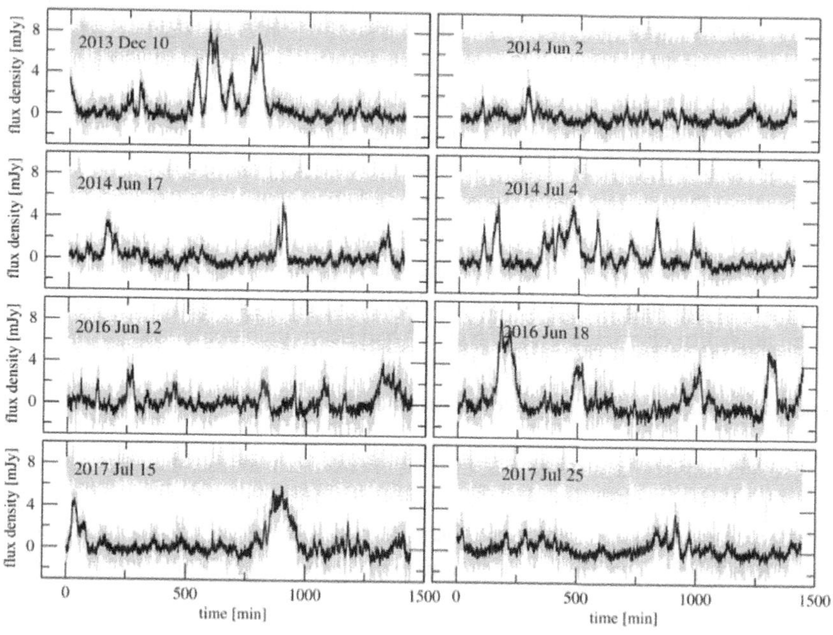

Figure 2.5. *Twenty-four-hour light curves of Sgr A* at 4.5 μm, as measured by the Spitzer/IRAC camera. In each panel, the gray lines show the flux density for each 6.4 s frame set, and the black lines show the data binned in one-minute intervals. The light-gray light curves at the top show the flux density of a reference pixel. (source: Hora et al. (2014); Witzel et al. (2018))*

In their statistical analysis, Witzel *et al.* (2018) identified a break frequency in the power spectral density of the infrared emission, indicating a characteristic coherence time scale of 243^{+82}_{-57} arcminutes[4]. This is the time over which the activity displayed in the light curve remains correlated, and can be interpreted as the lifetime of an active region, or the duration of the consequences of an energy release, perhaps by magnetic reconnection (Dexter *et al.* 2020) or of the sudden introduction of a mass concentration into the accretion flow. A similar characteristic time scale was found for submillimeter light curves by Dexter *et al.* (2014), although the fractional amplitude variability is less at those wavelengths than in the NIR.

So far, there is no evidence that the statistical properties of the infrared emission have changed during the time that the infrared emission from Sgr A* has been observed. Following a suggestion that the infrared flux density distribution of Sgr A* switches between two distinct states (Dodds-Eden *et al.* 2011), Meyer *et al.* (2014) investigated the data available at that time using a formal methodology – a "Hidden Markov Model" – known in economics research as the "regime switching model", and concluded that the emission from Sgr A* is sufficiently described by a stochastic process in a single intrinsic state. Later, Chen *et al.* (2019) extended the time frame over which Sgr A* had been detected, by reanalyzing data that had been obtained prior to 2005, using a speckle imaging technique. Using a speckle holography analysis, Chen *et al.* extended the depth of the images from that time period by a few magnitudes compared to the previously used shift-and-add technique, and they were able to extract a number of detections of Sgr A* between 1998 and 2005. The resulting source fluxes were found to be consistent with the flux density distributions obtained during the later ($\geq$ 2005) adaptive optics epochs, indicating that no obvious state changes had taken place in the earlier period, and that the variability characteristics of Sgr A* had been unchanged over 22 years. This conclusion is salient when we consider the evidence from X-rays (discussed below) that major outbursts of Sgr A* have occurred on time scales of hundreds of years.

However, a hint that anomalous outbursts can occasionally happen occurred in 2019, when a brightness excursion that was twice as bright as anything previously observed was captured at the Keck Observatory by Do *et al.* (2019b). The authors also presented light curves for other unusually bright events that occurred within weeks of the giant flare, and concluded that the probability of the flux levels in the ensemble of their light curves is less than 0.05%. This raises the possibility that Sgr A* was going through a phase of unusual activity in 2019. Do *et al.* considered the possibility that such activity might have been the delayed result of enhanced accretion caused by the 2018 periapse passage of the bright S star, S0-2, or even the

4. This verifies and refines an earlier estimation by Meyer *et al.* (2009).

earlier periapse passage of the dusty, extended G2 object in 2014, which underwent a clear tidal interaction with the GBH. The latter possibility was investigated further by Murchikova (2020), who concluded that the material pulled off of G2 by the tidal interaction could rain down later on the GBH and produce enhanced accretion luminosity. Indeed, Kawashima *et al.* (2017) predicted that a radio and infrared brightening would occur around 2020 as a result of the G2 passage. The 2019 data assembled by the GRAVITY Collaboration (2020b) provides additional evidence that Sgr A* was unusually active during 2019; it shows several near-IR intensity peaks that are markedly stronger than anything that had been seen in previous years, even though the median intensity was unchanged from year to year.

The infrared emission from Sgr A* displays strong linear polarization (Eckart *et al.* 2006, 2008; Trippe *et al.* 2007; Zamaninasab *et al.* 2010; Shahzamanian *et al.* 2015), which confirms that the infrared emission, like the radio, is optically thin synchrotron emission. In addition, the variability of the polarization adds a new dimension to the study of the infrared emission. Not only does the polarized signal vary largely in concert with the total intensity, but the polarized light curve is also more complex, consisting of "subflares", as the polarization fraction, typically ranging between 10% and 30%, undergoes marked changes through the total intensity maxima. Furthermore, in the course of an infrared flare, the orientation of the polarized E-vector often shows large and rapid changes on time scales of several minutes. These "swings" in the polarization angle have inspired the consideration of two classes of models in the references listed above. The most emphasized model involves a "hot spot" orbiting near the ISCO, perhaps as part of a transitory disk. The hot spot could be a density enhancement of emitting matter or a location in the accretion flow where energy injection has taken place as a result of some local instability. The swings in the polarization angle are then attributed to the changing projection of the ambient magnetic field within or around the hot spot as it orbits. The appeal of this model has been considerably enhanced by the direct observation of orbiting hot spots near the ISCO by the GRAVITY Collaboration (2018b), accompanied by polarization angle changes that varied continuously with orbital phase and with about the same 45 ± 15-minute period.

The second model for the polarization variations invokes emission from a jet or a collimated wind. The polarization angle from an outgoing plasma blob in a helically structured jet, for example, could change with location in the jet, as the projected direction of the magnetic field changes along the helix. In general, however, the polarization angle should reflect the jet direction, and thereby vary around a particular angle. Eckart *et al.* (2006) discussed the potential relationships of their observed polarization angles to possible outflow or jet signatures in X-ray and infrared images, and Shahzamanian *et al.* (2015) identified a preferred polarization angle – 13 ± 15 degrees – and suggested that it could be linked to either an outflow direction, be it that of a jet or a wind, or to the orientation of a temporary accretion disk. Continuing observations with GRAVITY will be of great value in refining the conclusions that might be drawn from polarization measurements.

The submillimeter emission from Sgr A* also displays a rather strong linear polarization, and like the NIR emission, it varies on similar time scales, with the polarization angle undergoing rapid occasional swings (Marrone *et al.* 2006).

2.4.2. *Variable X-ray emission from Sgr A**

X-ray emission from Sgr A*, discovered by Baganoff *et al.* (2001, 2003), consists of two components: a relatively weak, measurably extended quiescent component that has been invariant for at least 14 years (Yuan and Wang 2016), and an unresolved component consisting of occasional dramatic "flares" that rise above the quiescent component by factors up to several hundred (Wang *et al.* 2013). The quiescent component (see Figure 2.6) has been attributed to the virialization of the accreting gas at the Bondi radius, where it shocks and thermalizes at a temperature of ~10^7 K (e.g. Corrales *et al.* 2020). The Bondi radius occurs at ~ GM/c_s^2, for sound speed c_s, corresponding to ~0.1 pc (Quataert 2002; Baganoff *et al.* 2003; Yuan 2011; Wang *et al.* 2013).

The detectable X-ray flares, which occur about once per day on average, have durations of 0.5 to several hours and often show considerable temporal substructure (Baganoff *et al.* 2001; Nowak *et al.* 2012; Ponti *et al.* 2017; Haggard *et al.* 2019), including an occasional precursor (see Figure 2.7). The time scales indicate that the flares arise from regions that are the size of several gravitational radii, and those regions are probably quite close to the GBH. However, the flare mechanism has not been reliably identified; there remain several possibilities: magnetic reconnection events (Ball *et al.* 2016, 2020), hydrodynamic or magnetohydrodynamic instabilities in the inner accretion disk (Tagger and Melia 2006), a strongly enhanced accretion rate associated with an inspiraling lump of material, shocks caused by the impact of a jet on material in the accretion flow, gravitational lensing of hot spots in the accreting plasma, or even tidal disruptions of comets, asteroids or planets (Zubovas *et al.* 2012).

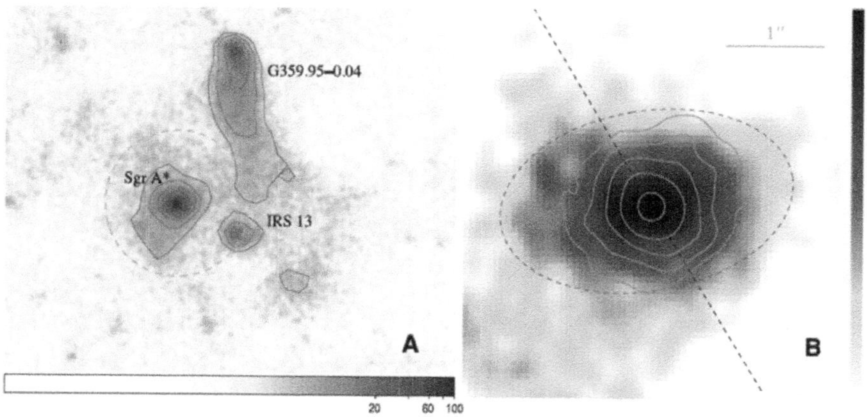

Figure 2.6. *X-ray image of Sgr A* from 1 to 9 keV, as measured by the Chandra X-Ray Observatory. Left: image of the quiescent emission, with flare intervals removed from the integration. The dashed circle surrounding Sgr A*, with a radius of 4 arcseconds, corresponds to the estimated Bondi capture radius. Note the nearby pulsar wind nebula, G359.95-0.04, which is a candidate source of the gamma-ray emission from this region. Right: a magnified version of the image on the left. The contours show the unresolved flare component. The dashed line indicates the orientation of the Galactic plane, and the dashed ellipse depicts the projection and orientation of the central disk of massive young stars, extending well beyond the ellipse. (source: Wang et al. (2013); see also Shcherbakov and Baganoff (2010))*

2.4.3. *Emission mechanisms, from radio to X-rays and frequency-dependent time lags*

There has been much discussion in the literature about the emission mechanism powering the X-ray flares, and the debate continues, primarily between synchrotron emission and synchrotron self-Compton (SSC) emission. The submillimeter and NIR emissions are readily explained as synchrotron emissions, and many have regarded the X-ray flares as a continuation of the direct synchrotron emission causing the infrared emission (e.g. Dodds-Eden *et al.* 2010; Barriere *et al.* 2014; Dibi *et al.* 2014, 2016; Ponti *et al.* 2017; Zhang *et al.* 2017; GRAVITY Collaboration 2021), albeit with a variable cooling break in the electron energy distribution. This cooling break is needed to account for the absence of detectable X-ray counterparts to most bright infrared maxima, while all sufficiently well-observed X-ray flares are coincident with a NIR maximum (Fazio *et al.* 2018), indicating that the emission mechanisms for NIR and X-rays are inherently linked.

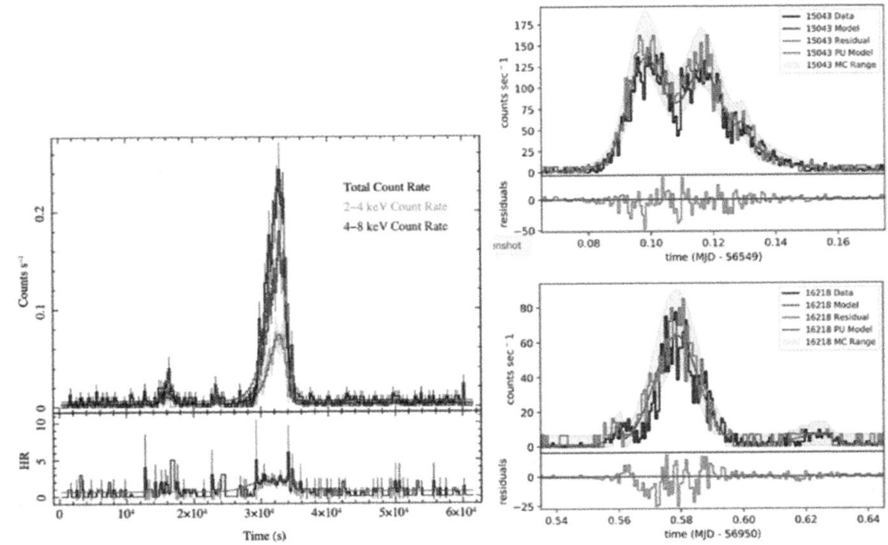

Figure 2.7. *Light curves of three of the brightest X-ray flares from Sgr A*, observed from the Chandra X-Ray Observatory. Left: February 9, 2012 event, in different bands, and the hardness ratio (HR) at the bottom (source: Nowak et al. (2012)). Right: September 14, 2013 and October 20, 2014 flare profiles and model fits, plus residuals (source: Haggard et al. (2019))*

The SSC model rests on the supposition that the electrons responsible for producing the low-energy photons by synchrotron radiation can also elevate the energies of those same photons to the X-ray regime, via inverse Compton scattering. This model was advanced by Falcke and Markoff (2000) and Markoff *et al.* (2001) in the context of their jet model for the radio and X-ray spectrum of Sgr A*. Many later authors have investigated the SSC model for emission from the accretion flow (Liu *et al.* 2006a, 2006b; Marrone *et al.* 2008; Trap *et al.* 2011; Eckart *et al.* 2012; Ma *et al.* 2019).

The relative merits of the synchrotron and SSC models have recently been summarized by Boyce *et al.* (2019), who analyzed all existing measurements of simultaneous observations of Sgr A* in the NIR and X-ray, in order to determine whether there is a significant time lag between the emission peaks in those two wavelength ranges. They found that X-ray peaks may have a lead on the NIR peaks by 10–20 minutes, but only with 68% confidence, and that both synchrotron and SSC emission processes are consistent with such a time lag. Additional cases of simultaneous X-ray/NIR observations are needed to obtain a more robust result, although it is quite challenging to arrange for simultaneous observations with an

X-ray satellite and a large-aperture ground-based telescope equipped with adaptive optics. The question of whether submm maxima can also be associated with NIR or X-ray peaks with a time lag is complicated by the fact that the time lag could become comparable to the range of typical intervals between maxima in the light curves. Tentative NIR-submm time lags from 20 to 200 minutes were listed by Morris *et al.* (2012), but Fazio *et al.* (2019) presented examples of simultaneous light curves that showed no such obvious trends. Unfortunately, what is needed are simultaneous light curves with durations exceeding what is usually feasible from a ground-based submm observatory.

The idea of an increasing time lag with increasing wavelength in the emissions from Sgr A* was first suggested by Yusef-Zadeh *et al.* (2006), who compared light curves at two radio frequencies (22 and 43 GHz), finding a 20–40 minute lag between the two frequencies. They interpreted their observation in terms of an expanding plasma blob, invoking a model originally suggested by van der Laan (1966). A similar result using much more extensive observations was obtained by Brinkerink *et al.* (2015), who observed seven frequencies from 19 to 100 GHz and found an increasing time lag with decreasing frequency.

With the expanding plasma blob paradigm, any significant time lags could be combined with source size measurements to estimate the expansion speed of the emitting region. The apparent source size, however, is broadened by interstellar scattering, which closely follows the theoretically expected wavelength-squared law (Shen *et al.* 2005; Bower *et al.* 2006, and references therein). Subsequently, Falcke *et al.* (2009), and later, Johnson *et al.* (2018) used VLBI measurements taken at several wavelengths between 1.3 and 13 mm, and from a variety of telescopes, to refine the relation for the major- and minor-axis diameters of the elliptical scatter-broadened disk of Sgr A*:

$$\phi_{scattering} = (1.38 \pm 0.013) \times (\lambda_{cm})^2 \text{ (major) mas and}$$

$$(0.703 \pm 0.013) \times (\lambda_{cm})^2 \text{ (minor) mas.}$$

Johnson *et al.* also found that Sgr A* has a nearly circular *intrinsic* structure, with a diameter roughly proportional to wavelengths between 1.3 and 13 mm: $\phi_{Sgr\,A*} = 0.4$ mas $\times\,\lambda_{cm}$, which corresponds to 3.2AU $\times\,\lambda_{cm}$, or $39\lambda_{cm}$ Schwarzschild radii (R_s). Issaoun *et al.* (2019, 2021) reported similar results, albeit with a slightly elongated intrinsic structure (major/minor axis ratio of ~1.5), with their 3 mm VLBI observations. This stands in contrast to the 7-mm observations by Bower *et al.* (2014), who reported a rather elongated intrinsic source shape of 35.4 × 12.6 R_S. Also, while observing at 1.3 mm, Lu *et al.* (2018) reported evidence for a compact intrinsic source structure on scales of ~3 R_S.

From the reported size distribution and their time lag measurements, Brinkerink *et al.* (2015) were able to constrain the expansion velocity of the emitting medium, and they concluded that the radio emission arises in a moderately relativistic, collimated outflow. It is not clear that significant time lags can be found at frequencies at or above 100 GHz; Miyazaki *et al.* (2013) found no time lag between 90 and 102 GHz. Furthermore, radio emission at millimeter wavelengths and longer likely arises at radii larger than the radii at which the variable infrared and X-ray emission are produced (e.g. Witzel *et al.* 2021), and this must be accounted for when extending time-lag arguments across the full electromagnetic spectrum. The bulk of the radio emission might not be physically linked to the shorter-wavelength emission.

In a recent study, Witzel *et al.* (2021) examined the variability characteristics of Sgr A* across the accessible electromagnetic spectrum, from submillimeter to X-rays. The authors engendered a model in which the correlations between NIR and X-ray emission were attributed to two strictly correlated stochastic processes, with the X-ray process being the low-pass-filtered version of the NIR process. The coupling between X-rays and NIR, indicated by the occurrence of a NIR peak coincident with every X-ray flare, evokes a model in which the NIR emission arises in a self-absorbed synchrotron volume, with a high-frequency exponential cooling cutoff within or just above NIR frequencies. The same volume generates SSC emission that occasionally contributes to the NIR flux, but is dominant at higher frequencies, accounting entirely for the X-ray emission. A simulation of the resulting spectral energy distribution is shown in Figure 2.8 for two chosen moments, one which produces a detectable X-ray flare (bottom) and one in which only the NIR is detectable (top). In this model, some random energy injection process causes an enhancement in the electron energy distribution that translates into enhanced synchrotron emission from the submm to the NIR, as well as into a higher cutoff in the electron energy distribution, and consequently a higher cutoff frequency in the synchrotron spectrum. As a result, the X-ray flux, while not strictly linked to the NIR flux, varies over a much broader intensity range than the NIR flux, and most of the NIR excursions are accompanied by X-ray flares whose fluence is undetectable because it is below the constant quiescent level arising much further out near the Bondi radius. This is consistent with the X-ray flare luminosity distribution extracted from the Chandra satellite data by Neilsen *et al.* (2013 and 2015): $dN/dL \propto L^{-1.9}$, from which they conclude that ~10% of the quiescent flux is owed to weak, individually undetectable X-ray flares. From his models, Witzel (personal communication) estimates that this variable fraction of the quiscent flux might be as large or larger than 50%.

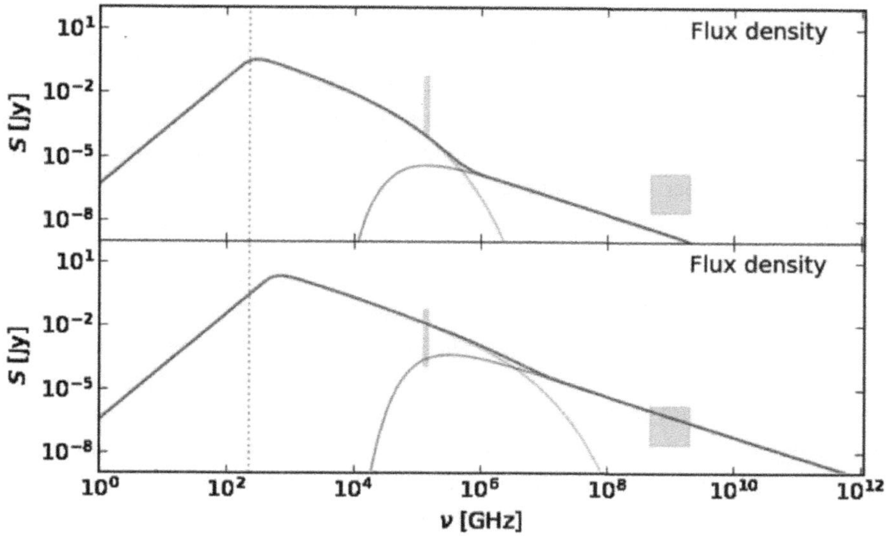

Figure 2.8. *Snapshots of the animation of the spectral energy distribution of Sgr A* at two different times, from Witzel et al. (2021). Orange lines show the synchrotron emission, blue lines show the SSC emission and red lines show their sum. The vertical dashed line marks the frequency of 230 GHz, which is near the turnover where the synchrotron emission becomes optically thick. The gray-shaded zones show the NIR and 2–8 keV X-ray bands where observations have been made. Their vertical extent corresponds to the typical range of observed NIR variability and the typical range of peak 2–8 keV flux densities above the detection limit. Non-variable radio emission at frequencies below the submillimeter bump located just above 230 GHz arises at larger radii than the variable emission, and is not depicted here, although it dominates the observed flux density. (source: Witzel et al. (2021))*

2.4.4. *Long-term variations in the accretion luminosity of Sgr A**

Evidence has grown over the past decade that multiple powerful pulses of relatively hard X-rays are presently propagating across the central molecular zone of the Galaxy, and the consensus is that those outgoing X-ray fronts originated in an energetic accretion event from the GBH within the past few hundred years. Observers using the Chandra, XMM-Newton and NuSTAR X-ray satellites have followed multiple "light echoes" consisting of a combination of fluorescent X-ray line emission and scattered hard X-rays, and that echo has been observed to move at apparent relativistic speeds (even superluminal speeds) in the general direction away from Sgr A* (Muno *et al.* 2007; Inui *et al.* 2009; Ponti *et al.* 2010; Terrier *et al.* 2010, 2018; Capelli *et al.* 2012; Nobukawa *et al.* 2012; Clavel *et al.* 2013;

Zhang *et al.* 2015; Chuard *et al.* 2018). The most prominent line is the 6.4 keV Kα line of neutral or singly ionized iron, which results from the photo-ejection of the most tightly bound electron by an X-ray with energy exceeding the 7.1 keV ionization edge of iron. In the ensuing radiative cascade, K lines are emitted, and the Kα line is the most intense. This line, and the scattered flux of hard X-rays, are produced in a number of molecular clouds in the central molecular zone as the flash of hard X-rays passes through them (see Figure 2.9).

The fluorescent iron line emission phenomenon was anticipated by Sunyaev *et al.* (1993) as a means to constrain the occurrence of luminous X-ray outbursts during the last several hundred years. The subsequent detection of bright 6.4 keV line emission towards two molecular complexes in the central molecular zone by Koyama *et al.* (1996), led those authors to suggest that the Galactic center could have undergone an outburst as strong as 2×10^{39} ergs s^{-1}. A similar luminosity has been inferred by most later studies. Even higher luminosities are likely implied by the fluorescent emission from the distant molecular cloud, Sgr D (see Terrier *et al.* 2018), if Sgr A* is indeed the illuminating source.

The duration of the X-ray outbursts from Sgr A* can be constrained by determining how long it takes for the intensity of the 6.4 keV line from a cloud to rise or fall substantially. Of course, that time is affected by the cloud geometry and its relation to the direction of the illuminating source. Taking this into account, Clavel *et al.* (2012) inferred the presence of at least two events, one lasting ~10 years and the other about two years, from observations of clouds at positive Galactic longitudes. Similarly, Chuard *et al.* (2018), using observations of the Sgr C complex on the negative-longitude side of Sgr A*, inferred from their long-term light curves that there were two illuminating events, one lasting 10 years or longer and the other lasting about two years. Figure 2.9 schematically illustrates, from a terrestrial perspective, the present observable locations of the X-ray fronts and the clouds with which they are interacting.

The occurrence date of the X-ray flashes can be determined if the line-of-sight distance to the fluorescing/scattering clouds can be determined (e.g. Churazov *et al.* 2017a, 2017b). This is possible because the cross-section for the scattering of hard X-rays is dependent on their energy and on the angle subtended at the scattering cloud between the illuminating source (presumed to be Sgr A*) and the line of sight to the observer. Furthermore, that angle, along with the column density of the cloud, also affects the strength of the Fe Kα line and the depth of the iron edge (Walls *et al.* 2016). The shape of the spectrum and the relative 6.4 keV line flux can therefore be used to infer the scattering angle. The metallicity of the scattering medium must also be taken into account (Chuard *et al.* 2018). Current estimations of the time of the flashes based on the Sgr B2, Sgr A and Sgr C clouds range between 110 and 250 years ago (Churazov *et al.* 2017; Chuard *et al.* 2018), and the variable emission from the more distant Sgr D cloud could imply an even earlier event between 300 and 1100 years ago (Terrier *et al.* 2018).

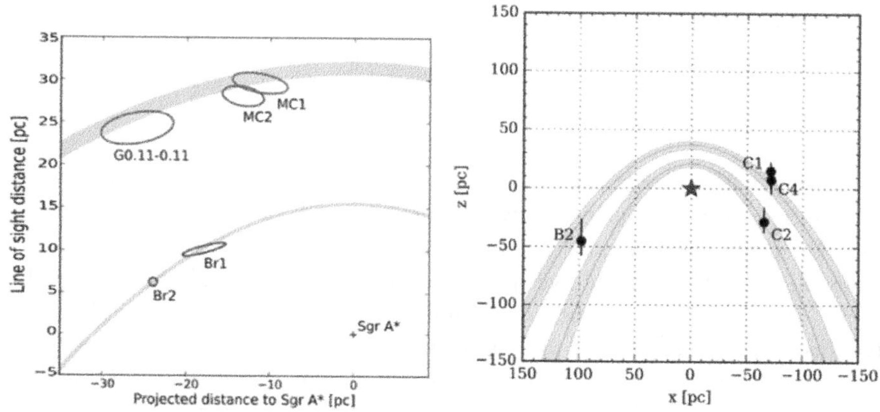

Figure 2.9. *Inferred present loci of molecular clouds recently experiencing the intrusion of the hard X-ray fronts presumed to have originated from Sgr A*. Left: clouds observed by Clavel et al. (2014). The parabolic curves show the presently observable regions of the planar central molecular zone and the durations of the flashes from two separate events, one launched ~100 years ago (red), and the other launched ~200 years ago (blue). Right: the same, but for a different set of clouds observed by Chuard et al. (2018) near Sgr C, plus the Sgr B2 cloud, in which the fluorescent iron line emission has recently subsided, indicating that an ~10-year X-ray front has largely passed through it (Terrier et al. 2010, 2018). Sgr A* is at the origin. The X-ray fronts in the two studies do not necessarily correspond to the same events, but their inferred durations more or less agree with each other*

The multiplicity of these X-ray flashes with a luminosity of five orders of magnitude greater than anything observed over the last 22 years of X-ray monitoring of Sgr A*, and their occurrence in the relatively recent past, make it all the more remarkable that Sgr A* has maintained such a statistically constant emission state during the last quarter century (Ponti *et al.* 2015b; Chen *et al.* 2019), with the possible minor exception of the excess activity observed in 2019 (Do *et al.* 2019b; GRAVITY Collaboration 2020b).

The occurrence of these 100-year events could be provoked by the accretion of a particularly large lump of gas from the rather inhomogeneous interstellar entourage of Sgr A*, by minor tidal disruption events of small planetary-size bodies, or by tidal stripping of the envelopes of giant stars or G objects that pass perilously close to Sgr A* as they move through their orbital periapse. Tidal disruption events involving stars should occur every 10^4–10^5 years (e.g. Phinney 1989), and would be much more energetic than the 100-year events. On even longer time scales, Sgr A* has clearly undergone extremely active stages. A detailed summary of the past activity of Sgr A* has been published by Ponti *et al.* (2013).

2.5. Dynamics of the accretion flow

The source of matter feeding the accretion flow appears to be the winds from the surrounding cluster of young, massive stars (Coker and Melia 1997; Quataert *et al.* 1999; Quataert 2004; Cuadra *et al.* 2008; Ressler *et al.* 2018, 2020a, 2020b; Calderon *et al.* 2020), which give rise to an accretion rate through the Bondi radius of ~10^{-5} $M_{\odot}$ yr^{-1}. Because there are numerous moving wind sources of varying strengths and velocities, the accretion flow is necessarily quite chaotic and time-variable (see Figure 2.10). Observationally, no concentrations of dust and gas have been identified within about 0.04 pc of Sgr A*[5], except for the accretion flow itself and matter contributed by the occasional orbital passage of objects from which material can be tidally removed, such as the mysterious G objects, which could be compact dust clouds or distended stellar envelopes (Gillessen *et al.* 2012; Witzel *et al.* 2014, 2017; Pfuhl *et al.* 2015; Ciurlo *et al.* 2020).

Inside the Bondi radius, the accretion flow eventually migrates inward to the radius at which its angular momentum imposes itself, and the flow eventually circularizes and forms a disk. That disk, however, is likely to be thick, inhomogeneous and time-variable in its orientation, given the variability of the locations and the angular momenta of the sources of the accreting material.

Submillimeter observations of the H30α hydrogen recombination line have been reported, arising from the inner accretion flow onto Sgr A*, by Murchikova *et al.* (2019; see also Yusef-Zadeh *et al.* (2020)); they interpret their very broad (2,200 km s^{-1}), double-peaked line profile and a positional displacement of ± 0.11" (± 0.005 pc) between the blue and red sides of the profile in terms of emission from an accretion disk. Except for the X-ray lines from highly ionized material much further out near the Bondi radius (1.5", Wang *et al.* 2013), the H30α line is the only spectral line that has been seen from the accretion flow so far. Indeed, the fact that the hydrogen

5. There are IR-emitting dust clouds superimposed on or near Sgr A* (e.g. Peissker *et al.* 2020), but their limited proper motions and small internal velocity gradients imply that they are well displaced from the black hole along the line of sight, and are therefore unlikely to be contributing to the accretion flow.

Brackett-γ line is absent at such a low level (Ciurlo *et al.* 2021) led Murchikova *et al.* (2019) to argue that the H30α line must be amplified by a factor of 80 via maser amplification[6]. The formation of a "disk-like structure" of radius ~0.01 pc at temperatures favoring recombination line emission (~10^4 K) emerges from the hydrodynamical models of Calderón *et al.* (2020), based on accretion from the surrounding stellar winds. However, the disk mass in their models (~5×10^{-3} $M_{\odot}$) is one or two orders of magnitude larger than the disk mass estimated by Murchikova *et al.* (2019).

Figure 2.10. *Simulation of the accretion flow onto the GBH (centered), and the stellar winds feeding that flow. The simulation is on a 0.5 pc × 0.5 pc scale. The stellar wind sources appear as circular rings that form bow shocks as their sources orbit the GBH (source: Ressler et al. (2018))*

6. Because Ciurlo *et al.* (2021) question this important result, it clearly warrants verification.

Other models (Ressler *et al.* 2020a, 2020b) indicate that circularization does not happen, except on much smaller scales where the concentrated magnetic field becomes strong enough to assert itself, forming a *magnetically arrested disk* with a radius of a few tens of gravitational radii. This is because the infall velocities at larger distances are comparable to the free-fall time, and because the radiative cooling efficiency is low on such time scales, so the infalling gas heats up and has a tendency to expand back outwards.

The low luminosity of Sgr A* implies that the radiative efficiency of the material that does accrete onto the GBH is very low, as many authors have noted, and have discussed in terms of various categories of a radiatively inefficient accretion flow (see Melia and Falcke (2001), and the references therein). The accretion flow onto the GBH has been interpreted as a hot, optically thin accretion disk, heated by the viscous dissipation of the accretion energy. In this plasma, the ion temperature is much higher than the electron temperature because of the much higher radiative efficiency of the electrons (Shapiro *et al.* 1976). The thermal energy is then advected inward by the accretion flow, giving rise to the generic term advection-dominated accretion flow, or ADAF (Narayan and Yi 1994, 1995; Yuan and Narayan 2014, and references therein). The ions remain hotter than the electrons because the thermal equilibration time scale far exceeds the accretion time.

Two principal variants of the ADAF model have been forwarded: the convection-dominated accretion flow (CDAF) and the adiabatic inflow–outflow solution (ADIOS). The CDAF model is based on the assumption that hot accretion flows are convectively unstable (Narayan *et al.* 2000; Quataert and Gruzinov 2000). The accreting fluid therefore becomes stalled in convective eddies, with the consequence that there is an inward decrease of the accretion rate. The ADIOS model posits that much of the accretion energy acquired by the disk drives a wind from the disk, so there, too, the accretion rate decreases inward (Blandford and Begelman 1999, 2004; Shcherbakov and Baganoff 2010; Begelman 2012; Yuan *et al.* 2012a, 2012b).

2.6. Outflows from the GBH

2.6.1. *An outflowing wind from the GBH accretion flow*

The vast majority of the gas falling through the Bondi radius does not make it to the event horizon. Measurements of the Faraday rotation of the millimeter and submillimeter synchrotron emission from Sgr A* indicate that the accretion rate must be less than or on the order of $1 - 3 \times 10^{-8}$ $M_{\odot}$ yr^{-1} (Bower *et al.* 2003; Marrone *et al.* 2007). Such a low rate is borne out by radiatively inefficient accretion models and MHD models of geometrically thick accretion flows (Sharma *et al.* 2007a, 2007b; Ressler *et al.* 2020b). As a consequence of the fact that the accretion rate onto the GBH is about three orders of magnitude smaller than the rate of material passing through the Bondi radius, there must be an outflowing wind, presumably somewhat collimated by the geometry of the accretion flow. Most of the outflow arises well outside the innermost regions, at distances exceeding ~10 gravitational radii (Yuan *et al.* 2012a).

Observationally, there is ample evidence for an outflow, perhaps quite variable or intermittent, from the GBH. Radio images, for example, show features such as bow shocks that can be attributed to an outflow (Yusef-Zadeh *et al.* 2015; Zhao *et al.* 2016). Infrared images also show bow-shock sources near the GBH (Muzic *et al.* 2010). However, in both the radio and infrared, the bow-shock sources could have resulted from the collective winds of the WR stars in the young nuclear cluster, so that evidence for wind from the GBH is not conclusive. Another radio feature for which a collimated wind from Sgr A* has been suggested is a pair of non-thermal radio structures located 5–10 pc away on opposite sides of the GBH, which are non-thermally emitting structures showing very strongly curved magnetic filaments (Morris *et al.* 2014). The center of curvature of these distorted filaments (named the northern and southern curls) is towards the GBH, consistent with the possibility that a collimated wind, or even a broad jet, propagated out in that direction in the recent past.

X-ray images also show features that can be attributed to collimated outflows from Sgr A*. On a scale of ~ ± 15 pc, there is a bipolar X-ray source straddling the location of the GBH and oriented perpendicular to the Galactic plane (Morris *et al.* 2003; Ponti *et al.* 2015a). The bipolar feature could have resulted from a large outburst at the GBH within the last 10^4 years – a tidal disruption event of a star, for example, although a supernova remnant channeled by a strong vertical magnetic field cannot be ruled out as the cause of the bipolar structure.

On scales of several hundred parsecs, extended X-ray and radio data show the presence of "chimneys" – quasi-cylindrical emission structures extending to both sides of the Galactic plane (Heywood *et al.* 2019; Nakashima *et al.* 2019; Ponti *et al.* 2019, 2021). X-ray emission arises from the interior of the chimneys, heralding a hot outflow arising from somewhere in the central ~100 pc, while radio emission arises from the periphery of the chimneys, presumably from an interface between their hot interiors and the ambient interstellar medium. Episodic ejections from the GBH, caused by occasional large accretion events or tidal disruptions of stars, have almost certainly contributed to the sculpting of the chimneys, but a quasi-continuous wind from the GBH could also be an important contributor. Again, however, relatively frequent supernovae occurring in the central ~100 pc could also be an important or even dominant contributor to the fabrication of the chimneys (Zhang *et al.* 2021).

On the largest scales (± 8 kpc), the γ-ray emitting Fermi bubbles (e.g. Yang *et al.* 2018) are another form of giant bipolar nebula that have possibly been energized by high-velocity emanations from the GBH. The Fermi bubbles could have been created by powerful collimated outflows in a single AGN-level event several million years ago (Zubovas *et al.* 2011; Zubovas and Nayaksin 2012), or by winds generated in a continuous series of tidal disruption events by the GBH extending over a much longer period of time (Ko *et al.* 2020). The energetic particles and hot plasma that are characteristic of the Fermi bubbles have likely been fed through the X-ray chimneys discussed above (Ponti *et al.* 2019). Once again, there are potential supernova alternatives for powering the outflows that have generated the Fermi bubbles. A model in which star formation in the central molecular zone leads to supernova-produced cosmic rays that diffuse upward into the Fermi bubbles, leading to hadronically produced γ-rays, has been proposed by Crocker and Aharonian (2011), Crocker (2012), and Crocker *et al.* (2015). Further work is under way to elucidate the roles of the GBH versus star formation in generating the Fermi bubbles. Similar questions have been raised by the discovery of the even larger (± 14 kpc) X-ray emitting bubbles found by the eROSITA satellite (Predehl *et al.* 2020). As with the Fermi Bubbles, the GBH is a strong candidate for the source of the tremendous energy powering the eROSITA bubbles.

2.6.2. *Does the GBH produce a jet?*

Accreting black holes of all masses produce jets when the accretion flow is sufficiently organized to form a well-defined disk, or when the accreting material carries magnetic flux into a black hole with a substantial spin, in which case the Blandford-Znajec (1977) mechanism operates to drive the Poynting flux out along the spin axis that can accelerate particles into a collimated beam. The degree of collimation depends on the geometry and the inertia of the collimating accretion disk, as well as on the black hole's spin. In the case of the GBH, no *radio* jet has yet been directly observed, although a considerable amount of work has been done to interpret the variable radio emission from Sgr A* in terms of emission from a jet

(Falcke *et al.* 1993, 2009; Falcke and Markoff 2000; Markoff *et al.* 2001, 2007; Yuan *et al.* 2002; Moscibrodzka and Falcke 2013).

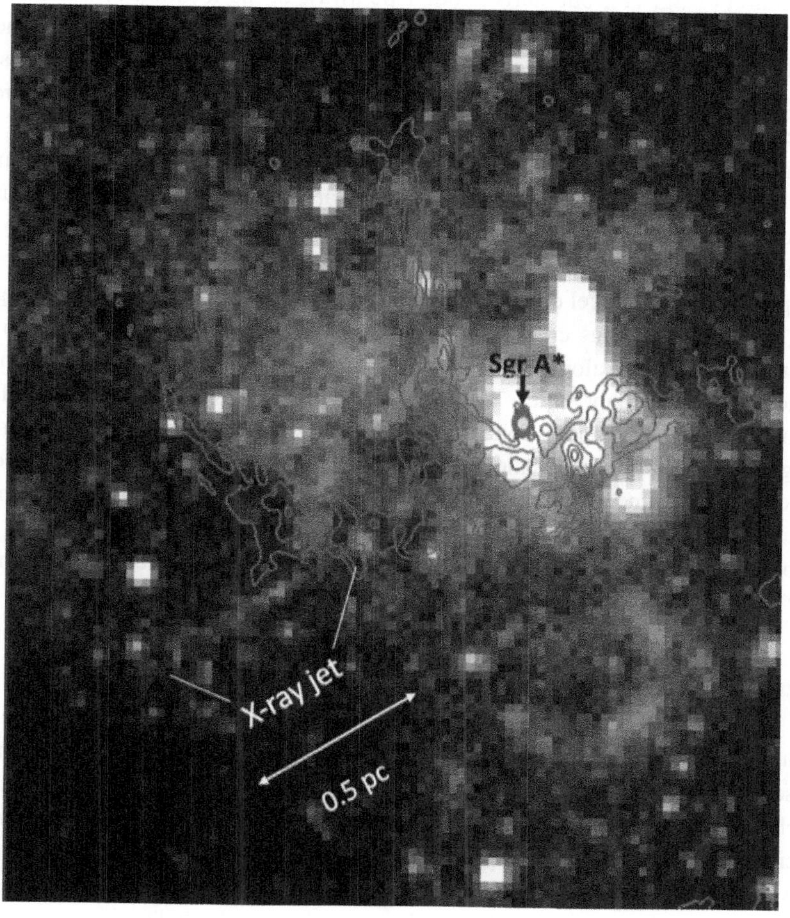

Figure 2.11. *2–8 keV X-ray image of the central few parsecs of the Galaxy, showing the location of the hypothetical X-ray jet from Sgr A* (data from Zhu et al. (2019)). Contours of 6 cm radio continuum emission from Zhao et al. (2016) are superimposed. In this image, equatorial north is up. The jet feature is aligned with the Galaxy's rotation axis. The observable portion of the jet begins at the location of a shock that appears here as an enhanced radio contour*

Various claims have been made for observable, large-scale morphological manifestations of a jet from Sgr A* (Yusef-Zadeh *et al.* 1986, 2012, 2016, 2020; Sofue *et al.* 1989; Su and Finkbeiner 2012; Li *et al.* 2013). The jet candidate discussed by Li *et al.* (2013), and originally considered as a jet by Morris *et al.* (2004) and Muno *et al.* (2008) is based on a ~1 pc linear X-ray feature that is aligned with the location of Sgr A*, but starts about 1 pc away, at the location of a shock front (see Figure 2.11). Li *et al.* (2013) argued that the shock front is created where a broadly collimated wind centered on the X-ray jet impacts an orbiting gaseous stream, known as the Eastern Arm of the Sgr A West HII region. The shock creates a discontinuity in the velocity field. Follow-up work by Zhu *et al.* (2019) with 5.6 megaseconds of X-ray data from the Chandra X-ray Observatory, spanning 18 years, shows that the spectrum of the linear feature is unusually hard at the location closest to Sgr A*, and that it undergoes significant spectral softening with increasing distance from Sgr A*. The hard spectrum is interpreted as resulting from the rapid cooling of the jet electrons in the immediate post-shock region, leading to a piling up in energy, thus creating a quasi-monoenergetic distribution that cools by synchrotron radiation along the jet path. The X-ray data also show that the jet luminosity is invariant over 18 years, with the possible exception of the measurements in 2002, about one year after the extended, dusty object G1 orbited through its periapse at a distance of only about 300 AU from the GBH, where it apparently had some material tidally removed (Witzel *et al.* 2017). The newly unbound material presumably joined the accretion flow onto the GBH and may have thereby been responsible for powering a factor-of-2 increase in the X-ray luminosity of the putative jet in 2002, which is significant at the 2-σ level.

2.7. Perspectives

Sagittarius A* is a prime target of the Event Horizon Telescope (EHT), and we will soon be learning much more about the region surrounding the event horizon. The challenge presented to such a very-long-baseline interferometer is in capturing a sufficient sampling of the UV-plane, that is, enough projected baselines, in the relatively brief time interval over which the source undergoes significant variations. Since Sgr A* varies on time scales as short as minutes, the EHT collaboration will likely be drawing their conclusions on a statistical basis, rather than producing single deep images, as was done for the much slower-varying black hole in M87. The anticipated EHT results on Sgr A*, combined with those to be expected from the next generation of large optical/IR telescopes, which will allow much greater precision on stellar orbits, will no doubt provide an exciting array of information on the accretion process near the event horizon of a spinning black hole, as well as more refined tests of general relativity in such an extreme environment, and perhaps some much-needed information on the launching of a jet.

2.8. References

Acero, F. *et al.* (HESS Collaboration) (2010). *MNRAS*, 402, 1877.

Baganoff, F.K., Bautz, M.W., Brandt, W.N., Chartas, G., Feigelson, E.D., Garmire, G.P., Maeda, Y., Morris, M., Ricker, G.R., Townsley, L.K., Walter, F. (2001). *Nature*, 413, 45.

Baganoff, F.K., Maeda, Y., Morris, M., Bautz, M.W., Brandt, W.N., Cui, W., Doty, J.P., Feigelson, E.D., Garmire, G.P., Pravdo, S.H., Ricker, G.R., Townsley, L.K. (2003). *ApJ*, 591, 891.

Bahcall, J.N. and Wolf, R.A. (1976). *ApJ*, 209, 214.

Ball, D., Özel, F., Christian, P., Chan, C.K., Psaltis, D. (2000). arXiv2005.14251.

Ball, D., Özel, F., Psaltis, D., Chan, C.K. (2016). *ApJ*, 826, 77.

Barriére, N.M., Tomsick, J.A., Baganoff, F.K., Boggs, S.E., Christensen, F.E., Craig, W.W., Dexter, J., Grefenstette, B., Hailey, C.J., Harrison, F.A. (2014). *ApJ*, 786, 46.

Becklin, E.E. and Neugebauer, G. (1968). *ApJ*, 151, 145.

Becklin, E.E. and Neugebauer, G. (1975). *ApJ*, 200, L71.

Begelman, M. (2012). *MNRAS*, 420, 2912.

Blandford, R.D. and Begelman, M. (1999). *MNRAS*, 303, L1.

Blandford, R.D. and Begelman, M. (2004). *MNRAS*, 349, 68.

Blandford, R.D. and Znajec, R.L. (1977). *MNRAS*, 179, 433.

Boehle, A., Ghez, A.M., Schödel, R. *et al.* (2016). *ApJ*, 830, 17.

Bower, G.C., Wright, M.C.H., Falcke, H., Backer, D.C. (2003). *ApJ*, 588, 331.

Bower, G.C., Goss, W.M., Falcke, H., Backer, D.C., Lithwick, Y. (2006). *ApJL*, 648, L127.

Bower, G.C., Markoff, S., Brunthaler, A., Law, C., Falke, H., Maitra, D., Clavel, M., Goldwurm, A., Morris, M.R., Witzel, G., Meyer, L., Ghez, A.M. (2014). *ApJ*, 790, 1.

Bower, G.C., Broderick, A., Dexter, J., Doelman, S., Falcke, H., Fish, V., Johnson, M.D., Marrone, D.P., Moran, J.M., Moscibrodzka, M., Peck, A. (2018). *ApJ*, 868, 101.

Boyce, H., Haggard, D., Witzel, G. *et al.* (2019). *ApJ*, 871, 161.

Brinkerink, C.D., Falcke, H., Law, C.J. *et al.* (2015). *A&A*, 576, A41.

Brown, R.L. and Lo, K.Y. (1982). *ApJ*, 253, 108.

Buchholz, R.M., Schödel, R., Eckart, A. (2009). *A&A*, 499, 483.

Calderón, D., Cuadra, J., Schartmann, M., Burkert, A., Russell, C.M.P. (2020). *ApJL*, 888, L2.

Capelli, R., Warwick, R.S., Porquet, D., Gillessen, S., Predehl, P. (2012). *A&A*, 545, A35.

Chen, Z., Gallego-Cano, E., Do, T. *et al.* (2019). *ApJL*, 882, L28.

Chuard, D., Terrier, R., Goldwurm, A., Clavel, M., Soldi, S., Morris, M.R., Ponti, G., Walls, M., Chernyakova, M. (2018). *A&A*, 610, A34.

Churazov, E., Khabibullin, I., Sunyaev, R., Ponti, G. (2017a). *MNRAS*, 465, 45.

Churazov, E., Khabibullin, I., Ponti, G., Sunyaev, R. (2017b). *MNRAS*, 468, 165.

Ciurlo, A., Campbell, R.D., Morris, M.R., Do, T., Ghez, A.M., Hees, A., Sitarski, B.N., Kosmo O'Neil, K., Chu, D.S., Martinez, G.D., Naoz, S., Stephan, A.P. (2020). *Nature*, 577, 377.

Ciurlo, A., Morris, M.R., Campbell, R.D., Ghez, A.M., Do, T., Chu, D.S. (2021). *ApJ*, 910, 143.

Clavel, M., Terrier, R., Goldwurm, A., Morris, M.R., Ponti, G., Soldi, S., Trap, G. (2013). *A&A*, 558, A32.

Clavel, M., Terrier, R., Goldwurm, A., Morris, M.R., Ponti, G., Soldi, S., Trap, G. (2014). *IAUS*, 303, 344.

Coker, R.F. and Melia, F. (1997). *ApJL*, 488, L149.

Corrales, L., Baganoff, F.K., Wang, Q.D., Nowak, M., Neilsen, J., Markoff, S., Haggard, D., Davis, J., Houck, J., Principe, D. (2020). *ApJ*, 891, 71.

Crocker, R.M. (2012). *MNRAS*, 423, 3512.

Crocker, R.M. and Aharonian, F. (2011). *PRL*, 106, 101102.

Crocker, R.M., Bicknell, G.V., Taylor, A.M., Carretti, E. (2015). *ApJ*, 808, 107.

Cuadra, S., Nayakshin, S., Martins, F. (2008). *MNRAS*, 383, 458.

Dexter, J., Kelly, B., Bower, G.C., Marrone, D.P., Stone, J., Plambeck, R. (2014). *MNRAS*, 442, 2797.

Dexter, J., Tchekhovskoy, A., Jiménez-Rosalez, A. *et al.* (2020). *MNRAS*, 497, 4999.

Do, T., Ghez, A.M., Morris, M.R., Yelda, S., Meyer, L., Lu, J.R., Hornstein, S.D., Matthews, K. (2009a). *ApJ*, 691, 1021.

Do, T., Ghez, A.M., Morris, M.R., Lu, J.R., Matthews, K., Yelda, S., Larkin, J. (2009b). *ApJ*, 703, 1323.

Do, T., Lu, J.R., Ghez, A.M., Morris, M.R., Yelda, S., Martinez, G.D., Wright, S.A., Matthews, K. (2013a). *ApJ*, 764, 154.

Do, T., Martinez, G.D., Yelda, S., Ghez, A., Bullock, J., Kaplinghat, M., Lu, J.R., Peter, A.H.G., Phifer, K. (2013b). *ApJL*, 779, L6.

Do, T., Hees, A., Ghez, A. *et al.* (2019a). *Science*, 365, 664.

Do, T., Witzel, G., Gautam, A.K. *et al.* (2019b), *ApJ*, 882, L27.

Dodds-Eden, K., Gillessen, S., Fritz, T.K. *et al.* (2011). *ApJ*, 728, 37.

Eckart, A. and Genzel, R. (1996). *Nature*, 383, 415.

Eckart, A. and Genzel, R. (1997). *MNRAS*, 284, 576.

Eckart, A., Schödel, R., Meyer, L., Trippe, S., Ott, T., Genzel, R. (2006). *A&A*, 455, 1.

Eckart, A., Baganoff, F.K., Zamaninasab, M. *et al.* (2008). *A&A*, 479, 625.

Eckart, A., García-Marín, M., Vogel, S.N. *et al.* (2012). *A&A*, 537, A52.

Eisenhauer, F., Schödel, R., Genzel, R., Ott, T., Tecza, M., Abuter, R., Eckart, A., Alexander, T. (2003). *ApJL*, 597, L121.

Eisenhauer, F., Genzel, R., Alexander, T. *et al.* (2005). *ApJ*, 628, 246.

Event Horizon Telescope Collaboration (2019). *ApJL*, 875, L1.

Falcke, H. and Markoff, S. (2000). *A&A*, 362, 113.

Falcke, H., Mannheim, K., Biermann, P.L. (1993). *A&A*, 278, L1.

Falcke, H., Markoff, S., Bower, G.C. (2009). *A&A*, 496, 77.

Fazio, G.G., Hora, J.L., Witzel, G. *et al.* (2018). *ApJ*, 864, 58.

von Fellenberg, S.D., Gillessen, S., Graciá-Carpio, J. *et al.* (2018). *ApJ*, 862, 129.

Figer, D.F., Gilmore, D., Kim, S.S., Morris, M., Becklin, E.E., McLean, I.S., Gilbert, A.M., Graham, J.R., Larkin, J.E., Levenson, N.A., Teplitz, H.I. (2003). *ApJ*, 599, 1139.

Fritz, T.K., Chatzopoulos, S., Gerhard, O., Gillessen, S., Genzel, R., Pfuhl, O., Tacchella, S., Eisenhauer, F., Ott, T. (2016). *ApJ*, 821, 44.

Gallego-Cano, E., Schödel, R., Dong, H., Nogueras-Lara, F., Gallego-Calvente, A.T., Amaro-Seoane, P., Baumgardt, H. (2018). *A&A*, 609, A26.

Genzel, R., Eckart, A., Ott, T., Eisenhauer, F. (1997). *MNRAS*, 291, 219.

Genzel, R., Pichon, C., Eckart, A., Gerhard, O.E., Ott, T. (2000). *MNRAS*, 317, 348.

Genzel, R., Schödel, R., Ott, T., Eckart, A., Alexander, T., LaCombe, F., Rouan, D., Aschenbach, B. (2003). *Nature*, 425, 934.

Genzel, R., Eisenhauer, F., Gillessen, S. (2010). *Rev. Mod. Phys.*, 82, 3121.

Ghez, A.M., Klein, B.L., Morris, M., Becklin, E.E. (1998). *ApJ*, 509, 678.

Ghez, A.M., Morris, M., Becklin, E.E., Tanner, A., Kremenek, T. (2000). *Nature*, 407, 349.

Ghez, A.M., Wright, S.A., Matthews, K., Thompson, D., Le Mignant, D., Tanner, A., Hornstein, S.D., Morris, M., Becklin, E.E., Soifer, B.T. (2004). *ApJL*, 601, L159.

Ghez, A.M., Salim, S., Hornstein, S.D., Tanner, A., Lu, J.R., Morris, M., Becklin, E.E., Duchêne, G. (2005). *ApJ*, 620, 744.

Ghez, A.M., Salim, S., Weinberg, N.N., Lu, J.R., Do, T., Dunn, J.K., Matthews, K., Morris, M., Yelda, S., Becklin, E.E., Kremenek, T., Milosavljevic, M., Naiman, J. (2008). *ApJ*, 689, 1044.

Gillessen, S., Eisenhauer, F., Trippe, S., Alexander, T., Genzel, R., Martins, F., Ott, T. (2009). *ApJ*, 692, 1075.

Gillessen, S., Genzel, R., Fritz, T.K., Quataert, E., Alig, C., Burkert, A., Cuadra, J., Eisenhauer, F., Pfuhl, O., Dodds-Eden, K., Gammie, C.F., Ott, T. (2012). *Nature*, 481, 51.

Gillessen, S., Plewa, P.M., Eisenhauer, F., Sari, R., Waisberg, I., Habibi, M., Pfuhl, O., George, E., Dexter, J., von Fellenberg, S., Ott, T., Genzel, R. (2017). *ApJ*, 837, 30.

Goss, W.M., Brown, R.L., Lo, K.Y. (2003). *ANS*, 324, 497.

GRAVITY Collaboration: Abuter, R. *et al.* (2018a). *A&A*, 615, L15.

GRAVITY Collaboration: Abuter, R. *et al.* (2018b). *A&A*, 618, L10.

GRAVITY Collaboration: Abuter, R. *et al.* (2019). *A&A*, 625, L10.

GRAVITY Collaboration: Abuter, R. *et al.* (2020a). *A&A*, 636, L5.

GRAVITY Collaboration: Abuter, R. *et al.* (2020b). *A&A*, 638, A2.

GRAVITY Collaboration: Abuter, R. *et al.* (2021). *A&A*, 654, A22.

Haggard, D., Nynka, M., Mon, B. *et al.* (2019). *ApJ*, 886, 96.

Haller, J.W., Rieke, M.J., Rieke, G.H., Tamblyn, P., Close, L., Melia, F. (1996). *ApJ*, 456, 194.

Heywood, I., Camilo, F., Cotton, W.D. *et al.* (2019). *Nature*, 573, 235.

Hora, J.L., Witzel, G., Ashby, M.L.N. *et al.* (2014). *ApJ*, 793, 120.

Hornstein, S.D., Matthews, K., Ghez, A.M., Lu, J.R., Morris, M., Becklin, E.E., Rafelski, M., Baganoff, F.K. (2007). *ApJ*, 667, 900.

Inui, T., Koyama, K., Matsumoto, H., Tsuru, T.G. (2009). *PASJ*, 61, 241.

Issaoun, S., Johnson, M.D., Blackburn, L. *et al.* (2019). *ApJ*, 871, 30.

Issaoun, S., Johnson, M.D., Blackburn, L. *et al.* (2021). *ApJ*, 915, 99.

Johnson, M.D., Narayan, R., Psaltis, D. *et al.* (2018). *ApJ*, 865, 104.

Kawashima, T., Matsumoto, Y., Matsumoto, R. (2017). *PASJ*, 69, 43.

Ko, C.M., Breitschwerdt, D., Chernyshov, D.O., Cheng, H., Dai, L., Dogiel, V.A. (2020). *ApJ*, 904, 46.

Koyama, K., Maeda, Y., Sonobe, T., Takeshima, T., Tanaka, Y., Yamauchi, S. (1996). *PASJ*, 48, 249.

van der Laan, H. (1966). *Nature*, 211, 1131.

Lacy, J.H., Baas, F., Townes, C.H., Geballe, T.R. (1979). *ApJ*, 227, L17.

Lacy, J.H., Townes, C.H., Geballe, T.R., Hollenbach, D.J. (1980). *ApJ*, 241, L132.

Liu, S., Melia, F., Petrosian, V. (2006a). *ApJ*, 636, 798.

Liu, S., Petrosian, V., Melia, F., Fryer, C.L. (2006b). *ApJ*, 648, 1020.

Lo, K.-Y., Schilizzi, R.T., Cohen, M.H., Ross, H.N. (1975). *ApJ*, 202, L63.

Lu, R.-S., Krichbaum, T.P., Roy, A.L. *et al.* (2018). *ApJ*, 859, 60.

Ma, R.-Y., Roberts, S.R., Li, Y.-P., Wang, Q.D. (2019). *MNRAS*, 483, 5614.

Macquart, J.P. and Bower, G.C. (2006). *ApJ*, 641, 302.

Maoz, E. (1998). *ApJL*, 494, L181.

Markoff, S., Falcke, H., Yuan, F., Biermann, P.L. (2001). *A&A*, 379, L13.

Markoff, S., Bower, G.C., Falcke, H. (2007). *MNRAS*, 379, 1519.

Marrone, D.P., Moran, J.M., Zhao, J.-H., Rao, R. (2006). *ApJ*, 640, 308.

Marrone, D.P., Moran, J.M., Zhao, J.-H., Rao, R. (2007). *ApJL*, 654, L57.

Marrone, D.P., Baganoff, F.K., Morris, M.R. *et al.* (2008). *ApJ*, 682, 373.

Mauerhan, J.C., Morris, M., Walter, F., Baganoff, F.K. (2005). *ApJL*, 623, L25.

McGinn, M.T., Sellgren, K., Becklin, E.E., Hall, D.N.B. (1989). *ApJ*, 338, 824.

Melia, F. and Falcke, H. (2001). *ARAA*, 39, 309.

Meyer, L., Do, T., Ghez, A., Morris, M.R., Yelda, S., Schödel, R., Eckart, A. (2009). *ApJL*, 694, L87.

Meyer, L., Witzel, G., Longstaff, F.A., Ghez, A.M. (2014). *ApJ*, 791, 24.

Miyazaki, A., Tsuboi, M., Tsutsumi, T. (2013). *PASJ*, 65, L6.

Morris, M.R., Baganoff, F., Muno, M., Howard, C., Maeda, Y., Feigelson, E., Bautz, M., Brandt, N., Chartas, G., Garmire, G., Townsley, L. (2003). *Astron. Nachr.*, 324(S1), 167.

Morris, M.R., Meyer, L., Ghez, A.M. (2012). *RAA*, 12, 995.

Morris, M.R., Zhao, J.-H., Goss, W.M. (2014). *IAUS*, 303, 369.

Moscibrodzka, M. and Falcke, H. (2013). *A&A*, 559, L3.

Muno, M.P., Baganoff, F.K., Brandt, W.N., Park S., Morris, M.R. (2007). *ApJL*, 656, L69.

Muno, M.P., Baganoff, F.K., Brandt, W.N., Morris, M.R. (2008). *ApJ*, 673, 251.

Murchikova, L. (2021). *ApJL*, 910, L1, 4 pp.

Murchikova, E.M., Phinney, E.S., Pancoast, A., Blandford, R.D. (2019). *Nature*, 570, 83.

Muzic, K., Eckart, A., Schödel, R., Buchholz, R., Zamaninasab, M., Witzel, G. (2010). *A&A*, 521, A13.

Nakashima, S., Koyama, K., Wang, Q.D., Enokiya, R. (2019). *ApJ*, 875, 32.

Narayan, R. and Yi, I. (1994). *ApJL*, 428, L13.

Narayan, R. and Yi, I. (1995). *ApJ*, 452, 710.

Narayan, R., Igumenshchev, I.V., Abramowicz, M.A. (2000). *ApJ*, 539, 798.

Neilsen, J., Nowak, M.A., Gammie, C. *et al.* (2013). *ApJ*, 774, 42.

Neilsen, J., Markoff, S., Nowak, M.A. *et al.* (2015). *ApJ*, 799, 199.

Nobukawa, M., Ryu, S.G., Tsuru, T.G., Koyama, K. (2011). *ApJL*, 739, L52.

Nowak, M.A., Neilsen, J., Markoff, S.B., Baganoff, F.K., Porquet, D., Grosso, N., Levin, Y., Houck, J., Eckart, A., Falcke, H., Ji, L., Miller, J.M., Wang, Q.D. (2012). *ApJ*, 759, 95.

Oort, J.H. (1977). *Ann. Rev. Astron. Ap.*, 15, 295.

Paumard, T., Genzel, R., Martins, F. *et al.* (2006). *ApJ*, 643, 1011.

Peissker, F., Eckart, A., Sabha, N.B., Zajacek, M., Bhat, H. (2020). arXiv:2006.03648.

Pfuhl, O., Gillessen, S., Eisenhauer, F., Genzel, R., Plewa, P.M., Ott, T., Ballone, A., Schartmann, M., Burkert, A., Fritz, T.K., Sari, R., Steinberg, E., Madigan, A.-M. (2015). *ApJ*, 798, 111.

Phinney, E.S. (1989). *IAUS*, 136, 543.

Ponti, G., Terrier, R., Goldwurm, A., Belanger, G., Trap, G. (2010). *ApJ*, 714, 732.

Ponti, G., Morris, M.R., Terrier, R., Goldwurm, A. (2013). Cosmic rays in star-forming environments. *ASSP*, 34, 331.

Ponti, G., Morris, M.R., Terrier, R. *et al.* (2015a). *MNRAS*, 453, 172.

Ponti, G., De Marco, B., Morris, M.R. *et al.* (2015b). *MNRAS*, 454, 1525.

Ponti, G., George, E., Scaringi, S. *et al.* (2017). *MNRAS*, 468, 2447.

Ponti, G., Hofmann, F., Churazov, E., Morris, M.R., Haberl, F., Nandra, K., Terrier, R., Clavel, M., Goldwurm, A. (2019). *Nature*, 567, 347.

Ponti, G., Morris, M.R., Churazov, E., Heywood, I., Fender, R.P. (2021). *A&A*, 646, A66.

Predehl, P., Sunyaev, R.A., Becker, W. *et al.* (2020). *Nature*, 588, 227.

Quataert, E. (2004). *ApJ*, 613, 322.

Quataert, E. and Gruzinov, A. (2000). *ApJ*, 539, 809.

Quataert, E., Narayan, M.J., Reid, M.J. (1999). *ApJL*, 517, L101.

Ressler, S.M., Quataert, E., Stone, J.M. (2018a). *MNRAS*, 478, 3544.

Ressler, S.M., White, C.J., Quataert, E., Stone, J.M. (2018b). *ApJL*, 896, L6.

Ressler, S.M., Quataert, E., Stone, J.M. (2020). *MNRAS*, 492, 3272.

Roberts, D.A., Yusef-Zadeh, F., Goss, W.M. (1996). *ApJ*, 459, 627.

Rubilar, G.F. and Eckart, A. (2001). *A&A*, 374, 95.

Schödel, R., Morris, M.R., Muzic, K., Alberdi, A., Meyer, L., Eckart, A., Gezari, D.Y. (2011). *A&A*, 532, A83.

Sellgren, K., McGinn, M.T., Becklin, E.E., Hall, D.N. (1990). *ApJ*, 359, 112.

Serabyn, E. and Lacy, J.H. (1975). *ApJ*, 293, 445.

Serabyn, E., Carlstrom, J., Lay, O., Lis, D.C., Hunter, T.R., Lacy, J.H., Hills, R.E. (1997). *ApJL*, 490, L77.

Shahzamanian, B., Eckart, A., Valencia-S., M. *et al.* (2015). *A&A*, 576, A20.

Shapiro, S.L., Lightman, A.P., Eardley, D.M. (1976). *ApJ*, 204, 187.

Sharma, P., Quataert, E., Hammett, G.W., Stone, J.M. (2007a). *ApJ*, 667, 714.

Sharma, P., Quataert, E., Stone, J.M. (2007b). *ApJ*, 671, 1696.

Shcherbakov, R.V. and Baganoff, F.K. (2010). *ApJ*, 716, 504.

Shen, Z.-Q., Lo, K.Y., Liang, M.-C., Ho, P.T.P., Zhao, J.-H. (2005). *Nature*, 438, 62.

Sofue, Y., Reich, W., Reich, P. (1989). *ApJ*, 341, L47.

Su, M. and Finkbeiner, D.P. (2012). *ApJ*, 753, 61.

Sunyaev, R., Markevitch, M., Pavlinsky, M. (1993). *ApJ*, 407, 606.

Tagger, M. and Melia, F. (2006). *ApJL*, 636, L33.

Terrier, R., Ponti, G., Bélanger, G., Decourchelle, A., Tatischeff, V., Goldwurm, A., Trap, G., Morris, M.R., Warwick, R. (2010). *ApJ*, 719, 143.

Terrier, R., Clavel, M., Soldi, S., Goldwurm, A., Ponti, G., Morris, M.R., Chuard, D. (2018). *A&A*, 612, A102.

Torres, D.F., Capozziello, S., Liambase, G. (2000). *Phys. Rev. D.*, 62, 104012.

Trap, G., Goldwurm, A., Dodds-Eden, K. *et al.* (2011). *A&A*, 528, A140.

Trippe, S., Paumard, T., Ott, T., Gillessen, S., Eisenhauer, F., Martins, F., Genzel, R. (2007). *MNRAS*, 375, 764.

Tsiklauri, D. and Viollier, R.D. (1998). *ApJ*, 500, 591.

Walls, M., Chernyakova, M., Terrier, R., Goldwurm, A. (2016). *MNRAS*, 463, 2893.

Wang, Q.D., Lu, F.J., Gotthelf, E.V. (2006). *MNRAS*, 367, 937.

Wang, Q.D., Nowak, M.A., Markoff, S.B. *et al.* (2013). *Science*, 341, 981.

Witzel, G., Eckart, A., Bremer, M. *et al.* (2012). *ApJS*, 203, 18.

Witzel, G., Ghez, A.M., Morris, M.R., Sitarski, B.N., Boehle, A., Naoz, S., Campbell, R., Becklin, E.E., Canalizo, G., Chappell, S., Do, T., Lu, J.R., Matthews, K., Meyer, L., Stockton, A., Wizinowich, P., Yelda, S. (2014). *ApJL*, 796, 8.

Witzel, G., Sitarski, B.N., Ghez, A.M., Morris, M.R., Hees, A., Do, T., Lu, J.R., Naoz, S., Boehle, A., Martinez, G., Chappell, S., Schödel, R., Meyer, L., Yelda, S., Becklin, E.E., Matthews, K. (2017). *ApJ*, 847, 80.

Witzel, G., Martinez, G., Hora, J. *et al.* (2018). *ApJ*, 863, 15.

Witzel, G., Martinez, G., Willner, S.P. *et al.* (2021). *ApJ*, 917, 73.

Wollman, E.R., Geballe, T.R., Lacy, J.H., Townes, C.H., Rank, D.M. (1976). *ApJ*, 205, L6.

Yang, H.-Y.K., Ruszkowski, M., Zweibel, E.G. (2018). *Galaxies*, 6, 29.

Yuan, F. (2011). *ASP Conf. Ser.*, 439, 346.

Yuan, F. and Narayan, R. (2014). *ARAA*, 52, 529.

Yuan, Q. and Wang, Q.D. (2016). *MNRAS*, 456, 1438.

Yuan, F., Markoff, S., Falcke, H. (2002). *A&A*, 383, 854.

Yuan, F., Quataert, E., Narayan, R. (2003). *ApJ*, 598, 301.

Yuan, F., Wu, M., Bu, D. (2012a). *ApJ*, 761, 129.

Yuan, F., Bu, D., Wu, M. (2012b). *ApJ*, 761, 130.

Yusef-Zadeh, F., Morris, M., Slee, O.B., Nelson, G.J. (1986). *ApJL*, 300, L47.

Yusef-Zadeh, F., Roberts, D., Wardle, M., Heinke, C.O., Bower, G.C. (2006). *ApJ*, 650, 189.

Yusef-Zadeh, F., Arendt, R., Bushouse, H., Cotton, W., Haggard, D., Pound, M.W., Roberts, D.A., Royster, M., Wardle, M. (2012). *ApJL*, 758, L11.

Yusef-Zadeh, F., Roberts, D., Wardle, M., Cotton, M., Schödel, R., Royster, M.J. (2015). *ApJL*, 801, L26.

Yusef-Zadeh, F., Wardle, M., Schödel, R., Roberts, D.A., Cotton, W., Bushouse, H., Arendt, R., Royster, M. (2016). *ApJ*, 819, 60.

Yusef-Zadeh, F., Royster, M., Wardle, M., Cotton, W., Kunneriath, D., Heywood, I., Michail, J. (2020). *MNRAS*, 499, 3909.

Zamaninasab, M., Eckart, A., Witzel, G. *et al.* (2010). *A&A*, 510, A3.

Zhang, S., Hailey, C.J., Mori, K. *et al.* (2015). *ApJ*, 815, 132.

Zhang, M., Li, Z., Morris, M.R. (2021). *ApJ*, Under review.

Zhao, J.-H., Ekers, R.D., Goss, W.M., Lo, K.-Y., Narayan, R. (1989). *IAUS*, 136, 535.

Zhao, J.-H., Young, K.H., Herrnstein, R.M., Ho, P.T.P., Tsutsumi, T., Lo, K-Y., Goss, W.M., Bower, G.C. (2003). *ApJ*, 586, L29.

Zhao, J.-H., Morris, M., Goss, W.M. (2016). *ApJ*, 817, 171.

Zhu, Z., Li, Z., Morris, M.R., Zhang, S., Liu, S. (2019). *ApJ*, 875, 44.

Zubovas, K. and Nayakshin, S. (2012). *MNRAS*, 424, 666.

Zubovas, K., King, A.R., Nayakshin, S. (2011). *MNRAS*, 415, L21.

3

AGN Accretion Disks

Jean-Pierre LASOTA

Institut d'Astrophysique de Paris, France, and Nicolaus Copernicus Astronomical Center, Warsaw, Poland

3.1. Introduction

Accretion disks are ubiquitous in the universe. The spectacular ALMA image of the protostellar disk in HL Tau is breathtaking and the M87* black hole silhouette observed by the Event Horizon Telescope (EHT) showed us the light emitted by matter as close to the event horizon as possible.

Understanding accretion disks around black holes is interesting in itself because of the fascinating and complex physics involved; however it is also fundamental for understanding the coupled evolution of galaxies and their nuclear black holes, that is, it is fundamental for understanding the growth of structures in the universe. The chances that inflows onto black holes are strictly radial, as assumed in many models, are slim.

Accretion disks in active galactic nuclei (AGNs) play a central role in the complex engine operating around the central supermassive black hole. They are the heart of the accretion–ejection processes, playing the role of a rugby-union fly-half, receiving and passing matter, energy and magnetic fields to various players of the inflow–outflow team. Accretion disks, in their standard, geometrically thin version, have the advantage of being the only team members for which a well-confirmed, though not perfect, physical model exists. This section concentrates on this model, its properties and observational consequences, as well as its possible generalizations to situations where it loses its validity.

Active Galactic Nuclei,
coordinated by Françoise COMBES.

3.2. Black holes

Black holes, which are at the center of all AGNs, are purely general-relativistic objects, and the description of the properties and behavior of matter and light in their vicinity requires the use of the formalism of Einstein's theory of gravitation. Unfortunately, this theory, now used on a daily basis in astronomy (e.g. observations of the universe in gravitational waves) and even in everyday life (GPS in our smartphones), has still not made it to the physics curriculum in most universities. We will therefore present here only the most important facts about the physics in the vicinity of black holes without derivation and with minimal justification.

3.2.1. *The horizon*

The Schwarzschild radius (radius of a non-rotating black hole) is

$$R_S = \frac{2GM}{c^2} = 2.95 \times 10^{13} \left(\frac{M}{10^8\,\mathrm{M}_\odot}\right) \mathrm{cm} = 9.58 \times 10^{-6} \left(\frac{M}{10^8\,\mathrm{M}_\odot}\right) \mathrm{pc}, \qquad [3.1]$$

where G is the gravitational constant, M is the mass of the gravitating body, and c is the speed of light.

Quite often in the literature, as well as in this book, we find the so-called *gravitational radius* $R_g = 0.5R_S$ used as unit of length. Here however, we prefer to use as unit of length the Schwarzschild radius R_S.

When the black hole is rotating, it is represented by the Kerr solution and the radius of its horizon[1] is equal to

$$R_H = \frac{GM}{c^2} + \left[\left(\frac{GM}{c^2}\right)^2 - \left(\frac{J}{Mc}\right)^2\right]^{1/2} = \frac{GM}{c^2}\left[1 + \left(1 - a^2\right)^{1/2}\right], \qquad [3.2]$$

where J is the black hole's angular momentum and its dimensionless angular momentum is defined as[2]

$$a := \frac{Jc}{GM^2}. \qquad [3.3]$$

Therefore, a horizon, that is, a black hole, exists only for $0 \leq a \leq 1$. For $a > 1$, the Kerr solution represents a space-time containing a naked (not covered by a horizon,

1. This is the radius of the external horizon. There exists also an inner horizon of no astrophysical interest.

2. We must be careful not to confuse this dimensionless "a" with the "a" often used in the general-relativistic literature defined as J/Mc which has the dimension of length.

i.e. observable) singularity and time-like trajectories that violate causality. The cosmic censorship hypothesis, according to which all singularities in the real universe (except for the Big Bang) are covered by an event horizon, remains to be proved. However, the third law of black hole thermodynamics guarantees that no black hole can be spun-up to $a = 1$.

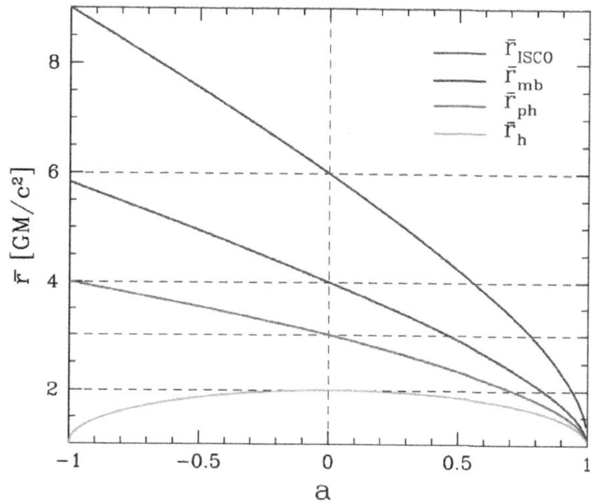

Figure 3.1. *Radii (in units of R_g) of the characteristic orbits in the Kerr metric as a function of a. The innermost stable circular orbit: $r_{\rm ISCO}$, the marginally bound orbit $r_{\rm IBCO}$ (marked $\bar{r}_{\rm mb}$), the photon orbit: $r_{\rm ph}$ and the black hole horizon: r_H (marked $\bar{r}_{\rm h}$). The negative values of a correspond to orbits counter-rotating with respect to the black hole (courtesy of Sądowski (2011)). For a color version of this figure, see www.iste.co.uk/combes/nuclei.zip*

The numerical value of the maximum angular momentum of a black hole is

$$J_{\rm max} = \frac{GM^2}{c} = 8.9 \times 10^{48} \left(\frac{M}{{\rm M}_\odot}\right)^2 {\rm g\ cm^2\ s^{-1}}. \qquad [3.4]$$

which, for a solar–mass black hole, is comparable to the angular momentum of the Sun: $J_\odot = 1.63 \times 10^{48}{\rm g\ cm^2\ s^{-1}}$, or $a_\odot = 0.185$.

For a black hole with the mass of the Galaxy ($M_G = 1.5 \times 10^{11}{\rm M}_\odot$), the maximum angular momentum would be $2.0 \times 10^{71}\,(M_G/{\rm M}_\odot)^2\,{\rm g\ cm^2\ s^{-1}}$, so the Galaxy whose angular momentum is $\sim 10^{74}{\rm g\ cm^2\ s^{-1}}$ would have to lose most of it, if it were to collapse into a black hole. The maximum observed mass of a black hole is $6 \times 10^{10}\,{\rm M}_\odot$.

3.2.2. *Characteristic orbits*

The space–time around a black hole is curved which implies that particles follow geodesics in a non-Euclidean geometry. Far from the black holes, the motion along these geodesics is well approximated by Newtonian orbits, but near the horizon the motion of bodies and light is strongly affected by general-relativistic effects. Whereas in Newtonian gravity, the angular momentum on Keplerian orbits decreases monotonically with the square-root of the distance from the center; in general relativity it has a minimum at the orbit called an ISCO, which stands for "innermost stable circular orbit". Closer to the horizon, free circular orbits are unstable, so that a Keplerian accretion disk (see below) has to end at the ISCO. "Keplerian" means that matter in the disk moves on approximately free-falling orbit, which is the case for low accretion rates. In the case of high accretion rates, the radial pressure gradients in the disk are no longer negligible and the inner disk's edge is pushed inwards, but cannot be closer to the horizon than the innermost bound circular orbit, which we refer to as an IBCO, but it should be noted that it is known under the name of "$R_{\rm mb}$"; "mb" corresponding to "marginally bound".

ISCO

The mathematical expression for the ISCO radius is complex and can be easily found in the literature. For the Schwarzschild solution, $R_{\rm ISCO} = 3R_S$. For a maximally rotating black hole, the radius of the counterrotating ISCO is at $R_{\rm ISCO} = 4.5R_S$, while the radius of a corotating ISCO is formally at $R_{ISCO} = 0.5R_S = R_H$, but this is the result of a degeneracy for $a = 1$ of the coordinate systems used.

For a Schwarzschild black hole, the frequency associated with the ISCO at $R_{\rm ISCO} = 3R_S$ is

$$\nu_K(R_{\rm ISCO}) = 2.2 \left(\frac{M}{10^8 {\rm M}_\odot}\right)^{-1} \mu{\rm Hz}. \qquad [3.5]$$

A quasi-periodic oscillation (QPO) of $\sim 27\mu$Hz that could be associated with the ISCO frequency has been observed in an active galaxy of type NLS1 (meaning Narrow Line Seyfert 1), whose mass is presumably $1 - 4 \times 10^6 {\rm M}_\odot$.

IBCO

The IBCO is at radius:

$$R^{\pm}_{\rm IBCO} = \frac{GM}{c^2}\left[2 \mp a + 2\sqrt{1 \mp a}\right], \qquad [3.6]$$

which for a Schwarzschild black hole gives $R_{\rm IBCO} = 2R_S$. When $a = 1$, for a counterrotating orbit $R_{\rm IBCO} \approx 2.9R_S$, in the corotating case, $R_{\rm IBCO} = 0.5R_S$ for the reasons mentioned above.

Photon orbit

The photon circular orbit around a black hole is also the innermost circular orbit (ICO), and its radius is at

$$R^{\pm}_{\text{phot}} = R_S \left(1 + \cos\left[\frac{2}{3}\cos^{-1}(\mp a)\right]\right). \quad [3.7]$$

The "photon orbit" is at $1.5R_S$ for a $a = 0$, at $4R_S$ for $a = 1$ retrograde orbit and, as all the other characteristic orbits, at $0.5R_S$ for prograde rotation with $a = 1$.

The photon circular orbit is unstable but in a suitable setting an observer could see a "photon ring" produced by photons "wrapping around" R_{phot} before reaching the detector. It has been claimed that such a ring has been observed by the EHT around the M87 central black hole, but this is far from being certain.

3.2.3. *Binding energy and accretion efficiency*

The binding energy per unit mass of an orbit is (assuming a unit system where c = 1),

$$E_{\text{bind}} = 1 - E_K, \quad [3.8]$$

where E_K is the kinetic energy per unit rest-mass of a Keplerian orbit (in the Newtonian case $E_K = (GM/2c^2R)$.

At the ISCO the binding energy per unit rest-mass is

– $1 - \sqrt{8/9} \approx 0.06$ for $a = 0$

– $1 - \sqrt{1/3} \approx 0.42$ for $a = 1$.

Therefore, this corresponds to the efficiencies of accretion in a Keplerian disk around a black hole.

By definition, the binding energy at the IBCO is equal to zero. Therefore, pressure gradients in an accretion flow that push the inner disk towards the black hole reduce accretion efficiency.

3.2.4. *Rotating space*

A purely general-relativistic effect, with no Newtonian equivalent or analogue is the rotation of space: the rotation of a gravitating body "drags" with it the surrounding space. The angular velocity of this dragging is:

$$\Omega_{\text{LT}} \approx \frac{2GJ}{c^2R^3}, \quad [3.9]$$

where J, as before is the black hole angular momentum. "LT" stands for "Lense–Thirring" – the names of the two discoverers of this effect in the weak-field approximation of Einstein's theory of gravitation. $\Omega_{\rm LT}$ corresponds to the angular velocity with which the orbits of misaligned (not in the equatorial plane) test particles precess around a black hole.

The corresponding precession time is

$$t_{\rm LT} \approx \frac{1}{|\Omega_{\rm LT}|} = \frac{c^2 R^3}{2GJ} = \frac{2}{a}\left(\frac{R}{R_{\rm S}}\right)^2 \frac{R}{c}, \qquad [3.10]$$

which, close to the black hole can be as short as the dynamical time $R/c \approx 980\left(\frac{M}{10^8 \rm M_\odot}\right)$ s, but since the precession is strongly differential, decreasing with radius, far from the black hole it is entirely negligible:

$$t_{\rm LT} \approx 7.1 \times 10^{10} a^{-1} \left(\frac{M}{10^8 \rm M_\odot}\right)^{-2} \left(\frac{R}{1 \rm pc}\right)^3 \text{ yrs}. \qquad [3.11]$$

However, dissipation in a misaligned disk can lead to the alignment of its innermost regions with the black hole spin, creating a warp, through the so-called Bardeen–Petterson effect (Bardeen and Petterson 1975) (see below).

3.2.5. *The ergoregion*

The surface of a black hole rotates with angular velocity

$$\Omega_{\rm H} = \frac{ac}{2R_H}, \qquad [3.12]$$

where $\Omega_{\rm H}$ is the angular velocity of the horizon, that is, the angular velocity of the horizon-forming light-rays with respect to infinity. The horizon rotates.

The fact that a rotating black hole induces rotation of space has two important astrophysical implications. First, it creates a torque that forces the matter in the tilted disk towards the equatorial plane in the so-called *Bardeen–Petterson effect*. This effect, strongest near the rotating body's surface, induces the disk precession. Second, in the case of a Kerr black hole, there exists an ergoregion between the horizon and the surface defined by

$$R_{\rm ergo}(\theta) = \frac{GM}{c^2}\left[1 + \left(1 - a^2 cos^2\theta\right)^{1/2}\right] \qquad [3.13]$$

where being at rest with respect to infinity is impossible: to resist the rotation of space would require local superluminal motion.

This implies that the ergoregion contains trajectories with negative mass energy as seen from infinity, although the energy measured in a frame rotating with the space is positive. If, for example, an ingoing particle with energy $E_1 > 0$ decays in the ergoregion into two particles, one with energy $E_2 > 0$, which escapes to infinity, the other with $\Delta E_H < 0$, which falls into the black hole (negative-energy particles cannot leave the ergoregion because this would require accelerating to speeds larger than the speed of light), $E_2 = E_1 - \Delta E_H > E_1$, so that the outgoing particle is more energetic than the ingoing one. This energy has been gained at the expense of the rotational energy of the black hole: the latter absorbed negative energy and negative angular momentum, since negative-energy particles counter-rotate with respect to the black hole spin. These particles carry negative angular momentum. The electromagnetic version of this *mechanical Penrose process* allows tapping the black hole rotational energy which is supposed to be the source of relativistic jets observed in some AGN (see section 3.8).

3.2.6. *Eddington accretion rate*

We will define the Eddington accretion rate as

$$\dot{M}_{\rm Edd} = \frac{L_{\rm Edd}}{\eta c^2} = \frac{1}{\eta}\frac{4\pi GM}{c\kappa_{es}}$$

$$= 1.6 \times 10^{26} \eta_{0.1}^{-1} \left(\frac{M}{10^8 {\rm M}_\odot}\right) {\rm g\,s^{-1}} \quad [3.14]$$

$$= 2.5\, \eta_{0.1}^{-1} \left(\frac{M}{10^8 {\rm M}_\odot}\right) {\rm M}_\odot\, {\rm yr}^{-1}, \quad [3.15]$$

where $\eta = 0.1\eta_{0.1}$ is the radiative efficiency of accretion, κ_{es} is the electron scattering (Thomson) opacity. We will often use accretion rate measured in units of the Eddington accretion rate:

$$\dot{m} = \frac{\dot{M}}{\dot{M}_{\rm Edd}} \quad \text{and use} \quad m_8 = \frac{M}{10^8\,{\rm M}_\odot}. \quad [3.16]$$

We can find in the literature definitions of $\dot{M}_{\rm Edd}$ implicitly using $\eta = 1.0$, or $\eta = 0.2$ so care is recommended when using published numerical values of the accretion rate in Eddington units.

3.3. Disk driving mechanism; viscosity

Very early in the studies of accretion disk physics, it became clear that, if their driving mechanism is viscous, an "anomalous", turbulent viscosity must be at work because astrophysical disks are too large (and their density too small) for the molecular viscosity to be efficient. After years of uncertainty, it is now obvious that the turbulence in ionized Keplerian disks is due to the presence of a magneto-rotational instability (MRI) also known as the Balbus–Hawley mechanism (Balbus and Hawley 1991), occurring in *weakly* magnetized, differentially rotating plasma. However, despite impressive developments, numerical simulations, even in their global 3D form, still suffer from weaknesses that, in most cases, make their direct application to real accretion flows almost infeasible.

One of the problems is related to the value of the ratio of the (vertically averaged) total stress to thermal (vertically averaged) pressure

$$\alpha = \frac{\langle \tau_{R\varphi} \rangle_z}{\langle P \rangle_z}, \qquad [3.17]$$

where the stress $\tau_{R\varphi}$ is the sum of the Maxwell and Reynolds stresses:

$$\tau_{R\varphi} = -\frac{B_R B_\varphi}{4\pi} + \rho v_R \delta v_\varphi, \qquad [3.18]$$

where indices R and φ denote, respectively, the radial and azimuthal components of the magnetic intensity $\mathbf{B}$, $\mathbf{v}$ and $\delta\mathbf{v}$ vectors, and ρ denotes the density. $\delta v_\varphi = v_\varphi + 3\Omega_K R/2$ is the difference between the azimuthal velocity and the mean rotational velocity in the disk, with Ω_K being the Keplerian orbital angular speed (equation [3.22]). $P = P_g + P_{rad}$ is the sum of the thermal pressures; magnetic pressure is not included in this definition. As we will see in the next section, and in the rest of this chapter, the α parameter has played a crucial role in the development of accretion disk astrophysics. Its value ($0 \leq \alpha \leq 1$) can be determined from observations of light variability of some accreting systems, as will be shown below.

According to most MRI simulations $\alpha \sim 10^{-3}$ whereas observations of ionized disks around white dwarfs in binary systems called cataclysmic variables, which are the best "life-size" realization of MRI disks, unambiguously show that $\alpha \approx 0.1 - 0.2$. Disk outbursts in X-ray binaries suggest even $\alpha \approx 0.2 - 2.0$. The effects of convection at temperatures $\sim 10^4$ K increase α to values ~ 0.1 and a net magnetic field in the region might bring its value up to the observed one. On the contrary, some observations of AGN variability give values of the viscosity parameter: $0.01 \leq \alpha \leq 0.03$ for $0.01 \leq L/L_{Edd} \leq 1.0$, still higher than the fiducial MRI-simulation value.

Another problem is related to cold disks. For the standard MRI to work, the degree of ionization in a weakly magnetized, quasi-Keplerian disk must be sufficiently high to produce the instability that leads to a breakdown of laminar flow into turbulence; the latter being the source of viscosity driving accretion onto the central body. In cold disks, the ionized fraction is very small and might be insufficient for the MRI to operate. In any case, in such a disk non-ideal MHD effects are always important. All these problems still await their solution even though a lot of progress has been made, especially in the context of protostellar accretion disks.

Finally, and very relevant to the subject of this chapter, there is the question of stability of disks in which the pressure is due to radiation and opacity to electron scattering. According to theory, such disks should be violently (thermally) unstable but observations of systems presumed to be in this regime totally infirm this prediction.

3.3.1. *The α-prescription*

The α-prescription is a rather simplistic description of the accretion disk physics but before one is offered better and physically more reliable options its simplicity makes it the best possible choice and has been the main source of progress in describing accretion disks in various astrophysical contexts.

We keep in mind that the accretion–driving viscosity is of magnetic origin; however, we use an effective hydrodynamical description of the accretion flow. The hydrodynamical stress tensor is

$$\tau_{R\varphi} = \rho\nu \frac{\partial v_\varphi}{\partial R} = \rho\nu \frac{d\Omega}{d\ln R}, \qquad [3.19]$$

where ρ is the density, ν is the kinematic viscosity coefficient, and v_φ is the azimuthal velocity ($v_\varphi = R\Omega$).

In 1973, Shakura and Sunyaev proposed the (now famous) prescription

$$\tau_{R\varphi} = \alpha P, \qquad [3.20]$$

where P is the total thermal pressure and $\alpha \le 1$. This leads to

$$\nu = \alpha c_s^2 \left[\frac{d\Omega}{d\ln R} \right]^{-1}, \qquad [3.21]$$

where $c_s = \sqrt{P/\rho}$ is the isothermal sound speed, and ρ is the density. For the Keplerian angular velocity

$$\Omega = \Omega_K = \left(\frac{GM}{R^3} \right)^{1/2} \qquad [3.22]$$

this becomes

$$\nu = \frac{2}{3} \alpha c_s^2 / \Omega_K. \qquad [3.23]$$

Using the approximate hydrostatic equilibrium (equation [3.34]), we can write this as

$$\nu \approx \frac{2}{3} \alpha c_s H. \qquad [3.24]$$

Multiplying the right-hand side of equation [3.19] by the ring length ($2\pi R$) and averaging over the (total = $2H$) disk height, we obtain the expression for the *total torque*

$$\mathfrak{T} = 2\pi R \Sigma \nu R \frac{d\Omega}{d\ln R}, \quad [3.25]$$

where

$$\Sigma = \int_{-\infty}^{+\infty} \rho\, dz. \quad [3.26]$$

For a Keplerian disk

$$\mathfrak{T} = 3\pi \Sigma \nu \ell_K, \quad [3.27]$$

($\ell_K = R^2 \Omega_K$ is the Keplerian *specific* (i.e. per unit mass) angular momentum).

The viscous heating is proportional to $\tau_{r\varphi}(d\Omega/dR)$. In particular, the viscous heating rate per unit volume is

$$q^+ = -\tau_{r\varphi} \frac{d\Omega}{d\ln R}, \quad [3.28]$$

which for a Keplerian disk, using equation [3.20], can be written as

$$q^+ = \frac{3}{2} \alpha \Omega_K P, \quad [3.29]$$

and the viscous heating rate per unit surface is therefore

$$Q^+ = \frac{\mathfrak{T}\Omega'}{4\pi R} = \frac{9}{8} \Sigma \nu \Omega_K^2. \quad [3.30]$$

(the denominator in the first right-hand side is $2 \times 2\pi R$ taking into account the existence of two disk surfaces).

$$\Omega' = \mathrm{d}\Omega/\mathrm{d}R. \quad [3.31]$$

3.4. Geometrically thin Keplerian disks

The 2D structure of geometrically thin, non-self-gravitating, axially symmetric accretion disks can be split into a 1D+1D structure corresponding to a hydrostatic vertical configuration and radial quasi-Keplerian viscous flow. These two 1D structures are coupled through the viscosity mechanism transporting angular momentum and providing the local release of gravitational energy.

3.4.1. *Disk vertical structure*

The vertical structure can be treated as a one-dimensional star with two essential differences, as follows:

1) the energy sources are distributed over the whole height of the disk, while in a star they are limited to the nucleus;

2) the gravitational acceleration *increases* with height because it is given by the tidal gravity of the accretor, while in stars, the (self)gravity decreases as the inverse square of the distance from the center.

Taking these differences into account, the standard stellar structure equations adapted to the description of the disk vertical structure are listed as follows:

– Hydrostatic equilibrium

The gravity force is counteracted by the force produced by the pressure gradient:

$$\frac{dP}{dz} = \rho g_z, \quad [3.32]$$

where g_z is the vertical component (tidal) of the accreting body gravitational acceleration:

$$g_z = \frac{\partial}{\partial z}\left[\frac{GM}{(R^2+z^2)^{1/2}}\right] \approx \frac{GM}{R^2}\frac{z}{R}. \quad [3.33]$$

The second equality follows from the assumption that $z \ll R$. Denoting the typical (pressure or density) scale-height by H, the condition of geometrical thinness of the disk is $H/R \ll 1$ and writing $dP/dz \sim P/H$, equation [3.32] can be written as

$$\frac{H}{R} \approx \frac{c_s}{v_K}, \quad [3.34]$$

where $v_K = \sqrt{GM/R}$ is the Keplerian velocity and we made use of equation [3.33]. From equation [3.34], it follows that

$$\frac{H}{c_s} \approx \frac{1}{\Omega_K} =: t_{\rm dyn}, \quad [3.35]$$

where $t_{\rm dyn}$ is the dynamical time.

For the parameters of interest

$$t_{\rm dyn} = 1.4 \times 10^3 m_8 r^{3/2}\ \mathrm{s}. \quad [3.36]$$

Here, and in what follows $r := R/R_S$ and $m_8 :=$ M/10^8 $M_\odot$. Therefore, for example, for $R \approx 10^{16}$ cm (or 330 R_S), and a 10^8 $M_\odot$ black hole, the dynamical time is about 3.3 months.

– Mass conservation

In 1D hydrostatic equilibrium, the mass conservation equation takes the simple form of

$$\frac{d\varsigma}{dz} = 2\rho, \qquad [3.37]$$

where ς is the surface density between $-z$ and $+z$.

– Energy transfer – temperature gradient

$$\frac{d\ln T}{dz} = \nabla \frac{d\ln P}{dz}. \qquad [3.38]$$

For radiative energy transport

$$\nabla_{\rm rad} = \frac{\kappa_{\rm R} P F_{\rm z}}{4 P_r c g_{\rm z}}, \qquad [3.39]$$

where P_r is the radiation pressure and $\kappa_{\rm R}$ is the Rosseland mean opacity. From equations [3.38] and [3.39], we recover the familiar expression for the radiative flux

$$F_z = -\frac{16}{3}\frac{\sigma T^3}{\kappa_{\rm R}\rho}\frac{\partial T}{\partial z} = -\frac{4\sigma}{3\kappa_{\rm R}\rho}\frac{\partial T^4}{\partial z} \qquad [3.40]$$

(F_z is positive because the temperature decreases with z, so $\partial T/\partial z < 0$).

The photosphere is at optical thickness $\tau \simeq 2/3$ (see equation [3.119]). The boundary conditions are as follows: $z = 0$, $F_z = 0$, $T = T_c$, $\varsigma = 0$ at the disk mid-plane; at the disk photosphere, $\varsigma = \Sigma$ and $T^4(\tau = 2/3) = T_{\rm eff}^4$ (for a detailed discussion of radiative transfer, temperature stratification and boundary conditions, see section 3.4.7).

In the same spirit as equation [3.34], we can write equation [3.40] as

$$F_z \approx \frac{4}{3}\frac{\sigma T_c^4}{\kappa_{\rm R}\rho H} = \frac{8}{3}\frac{\sigma T_c^4}{\kappa_{\rm R}\Sigma}, \qquad [3.41]$$

where T_c is the mid-plane ("central") disk temperature. Using the optical depth $\tau = \kappa_{\rm R}\rho H = (1/2)\kappa_{\rm R}\Sigma$, this can be written as

$$F_z(H) \approx \frac{8}{3}\frac{\sigma T_c^4}{\tau} = Q^-, \qquad [3.42]$$

(see equation [3.120] for a rigorous derivation of this formula).

In the case of convective energy transport $\nabla = \nabla_{\rm conv}$. Because convection in disks is still not well understood, there is no obvious choice for $\nabla_{\rm conv}$, so in practice, prescriptions based on stellar physics, such as the mixing-length approximation, are used, even though it is not obvious that they apply to accretion disks. Some evidence provided by MRI simulations suggests that they do not.

– Energy conservation

Vertical energy conservation should have the form

$$\frac{dF_z}{dz} = q^+(z), \qquad [3.43]$$

where $q^+(z)$ corresponds to the viscous energy dissipation per unit volume. Note that, in contrast with accretion disks, stellar envelopes are in *radiative equilibrium* $dF_z/dz = 0$.

The α prescription does not allow one to deduce the viscous dissipation stratification (z dependence), it just lets us know that the vertically averaged viscous torque is proportional to pressure. Most often, we assume therefore that

$$q^+(z) = \frac{3}{2}\alpha\Omega_K P(z), \qquad [3.44]$$

by analogy with equation [3.29] but such an assumption is chosen because of its simplicity and not because of some physical motivation. In fact, MRI numerical simulations suggest that dissipation is not stratified in the same way as pressure.

– The vertical structure equations have to be completed by the equation of state (EOS):

$$P = P_r + P_g = \frac{4\sigma}{3c}T^4 + \frac{\mathcal{R}}{\mu}\rho T, \qquad [3.45]$$

where $\mathcal{R}$ is the gas constant and μ is the mean molecular weight, and an equation describing the mean opacity dependence on density and temperature.

3.4.2. *Disk radial structure*

– Continuity (mass conservation) equation has the form

$$\frac{\partial \Sigma}{\partial t} = -\frac{1}{R}\frac{\partial}{\partial R}(R\Sigma v_r) + \frac{S(R,t)}{2\pi R}, \qquad [3.46]$$

where $S(R,t)$ is the matter source (sink) term.

– Angular momentum conservation

$$\frac{\partial \Sigma \ell}{\partial t} = -\frac{1}{R}\frac{\partial}{\partial R}(R\Sigma \ell v_r) + \frac{1}{R}\frac{\partial}{\partial R}\left(R^3\Sigma\nu\frac{d\Omega}{dR}\right) + \frac{S_\ell(R,t)}{2\pi R}, \qquad [3.47]$$

ℓ is the specific (per unit mass) angular momentum.

This conservation equation reflects the fact that angular momentum is transported through the disk by a viscous stress $\tau_{r\varphi} = R\Sigma\nu d\Omega/dR$. Therefore, if the disk is not considered infinite (recommended in application to real processes and systems), then somewhere there must be a sink of this transported angular momentum $S_\ell(R,t)$.

For semi-detached binary systems, there is both a source (angular momentum brought in by the mass transfer from the stellar companion) and a sink (tidal interaction taking angular momentum back to the orbit). In the case of accretion disks in AGN, neither the processes through which matter is fed to the disk, nor the mechanism removing angular momentum are well established so the form of both $S(R,t)$ and $S_\ell(R,t)$ are unknown. We can only safely assume that the equations used in this section will not apply beyond the radius at which the disk becomes self-gravitating (see section 3.4.3), which defines a natural outer boundary of the disks considered here.

Assuming $\Omega = \Omega_K$, from equations [3.46] and [3.47], we can obtain a diffusion equation for the surface density Σ:

$$\frac{\partial \Sigma}{\partial t} = \frac{3}{R}\frac{\partial}{\partial R}\left\{R^{1/2}\frac{\partial}{\partial R}\left[\nu \Sigma R^{1/2}\right]\right\}. \qquad [3.48]$$

Comparing this with equations [3.46], we can see that the radial velocity induced by the viscous torque is

$$v_r = -\frac{3}{\Sigma R^{1/2}}\frac{\partial}{\partial R}\left[\nu \Sigma R^{1/2}\right], \qquad [3.49]$$

which is an example of the general relation

$$v_{\rm visc} \sim \frac{\nu}{R}. \qquad [3.50]$$

Using equation [3.24], we can write

$$t_{\rm vis} := \frac{R}{v_{\rm visc}} \approx \frac{R^2}{\nu} \approx \alpha^{-1}\frac{H}{c_s}\left(\frac{H}{R}\right)^{-2}. \qquad [3.51]$$

The relation between the viscous and the dynamical times is

$$t_{\rm vis} \approx \alpha^{-1}\left(\frac{H}{R}\right)^{-2} t_{\rm dyn}. \qquad [3.52]$$

In thin ($H/R \ll 1$) accretion disks, the viscous time is much longer than the dynamical time. In other words, during viscous processes, the vertical disk structure can be considered to be in hydrostatic equilibrium.

– Energy conservation

The general form of an energy conservation (thermal) equation can be written as:

$$\rho T\frac{ds}{dR} := \rho T\left(\frac{\partial s}{\partial t} + v_r\frac{\partial s}{\partial R}\right) = q^+ - q^- + \widetilde{q}, \qquad [3.53]$$

where s is the entropy density, q^+ and q^- are, respectively, the viscous and radiative energy density, and $\widetilde{q}$ is the density of external and/or radially transported energy densities. The term

$$\rho T v_r \frac{\partial s}{\partial R} =: q^{\rm adv} \quad [3.54]$$

describes the radial advection of energy.

Using the first law of thermodynamics $Tds = dU + PdV$, we can write

$$\rho T \frac{ds}{dt} = \rho \frac{dU}{dt} + P \frac{\partial v_r}{\partial r}, \quad [3.55]$$

where $U = \Re T_{\rm c}/\mu(\gamma - 1)$.

Vertically averaging, but taking $T = T_c$, using equation [3.46] and assuming a gas-pressure dominated disk ($P = P_g$), we obtain

$$\frac{\partial T_{\rm c}}{\partial t} + v_{\rm r} \frac{\partial T_{\rm c}}{\partial R} + \frac{\Re T_{\rm c}}{\mu c_P} \frac{1}{R} \frac{\partial (R v_{\rm r})}{\partial R} = 2 \frac{Q^+ - Q^-}{c_P \Sigma} + \frac{\widetilde{Q}}{c_P \Sigma}, \quad [3.56]$$

where Q^+ and Q^- are, respectively, the heating and cooling rates per unit surface. $\widetilde{Q} = Q_{\rm out} + J$ with $Q_{\rm out}$ corresponding to energy contributions by the mass-transfer stream and tidal torques; $J(T, \Sigma)$ represent the radial energy fluxes that are a, more or less, *ad hoc* addition to the 1D+1D scheme to which they do not belong: indeed, such a scheme assumes that radial gradients ($\partial / \partial R$) of physical quantities can be neglected when compared with gradients in the vertical direction. That is also the reason that as long as $H/R \ll 1$, $q^{\rm adv}$ is negligible compared to q^+ and q^-.

The viscous heating rate per unit surface can be written as (see equation [3.30])

$$Q^+ = \frac{9}{8} \nu \Sigma \Omega_{\rm K}^2 \quad [3.57]$$

while the cooling rate over unit surface (the radiative flux) is obviously

$$Q^- = \sigma T_{\rm eff}^4. \quad [3.58]$$

In thermal equilibrium, we have

$$Q^+ = Q^-. \quad [3.59]$$

The cooling time can be easily estimated from equation [3.59]. The energy density to be radiated away is $\rho U \sim \rho c_s^2$, so the energy per unit surface is $\sim \Sigma c_s^2$ and the cooling (thermal) time is

$$t_{\rm th} = \frac{\Sigma c_s^2}{Q^-} = \frac{\Sigma c_s^2}{Q^+} \sim \alpha^{-1} \Omega_K^{-1} = \alpha^{-1} t_{\rm dyn}. \quad [3.60]$$

Since $\alpha < 1, t_{\rm th} > t_{\rm dyn}$ and during thermal processes, the disk can be assumed to be in (vertical) hydrostatic equilibrium.

For geometrical thin ($H/R \ll 1$) accretion disks, we have the following hierarchy of the characteristic times:

$$t_{\rm dyn} < t_{\rm th} \ll t_{\rm vis}. \qquad [3.61]$$

(This hierarchy is similar to that of characteristic times in stars where the dynamical time is shorter than the thermal (Kelvin–Helmholtz) time, and the thermal time is much shorter than the thermonuclear timescale.)

3.4.3. *Self-gravity*

In this chapter, we are interested in disks that are not self-gravitating, that is, in disks where the vertical hydrostatic equilibrium is maintained against the pull of the accreting body's tidal gravity whereas the disk's self-gravity can be neglected. We will see now under what conditions this assumption is satisfied.

The equation of vertical hydrostatic equilibrium can be written as

$$\frac{1}{\rho}\frac{dP}{dz} = -g = (-g_z - g_s) = -g_z\left(1 + \frac{g_s}{g_z}\right) =: -g_z\,(1 + A)\,, \qquad [3.62]$$

therefore self-gravity is negligible when $A \ll 1$. Treating the disk as an infinite uniform plane (i.e. assuming the surface density does not vary too much with radius), we can write its self-gravity as $g_s = 2\pi G\Sigma$, whereas the z-component of the gravity provided by the central body is $g_z = \Omega_K^2\, z$ (equation [3.33]). Therefore, evaluating A at $z = H$, we get

$$A_H := \left.\frac{g_s}{g_z}\right|_H = \frac{2\pi G\Sigma}{\Omega_K^2 H}. \qquad [3.63]$$

A_H is related to the so-called Toomre parameter (Toomre 1964)

$$Q_T := \frac{c_s\Omega}{\pi G\Sigma}, \qquad [3.64]$$

widely used in the studies of gravitational stability of rotating systems, through $A_H \approx Q_T^{-1}$. We will therefore express the condition of negligible self-gravity (gravitational stability) as

$$Q_T > 1. \qquad [3.65]$$

Note that the condition of negligible disk self-gravity (equation [3.65]) is equivalent to $M_D < (H/R)\, M_{\rm BH}$ and not just to $M_D < M_{\rm BH}$, as sometimes claimed, where M_D is the disk mass, and $M_{\rm BH}$ is the mass of the accreting body, a black hole in our case.

Using equations [3.34], [3.24] and [3.75], we can write the Toomre parameter as

$$Q_T = \frac{3\alpha c_s^3}{G\dot{M}}, \qquad [3.66]$$

where $\dot{M}$ is the accretion rate (mass per unit time, see equation [3.70]), or as a function of the mid-plane temperature $T = 10^4\, T_4 \mathrm{K}$

$$Q_T \approx 0.5\, \frac{\alpha\, T_4^{3/2}}{m_8\, \dot{m}}. \qquad [3.67]$$

This shows that hot ionized ($T \gtrsim 10^4$K) disks become self-gravitating for high accretor masses and high accretion rates. Disks in close binary systems ($m_8 \lesssim 3 \times 10^{-7}$) are never self-gravitating for realistic accretion rates ($\dot{m} < 1000$, say), and even in intermediate-mass black holes (IMBH) binaries (if they exist) (hot) disks would also be free of the gravitational instability. Around a supermassive black hole, however, disks can become self-gravitating quite close to the black hole. For example, when the black hole mass is $m_8 = 1$, a hot disk will become self-gravitating at $r \approx 100$, for $\dot{m} \sim 10^{-2}$. In general, geometrically thin, non-self-gravitating accretion disks around supermassive black holes have a very limited radial extent. From equation [3.124] we obtain

$$Q_T \approx 10^4 \alpha^{7/10} m_8^{-13/10} \dot{m}^{-11/20} r^{-9/8}, \qquad [3.68]$$

which gives

$$r < 4.0 \times 10^3 \alpha^{28/45} m_8^{-52/45} \dot{m}^{-22/45}, \qquad [3.69]$$

or, for example, for $m_8 = 1$ and $\dot{m} = 0.01$, $R < 0.4$ pc, r being the distance in units of the Schwrzchild radius.

3.4.4. *Stationary disks*

In the case of stationary ($\partial/\partial t = 0$) disks, equation [3.46] can be easily integrated giving

$$\dot{M} := 2\pi R \Sigma v_{\mathrm{r}}, \qquad [3.70]$$

where the integration constant $\dot{M}$ (mass/time) is the *accretion rate*.

Also the angular momentum equation [3.47] can be integrated to give

$$-2\pi R \Sigma v_r \ell + 2\pi R^3 \Sigma \nu \frac{d\Omega}{dR} = const. \qquad [3.71]$$

Or, using equation [3.70]

$$-\dot{M}\ell + \mathfrak{T} = const., \qquad [3.72]$$

where the torque

$$\mathfrak{T} := 2\pi R^3 \Sigma\nu d\Omega/dR; \qquad [3.73]$$

(for a Keplerian disk $\mathfrak{T} = 3\pi R^2 \Sigma\nu\Omega_K$).

Assuming that the torque vanishes at the inner disk radius, we get $const. = -\dot{M}\ell_{\text{in}}$, where ℓ_{in} is the specific angular momentum at the disk inner edge. Therefore

$$\dot{M}(\ell - \ell_{\text{in}}) = \mathfrak{T} \qquad [3.74]$$

which is a simple expression of angular momentum conservation.

For Keplerian disks, we obtain an important relation between viscosity and accretion rate

$$\nu\Sigma = \frac{\dot{M}}{3\pi}\left[1 - \left(\frac{R_{\text{in}}}{R}\right)^{1/2}\right]. \qquad [3.75]$$

From equations [3.75], [3.57], [3.58], and the thermal equilibrium equation [3.59], it follows that

$$\sigma T_{\text{eff}}^4 = \frac{8}{3}\frac{\sigma T_c^4}{\tau} = \frac{3}{8\pi}\frac{GM\dot{M}}{R^3}\left[1 - \left(\frac{R_{\text{in}}}{R}\right)^{1/2}\right]. \qquad [3.76]$$

This relation assumes only a Keplerian disk in thermal ($Q^+ = Q^-$) and viscous ($\dot{M} = const.$) equilibrium. Since the disk is in thermal equilibrium, the emitted radiation flux cannot contain information about the heating mechanism which explains why the viscosity coefficient is absent from equation [3.76]. Steady disks do not provide information about the viscosity operating in disks or the viscosity parameter α. To get this information, we must consider (and observe) time-dependent states of accretion disks. It is often incorrectly claimed that the $R^{-3/4}$ temperature profile is a signature of the Shakura–Sunyaev solution, but obviously this profile is much more general and independent of the viscosity prescription assumed to obtain this celebrated solution (see below).

Equation [3.76] determines indeed a universal radial temperature profile for *stationary Keplerian accretion disks*

$$T_{\text{eff}} \sim R^{-3/4}. \qquad [3.77]$$

For an optically thick disk, the observed temperature $T \sim T_{\text{eff}}$ and $T \sim R^{-3/4}$ should be observed if stationary; optically thick Keplerian disks do exist in the universe. And vice versa, if they are observed, this proves that such disks exist not only on paper. Such $R^{-3/4}$ disk–temperature profiles have been clearly observed in bright, eclipsing cataclysmic variables. It seems, however, that in AGN disks, these profiles are a bit steeper, as will be discussed below.

The temperature profile of a stationary Keplerian accretion disk is given therefore by

$$T_{\text{eff}} = T_{\text{in}} \left(\frac{r}{3}\right)^{-3/4}, \qquad [3.78]$$

where

$$T_{\text{in}} = \left(\frac{3GM\dot{M}}{8\pi\sigma(3R_S)^3}\right)^{1/4} \approx 3.0 \times 10^5\, m_8^{-1/2} \dot{m}^{1/4}\text{K}, \qquad [3.79]$$

where we assumed that $R_{\text{in}} = 3R_S$, that is, the ISCO for a non-rotating black hole. Therefore, the maximum effective disk temperature $T_{\text{eff}}^{\max} = 0.488\, T_{\text{in}}$ is located at $r = 49/12 \approx 4.1$ and in AGN corresponds typically to UV radiation.

3.4.4.1. *The "no-torque condition"*

There has been a lot of discussion about the inner boundary condition in an accretion disk around a black hole. The usual reasoning is that for a thin disk, the inner boundary is at ISCO and since circular orbits end there, the boundary condition should be simply that the "viscous" torque vanishes there (there is no orbit below the ISCO to interact with). In fact, for geometrically thin accretion disks, the no-torque condition is a simple consequence of conservation of angular momentum. Numerical simulations of thin accretion disks that do not satisfy this condition do not conserve angular momentum.

In general, if we do not assume that the torque vanishes at the inner disk edge, equation [3.72] will take the form

$$\dot{M}(\ell - \ell_H) = \mathfrak{T}, \qquad [3.80]$$

where l_H is the specific angular momentum of the accretion flow at the black hole surface: torques must vanish on the horizon, since the horizon is a hypersurface causally detached from the rest of the universe.

From equations [3.80], [3.73] and the viscosity prescription $\nu \approx \alpha H^2 \Omega$, we can obtain

$$v_r \approx \alpha\, H^2 \frac{\ell}{\ell - \ell_H} \frac{d\Omega}{dR} \approx \alpha\, H^2 \frac{\ell}{\ell - \ell_H} \frac{\Omega}{R} \approx \alpha\, v_\varphi \left(\frac{H}{R}\right)^2 \frac{\ell}{\ell - \ell_H}, \qquad [3.81]$$

where $v_\varphi = R\Omega$. Equation [3.81] does not assume that the radial velocity is small, that is, this equation holds within the disk as well as within the stream below the ISCO.

Far out in the disk, where $\ell \gg \ell_H$, we obtain the standard formula (see equation [3.50])

$$v_r \approx \alpha\, v_\varphi \left(\frac{H}{R}\right)^2, \qquad R \gg R_{in}. \qquad [3.82]$$

The flow crosses the black hole surface at the speed of light and since it is subsonic in the disk, it must somewhere become transonic, that is, go through a sonic point, which has been shown to be close to the disk's inner edge.

At the sonic point $v_r = c_s \approx (H/R)v_\varphi$, and equation [3.81] becomes:

$$\frac{v_r}{c_s} = 1 \approx \alpha \frac{H_{\rm in}}{R_{\rm in}} \frac{\ell_{\rm in}}{\ell_{\rm in} - \ell_H}, \qquad R = R_{\rm in}. \qquad [3.83]$$

If the disk is thin, that is, $H_{\rm in}/R_{\rm in} \ll 1$, and the viscosity is small, that is, $\alpha \ll 1$, then equation [3.83] implies that $(\ell_{\rm in} - \ell_H)/\ell_{\rm in} \ll 1$, that is, the specific angular momentum at the sonic point is almost equal to the asymptotic angular momentum at the horizon.

In a steady-state disk, the torque $\mathfrak{T}$ has to satisfy the equation of angular momentum conservation [3.72], which can be written as

$$\mathfrak{T} = \dot{M}\,(\ell - \ell_H), \qquad \mathfrak{T}_{in} = \dot{M}\,(\ell_{in} - \ell_H)\,. \qquad [3.84]$$

Thus, it is clear that for a thin, low viscosity disk, the "no torque inner boundary condition" ($\mathfrak{T}_{in} \approx 0$) is an excellent approximation *following from angular momentum conservation*.

However, if the disk and the stream are thick, that is, $H/r \lesssim 1$, and the viscosity is high, that is, $\alpha \sim 1$, then the angular momentum can vary also in the stream in accordance with the simple reasoning presented above. In such a case, the no-stress condition at the disk inner edge might be unsatisfied.

3.4.4.2. *Total luminosity and spectrum*

The total luminosity of a stationary, geometrically thin accretion disk, that is, the sum of luminosities of its two surfaces, is

$$2\int_{R_{\rm in}}^{R_{\rm out}} \sigma T_{\rm eff}^4\, 2\pi R dR = \frac{3GM\dot{M}}{2}\int_{R_{\rm in}}^{R_{\rm out}} \left[1 - \left(\frac{R_{\rm in}}{R}\right)^{1/2}\right]\frac{dR}{R^2}. \qquad [3.85]$$

For $R_{\rm out} \to \infty$, this becomes

$$L_{\rm disc} = \frac{1}{2}\frac{GM\dot{M}}{R_{\rm in}} = \frac{1}{2}L_{\rm acc}. \qquad [3.86]$$

In the disk, the radiating particles move on Keplerian orbits; hence, they retain half of the potential energy. When the accreting body is a black hole, this leftover energy will be lost. In this case, however, the non-relativistic formula of equation [3.86] does not apply – see equation [3.8].

For an optically thick disk, we can assume that each of its rings radiate as black body emitter:

$$I_\nu = B_\nu\left[T_{\rm eff}(R)\right] = \frac{2h\nu^3}{c^2(e^{h\nu/kT(R)}-1)}\ \left(\mathrm{erg\,s^{-1}\,cm^{-2}\,Hz^{-1}\,sr^{-1}}\right). \quad [3.87]$$

This is a rather crude but useful approximation. We should keep in mind, however, that it does not represent real observed disk spectra. Stars are very optically thick, but their spectra do not look like black bodies, because the observed light had to cross the stellar atmosphere on its way to the detector.

With such an assumption, the flux at a given frequency, detected by an observer at a distance D, is equal to

$$F_\nu = \frac{2\pi\cos i}{D^2}\int_{3R_S}^{R_{\rm out}} I_\nu R\,dR = \frac{4\pi h\nu^3\cos i}{c^2D^2}\int_{3R_S}^{R_{\rm out}}\frac{R\,dR}{e^{h\nu/kT(R)}-1}, \quad [3.88]$$

where i is the disk inclination.

For frequencies

$$\frac{kT(R_{\rm out})}{h} \ll \nu \ll \frac{kT_{\rm in}}{h}, \quad [3.89]$$

we obtain from equation [3.88] $F_\nu \propto \nu^{1/3}$, known sometimes as the "disk" spectrum, but such a $\nu^{1/3}$ feature is prominent only if $T(R_{\rm out}) \ll T_{\rm in}$, that is, if the disk is sufficiently large. For an AGN disk with $R_{\rm out}\sim 100R_{\rm in}$ and $\dot{m}=0.01$, the optical and UV spectrum will indeed have a 1/3 slope.

3.4.5. *Size of the disk*

Assuming that thin disks in AGNs emit locally as black bodies allows us to determine the disk's size and compare it through observation. Defining the disk size as corresponding to the radius at which the disk temperature matches the wavelength R_λ:

$$kT(R_\lambda) = hc/\lambda, \quad [3.90]$$

and by using equation [3.87] in the form of

$$I_\nu = \frac{2h_pc}{\lambda^3}\left[\exp\left(\frac{R}{R_{\lambda_{\rm rest}}}\right)^{3/4} - 1\right]^{-1}, \quad [3.91]$$

we obtain from equations [3.78] and [3.79]

$$R_\lambda = \left[\frac{45G\lambda^4 M\dot{M}}{16\pi^6 hc^2}\right]^{1/3} = 2.1\times 10^{15}\left(\frac{\lambda}{\mu\text{m}}\right)^{4/3} m_8^{2/3}\left(\frac{L}{\eta L_E}\right)^{1/3}\ \text{cm}. \qquad [3.92]$$

λ is the wavelength in the rest-frame of the AGN. Thus, the prediction of the thin-disk model is that the size of the disk $R \sim M^{2/3}$.

Equation [3.90] assumes that emission at wavelength λ originates solely at radius R_λ, while in reality, this emission comes also from other radii. Therefore, a more appropriated size for comparison with observations would be a flux-weighted mean radius $\langle R_\lambda\rangle = XR_\lambda$. $X \approx 2-3$; however, if disk variability is taken into account, this factor can be even ~ 5. R_λ itself depends on the black hole mass and the accretion rate (through η) so that uncertainties in the observed values of these quantities impact the comparison of the model size with observations. AGN disk sizes are measured through reverberation mapping and microlensing. Observations are in agreement with the predicted $\sim M^{2/3}$ slope. There is no agreement about the actual size. While microlensing observations and some reverberation estimates claim values 2 – 3 larger than those predicted by the model, according to other reverberation observations, most measured sizes of AGN disks agree with the thin-disk value. The reasons for the discrepancies (if real) are not clear; however, it is possible that the real temperature profile is more shallow than the thin-disk $R^{-3/4}$ which should be the case when the disk is irradiated.

3.4.6. *Spectral lines from Keplerian accretion disks*

Observations of spectral lines from disks around black holes play an important role as tools allowing us to investigate the disk structure itself, but their shapes encode the properties of the accretor, in this case its mass and spin. In practice, this means taking into account the relativistic effects (both special and general) on the propagation of light rays which might obfuscate the basic properties of line shapes emitted by an accretion disk. We will therefore begin with the Newtonian case.

Emission lines from a Keplerian accretion disk have a characteristic double-peak structure. The reason is easy to understand. For a distant observer, the frequency of the lines emitted by the disk is shifted by the Doppler effect, corresponding to velocity

$$v_{\rm D} = v_{\rm K}\sin i\sin\theta, \qquad [3.93]$$

where $v_K = R\Omega_K$, i is the disk inclination ($i = 90°$ for a disk seen edge-on), and θ is the azimuth angle relative to the line of sight. The total disk emission flux is obtained by integrating the Doppler-shifted intensities over the whole disk's surface:

$$F_\nu = \frac{1}{D^2}\int I_\nu R\, dRd\theta. \qquad [3.94]$$

Here, the line specific intensity is $I_\nu = j(R)\phi_\nu$, with $\int \phi_\nu \, d\nu = 1$, ϕ_ν being the emission profile of the line. $j(R)$ is assumed to be isotropic. Each local line profile I_ν is shifted from the rest frequency ν' to the local Doppler frequency

$$\nu_{\mathrm{D}} = \nu' \left(1 - \frac{v_{\mathrm{D}}}{c}\right). \qquad [3.95]$$

As shown in Figure 3.2, the loci of constant radial velocity form a "dipole" pattern on the disk's surface. The disk's emission-line profile, shown in the right part of this figure, is divided into velocity bins corresponding to the disk regions between consecutive dipole field lines. The emission in each velocity bin arises from a different region of the disk surface. Assuming circular orbits, matter crossing the $\theta = 0, \pi$ line moves perpendicularly to the line of sight, so $v_D = 0$ and its emission correspond to the line center.

The crescents that are complete near the accretor become truncated at the outer disk radius R_D which produces cusps in the line profile at $v_D = \pm v_K(R_D) \sin i$. Within the assumptions of a Keplerian potential, and an almost constant disk surface density, *double-peaked line profiles thus correspond to the presence of an outer boundary to the emission-line region.*

The wings of the lines are obviously produced by the fastest moving disk matter, that is, close to the accretor. However, in a Keplerian disk, the area of the corresponding crescent $R\Delta R \sim v^{-5}\Delta v$ which explains its rapid decline with increasing velocity.

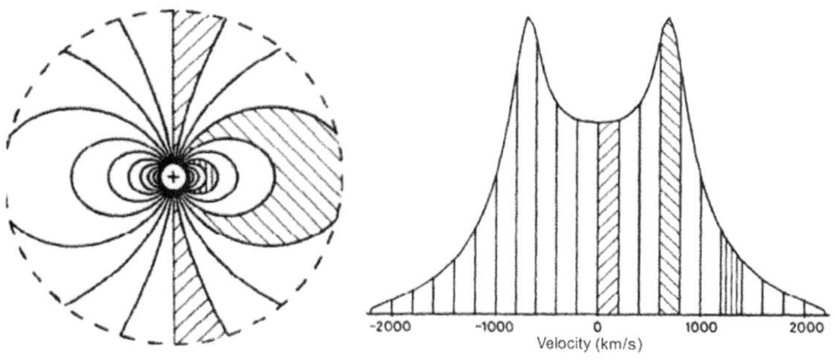

Figure 3.2. *The loci of constant radial velocity form a dipole field pattern on the surface of a Keplerian disk (left figure). The velocity profile of emission lines from the disk (right). Emission in the shaded velocity bins arises from the corresponding regions of the disk. This example corresponds to a stellar-mass accretor (source: a dapted from Horne and Marsh (1986))*

3.4.6.1. *General-relativistic spectral line description*

Equations [3.93], [3.94] and [3.95] cannot correctly describe the spectral line close to a black hole because in their derivation, higher-order terms in v/c and gravitational light-bending have been neglected. The description of such line profiles requires the use of the full formalism of general relativity and usually requires numerical calculations. In the following, we will show how the problem can be simplified even in the case of a Kerr black hole. Exceptionally in this section, we will use units in which c = G = 1 in the description of the motion of the emitter.

The specific intensity [erg $\mathrm{cm}^{-2}\,\mathrm{s}^{-1}\,\mathrm{Hz}^{-1}\,\mathrm{Sr}^{-1}$] is now written as

$$I_\nu = I_{\nu'}(1+z)^{-3}, \qquad [3.96]$$

where ν' is the rest-frame frequency, ν is the frequency measured at infinity, and $1+z = \nu'/\nu$ is the redshift factor. One of the redshift factors in the denominator accounts for the time dilation, while the other two correspond to the angular diameter correction to the solid angle. Therefore, the observed spectral flux is

$$F_\nu = \int_{\text{image}} I_\nu d\Omega = \int_{\text{image}} I_{\nu'}(1+z)^{-3} d\Omega, \qquad [3.97]$$

where $d\Omega$ is the solid angle and the integration is over the observed accretion disk image.

Therefore, to calculate the observed line profiles, we have to calculate the redshift factor $1+z$ which is given by the motion of the emitter in the gravitational field of the attracting body. In the case of a Keplerian motion around a Kerr black hole, with orbital frequency

$$\Omega_K = \frac{M^{1/2}}{R^{3/2} + aM^{3/2}}, \qquad [3.98]$$

the result is

$$1 + z = \qquad [3.99]$$

$$\frac{1 - L\Omega_K}{\sqrt{(1 - 2M/R - R^2\Omega_K^2) + 4a\Omega_K M^2/R - (1+2M/R)(aM)^2\Omega_K^2}}, \qquad [3.100]$$

where L is the component of the angular momentum of the emitter with respect to the black hole rotation axis. In general, equation [3.97] with [3.99] has to be solved numerically. However, for distances $R > 10$ ($r > 5$) one can neglect the effects of black hole rotation (they diminish as R^{-3}; see equation [3.9]) and equations [3.98] and [3.99] simplify to its Schwarzschild form

$$\Omega_K = \sqrt{\frac{M}{R^3}} \quad 1 + z = \frac{1 - \Phi\Omega_K}{\sqrt{1 - 3M/R}}. \qquad [3.101]$$

Then, for low inclinations ($i \sim 0$) and large radii, we can easily integrate equation [3.97], with $d\Omega \approx \cos i\, r dr\, d\varphi / 4\pi D^2$. Assuming a Dirac-delta intrinsic line profile $I_{\nu'} = j(r)\delta(\nu' - \nu_0)$ and defining

$$\alpha := \sqrt{1 - 3/\tilde{r}}, \quad \beta := \sin i / \sqrt{\tilde{r}}, \quad \tilde{r} := R/M, \qquad [3.102]$$

we obtain

$$F_\nu = \frac{\cos i}{2\pi D^2 \nu_0} \int_{R_{\rm in}}^{R_{\rm out}} \frac{\alpha \left(\nu/\nu_0\right)^3 \phi(R) R\, dR}{\left(\beta^2 \left(\nu/\nu_0\right)^2 - \left(\alpha - \nu/\nu_0\right)^2\right)^{1/2}}, \qquad [3.103]$$

which is an asymptotic formula for the spectrum observed at large distances from the black hole. Close to the black hole, this formula is inappropriate even for low inclinations, since it neglects both the effects of black hole rotation and terms of the order of $(v/c)^3$. However, it happens to provide an excellent approximation to the observed line profiles if α and β are redefined as

$$\alpha = \alpha_0$$

$$\beta = \beta_0 \left[1 + \frac{1.25}{\tilde{r}} + \frac{2.5 - 3.4\cos i}{\tilde{r}^2} + \frac{3 \sin i}{\tilde{r}^3}\right], \qquad [3.104]$$

for a Schwarzschild metric ($a = 0$), and as

$$\alpha = \alpha_0 \left[1 + \frac{5(\cos i - 1)}{\tilde{r}^{5/2}} + \frac{6.5 - 8.5\cos i}{\tilde{r}^{7/2}}\right]$$

$$\beta = \beta_0 \left[1 + \frac{1.25}{\tilde{r}} + \frac{3 - 3\cos i}{\tilde{r}^2} - \frac{3.5 \sin i}{\tilde{r}^3}\right], \qquad [3.105]$$

for a Kerr black hole with $a = 0.99$. $\alpha_0 = (1 - 3v^2)$, $\beta_0 = v \sin i\ v = R\Omega_K$, and $\tilde{r} = 1/2r$ with Ω_K given by equation [3.98]. α and β are related to the minimum and maximum frequencies of the observed line through

$$\alpha = \frac{2\nu_{\rm min}\nu_{\rm max}}{\left(\nu_{\rm min} + \nu_{\rm max}\right)/\nu_0}$$

$$\beta = \frac{\nu_{\rm max} - \nu_{\rm min}}{\left(\nu_{\rm min} + \nu_{\rm max}\right)/\nu_0}. \qquad [3.106]$$

The fit to α and β is accurate to better than 1% for $r > 1.5$ in the Schwarzschild metric and $r > 2.5$ for a Kerr black hole, if $i < 80°$. Finally, the flux as given by equation [3.97] has to be renormalized through factors that depend on the radius and inclination. These factors are

$$C = 1 + \frac{1.25(\cos i - \sin i) + 1.6\tan i}{\tilde{r}^{1 - 0.19\sin i}} + \frac{10 - 10\cos i - 3.5\tan i}{\tilde{r}^2}, \qquad [3.107]$$

for $a = 0$

$$C = 1 + \frac{1.5 - 2\sin i + 1.2\tan i}{\tilde{r}^{1-0.25\sin i}} + \frac{7 - 7.6\cos i + 2\sin i}{\tilde{r}^2} - \frac{3.5\sin i}{\tilde{r}^3}, \quad [3.108]$$

for $a = 0.99$. Multiplying by these factors gives the total line fluxes correct within less than 1% for $i < 80°$ when $r > 1.5$ in the Schwarzschild metric and $r > 2.5$ for a Kerr black hole. $\tan i$ in these formulae represents the light bending effect.

Figures 3.3 and 3.4 show two examples of "exact" (numerical) and approximated fitting of the profile of a line emitted by a ring around an $a = 0.99$ black hole. For a low inclination ($i = 30°$) even close to the black hole ($r = 2.5$), the numerical and approximate profiles are practically indistinguishable. However, even at a relatively large distance from the hole ($r = 12.5$) but a high inclination ($i = 80°$), the imperfections of the approximate version of the flux are clearly visible.

The double-peaked line shape is, as in the Newtonian case, due to the finite extent of the emission region, in this case a ring. The often used assumption $j(R) \sim R^{-b}$ will not produce double-peak lines if the outer emission region is not spatially limited.

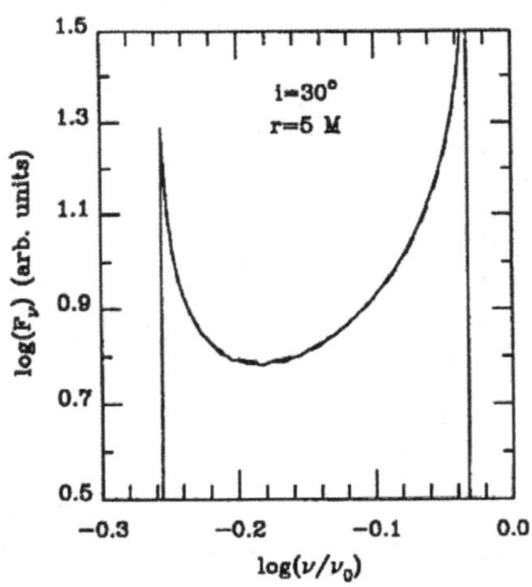

Figure 3.3. *Exact (solid curve) and approximate (dashed line) line profiles emitted by a ring of matter at* $r = 2.5$ $(R = 5GM/c^2)$. *The disk inclination is* $30°$ *(source: adapted from Hameury et al. (1994))*

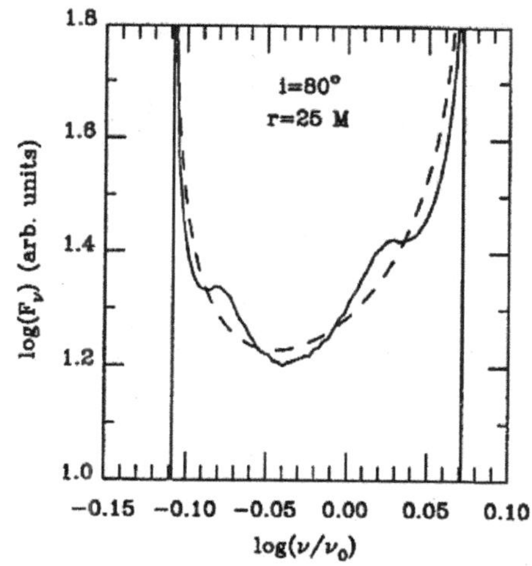

Figure 3.4. *Same as in Figure 3.3, but for $r = 12.5$ ($R = 25GM/c^2$) and with an inclination of $80°$. For such high disk inclinations, the effects of light-bending (producing the two additional inner peaks) cannot be neglected even for relatively large radii (source: adapted from Hameury et al. (1994))*

3.4.7. *Radiative structure*

Here, we will show an example of the solution for the vertical thin disk structure which exhibits properties impossible to identify when the structure is vertically averaged. We will also consider here irradiated disks – such as AGN disks.

We write the energy conservation as:

$$\frac{dF}{dz} = q^+(R, z), \qquad [3.109]$$

where F is the vertical (in the z direction) radiative flux and $q^+(R, z)$ is the viscous heating rate per unit volume. Equation [3.109] states that an accretion disk is not in radiative equilibrium ($dF/dz \neq 0$), contrary to a stellar atmosphere. For this equation to be solved, the function $q^+(R, z)$ must be known. As explained and discussed in section 3.4.1, the viscous dissipation is often written as

$$q^+(R, z) = \frac{3}{2}\alpha\Omega_{\rm K} P(z). \qquad [3.110]$$

Viscous heating of this form has important implications for the structure of optically thin layers of accretion disks and may lead to the creation of coronae and winds. In reality, it is an *ad hoc* formula inspired by equation [3.29]. We do not know yet how to describe the viscous heating stratification in a real *geometrically thin* accretion disk and equation [3.110] just *assumes* that it is proportional to pressure. It is simple and convenient, but it is not necessarily true.

When integrated over z, the right-hand side of equation [3.109] using equation [3.110] is equal to viscous dissipation per unit surface:

$$F^+ = \frac{3}{2}\alpha\Omega_\mathrm{K} \int_0^{+\infty} P dz, \qquad [3.111]$$

where $F^+ = (1/2)Q^+$ because of the integration from 0 to $+\infty$ while Q^+ contains Σ which is integrated from $-\infty$ to $+\infty$ (equation [3.26]).

We can rewrite equation [3.109] as

$$\frac{dF}{d\tau} = -f(\tau)\frac{F_\mathrm{vis}}{\tau_\mathrm{tot}}, \qquad [3.112]$$

where we introduced a new variable, the optical depth $d\tau = -\kappa_\mathrm{R}\rho dz$, κ_R being the Rosseland mean opacity and $\tau_\mathrm{tot} = \int_0^{+\infty} \kappa_\mathrm{R}\rho dz$ is the total optical depth. $f(\tau)$ is of order unity.

At the disk mid-plane, by symmetry, the flux must vanish: $F(\tau_\mathrm{tot}) = 0$, whereas at the surface, ($\tau = 0$)

$$F(0) \equiv \sigma T_\mathrm{eff}^4 = F^+. \qquad [3.113]$$

Equation [3.113] states that the total flux at the surface is equal to the energy dissipated by viscosity (per unit time and unit surface). The solution of equation [3.112] is thus

$$F(\tau) = F^+\left(1 - \frac{\int_0^\tau f(\tau)d\tau}{\tau_\mathrm{tot}}\right), \qquad [3.114]$$

where $\int_0^{\tau_\mathrm{tot}} f(\tau)d\tau = \tau_\mathrm{tot}$. Since $f \approx 1$, we have

$$F(\tau) \approx F^+\left(1 - \frac{\tau}{\tau_\mathrm{tot}}\right). \qquad [3.115]$$

To obtain the temperature stratification, we have to solve the transfer equation. Here, we use the diffusion approximation

$$F(\tau) = \frac{4}{3}\frac{\sigma dT^4}{d\tau}, \qquad [3.116]$$

appropriate for the optically thick disks we are dealing with. The integration of equation [3.116] is straightforward and gives:

$$T^4(\tau) - T^4(0) = \frac{3}{4}\tau\left(1 - \frac{\tau}{2\tau_\mathrm{tot}}\right)T_\mathrm{eff}^4. \qquad [3.117]$$

The upper (surface) boundary condition is:

$$T^4(0) = \frac{1}{2}T_{\text{eff}}^4 + T_{\text{irr}}^4, \qquad [3.118]$$

where T_{irr}^4 is the irradiation temperature, which depends on R, the albedo, the height at which the energy is deposited and on the shape of the disk. In equation [3.118], $T(0)$ corresponds to the *emergent* flux and, as mentioned above, T_{eff} corresponds to the *total* flux ($\sigma T_{\text{eff}}^4 = Q^+$) which explains the factor 1/2 in equation [3.118]. The temperature stratification is thus:

$$T^4(\tau) = \frac{3}{4}T_{\text{eff}}^4\left[\tau\left(1 - \frac{\tau}{2\tau_{\text{tot}}}\right) + \frac{2}{3}\right] + T_{\text{irr}}^4. \qquad [3.119]$$

For $\tau_{\text{tot}} \gg 1$, the first term on the right-hand side has the form familiar from the stellar atmosphere models in the Eddington approximation.

In this case at $\tau = 2/3$, *we have* $T(2/3) = T_{\text{eff}}$.

Also for $\tau_{\text{tot}} \gg 1$, the temperature at the disk mid-plane is

$$T_c^4 \equiv T^4(\tau_{\text{tot}}) = \frac{3}{8}\tau_{\text{tot}}T_{\text{eff}}^4 + T_{\text{irr}}^4. \qquad [3.120]$$

It is clear, therefore, that for the disk inner structure to be dominated by irradiation and the disk to be isothermal, we must have

$$\frac{F_{\text{irr}}}{\tau_{\text{tot}}} \equiv \frac{\sigma T_{\text{irr}}^4}{\tau_{\text{tot}}} \gg F^+ \qquad [3.121]$$

and not just $F_{\text{irr}} \gg F^+$ as is sometimes assumed. The difference between the two criteria is important both in low-mass X-ray binaries (LMXBs), where $\tau_{\text{tot}} \gtrsim 10^2 - 10^3$ in the outer disk, and in AGNs where τ can reach 10^4.

3.4.8. *Shakura–Sunyaev thin disk solution*

In their seminal and famous 1973 paper, Shakura and Sunyaev found power-law stationary solutions of the simplified version of the thin disk equations presented in sections 3.4.1, 3.4.2 and 3.4.4. The eight equations for the eight unknowns T_c, ρ, P, Σ, H, ν, τ and c_s can be written as

$$\Sigma = 2H\rho \qquad (\text{I})$$

$$H = \frac{c_s R^{3/2}}{(GM)^{1/2}} \qquad (\text{II})$$

$$c_s = \sqrt{\frac{P}{\rho}} \tag{iii}$$

$$P = \frac{\mathcal{R}\rho T}{\mu} + \frac{4\sigma}{3c}T^4 \tag{iv}$$

$$\tau(\rho, \Sigma, T_c) = \kappa_R(\rho, T_c)\Sigma \tag{v}$$

$$\nu(\rho, \Sigma, T_c, \alpha) = \frac{2}{3}\alpha c_s H \tag{vi}$$

$$\nu\Sigma = \frac{\dot{M}}{3\pi}\left[1 - \left(\frac{R_{\text{in}}}{R}\right)^{1/2}\right] \tag{vii}$$

$$\frac{8}{3}\frac{\sigma T_c^4}{\tau} = \frac{3}{8\pi}\frac{GM\dot{M}}{R^3}\left[1 - \left(\frac{R_{\text{in}}}{R}\right)^{1/2}\right]. \tag{viii}$$

Equations (i) and (ii) correspond to vertical structure equations [3.37] and [3.34], equation (vii) is the radial equation [3.75], while equation (viii) connects vertically to radial equations. Equation (iii) defines the sound speed, equation (iv) is the equation of state and equation (vi) contains the information about opacities. The viscosity α parametrization introduced in Shakura and Sunyaev (1973) provides the closure of the eight disk equations. Therefore, they can be solved for a given set of α, M, R and $\dot{M}$.

Power-law solutions of these equations exist in the physical regimes where the opacity can be represented in the Kramers form $\kappa = \kappa_0 \rho^n T^m$ and one of the two pressures, gas or radiation, dominates over the other. There are three regimes to be considered:

$a)$ $P_r \gg P_g$ and $\kappa_{\text{es}} \gg \kappa_{\text{ff}}$

$b)$ $P_g \gg P_r$ and $\kappa_{\text{es}} \gg \kappa_{\text{ff}}$

$c)$ $P_g \gg P_r$ and $\kappa_{\text{ff}} \gg \kappa_{\text{es}}$.

Regimes (a) and (b) in which opacity is dominated by electron scattering will be discussed in section 3.6. Here, we will present the solutions of regime (c), that is, we will assume that

$$P_r = 0 \quad \text{and} \quad \kappa_R = \kappa_{\text{ff}} = 5 \times 10^{24} \rho T_c^{-7/2}\, \text{cm}^2\text{g}^{-1}. \tag{3.122}$$

The solutions for the surface density Σ, the central temperature T_c, and the disk relative height (aspect ratio) are respectively

$$\Sigma = 1.6 \times 10^7\, \alpha^{-4/5} m_8^{+1/5} r^{-3/4} \dot{m}^{7/10} f^{7/10}\, \text{g}\,\text{cm}^{-2}, \tag{3.123}$$

$$T_c = 8.4 \times 10^6\, \alpha^{-1/5} m_8^{-1/5} r^{-3/4} \dot{m}^{3/10} f^{3/10}\, \text{K}, \tag{3.124}$$

$$\frac{H}{R} = 1.6 \times 10^{-3}\alpha^{-1/10}m_8^{-1/10}r^{1/8}\dot{m}^{3/20}f^{3/20}, \qquad [3.125]$$

where $m_8 = M/10^8\mathrm{M}_\odot$, $r = R/R_S$ and $f = 1 - (R_{\mathrm{in}}/R)^{1/2}$.

Since

$$\frac{P_g}{P_r} = 1.6 \times 10^{-2}\alpha^{-1/10}m_8^{1/2}r^{3/8}\dot{m}^{-7/20}f^{-7/20}, \qquad [3.126]$$

the regime (c) solution exists only for radii

$$r > 6 \times 10^4\alpha^{4/15}m_8^{4/3}\dot{m}^{14/15}f^{14/15}. \qquad [3.127]$$

This solution, the viscous time is

$$t_{\mathrm{vis}} \approx 26\,\alpha^{-6/5}\dot{m}_8^{6/5}\dot{m}^{-3/10}r^{5/4}\mathrm{yr}, \qquad [3.128]$$

so for the viscous time to be less than the age of the universe (13.8 Gyr), for an Eddington accretion rate onto a $10^8\,\mathrm{M}_\odot$ black hole, the outer disk radius must be $r_{\mathrm{out}} < 10^7\alpha^{24/25}$, that is, less than $\approx 100\,\mathrm{pc}$, or less than $1\,\mathrm{pc}$ if α in AGNs is $\sim 10^{-2}$.

In this solution, it was assumed that the opacity is given by the formula

$$\kappa_{\mathrm{R}} = 5 \times 10^{-4}\dot{m}^{-1/2}r^{3/4}\mathrm{cm}^2\mathrm{g}^{-1}, \qquad [3.129]$$

so that $\kappa_{\mathrm{R}} > \kappa_{\mathrm{es}} = 0.4\,\mathrm{cm}^2\mathrm{g}^{-1}$, for

$$r_{\mathrm{Res}} > 7400\dot{m}^{2/3}. \qquad [3.130]$$

Note that r_{Res} is independent on the accretor's mass.

It is characteristic of the Shakura–Sunyaev solution in this regime that the three Σ, T_c and T_{eff} radial profiles vary as $R^{-3/4}$ (this implies that the optical depth τ is constant with radius – see equation (VIII)). For high accretion rates and small radii, the assumption of opacity dominated by free-free and bound-free absorption will break down, and the solution will cease to be valid. We will first consider the other disk end: large radii.

There, the temperature will finally drop below 10^4 K, the disk plasma will recombine, leading to a drastic change in opacities which triggers a thermal instability.

3.5. Disk instabilities

In this section, we will present and discuss the disk thermal and the (related) viscous instabilities. In AGN disks, two instabilities are of interest: one is related to the hydrogen ionization/recombination, and the other occurs in disks where pressure is dominated by radiation. The instability is thermal when it grows on a thermal timescale, and in a geometrically thin disk, it means that the timescale is $\sim (\alpha\,\Omega_K)^{-1}$ (see equation [3.60]).

3.5.1. *The thermal instability*

A disk is thermally stable if the radiative cooling varies with temperature faster than the viscous heating. In other words, a disk is thermally stable if

$$\frac{d\ln \sigma T_{\rm eff}^4}{d\ln T_{\rm c}} > \frac{d\ln Q^+}{d\ln T_{\rm c}}. \qquad [3.131]$$

Using equation [3.120], we obtain

$$\frac{d\ln T_{\rm eff}^4}{d\ln T_{\rm c}} = 4\left[1-\left(\frac{T_{\rm irr}}{T_{\rm c}}\right)^4\right]^{-1} - \frac{d\ln \kappa}{d\ln T_{\rm c}}. \qquad [3.132]$$

We can see from these equations that disk stability is determined by the temperature dependence of the opacities and that irradiation stabilizes accretion disks.

3.5.1.1. *Thermal instability due to hydrogen ionization/recombination*

In a gas–pressure dominated disk, $Q^+ \sim \rho T H \sim \Sigma T \sim T_{\rm c}$. At high temperatures, $d\ln\kappa/d\ln T_{\rm c} \approx -4$ (see equation [3.122]). A thermal instability arises due to a rapid change of opacities with temperature when hydrogen begins to recombine ($T_c \sim 10^4$K) and this opacity dependence on temperature is no longer valid. In the instability region, the temperature exponent becomes large and positive: $d\ln\kappa/d\ln T_{\rm c} \approx 7-10$, and in the end, cooling decreases with temperature.

Thus, this thermal instability should be present in accretion disks with the mid-plane temperature $T_{\rm c}$ equal to few times 10^4K and below. It has been established that it is at the origin of outbursts observed in disks around stellar-mass black holes, neutron stars and white dwarfs. Systems containing the first two classes of objects are known as soft X-ray transients (SXTs, where "soft" relates to their X-ray spectrum), while those containing white-dwarfs are called dwarf-novae (despite the name that could suggest otherwise, nova and supernova outbursts have nothing to do with accretion disk outbursts). This range of temperatures is also present in disks around AGNs and we will present its consequences in section 3.5.3.

3.5.1.2. *Thermal instability of radiation–pressure-dominated disks*

Radiation–pressure-dominated ($P = P_{\rm rad} \propto T^4$) accretion disks are thermally unstable when the opacity is due to scattering on electrons. Indeed, in this case

$$\frac{d\ln T_{\rm eff}^4}{d\ln T_{\rm c}} = 4 \qquad [3.133]$$

because $\kappa_R = \kappa_{\rm es} = const.$, while in a radiation–pressure-dominated disk

$$Q^+ \sim HT^4 \sim T^8/\Sigma \qquad [3.134]$$

so that

$$\frac{d\ln Q^+}{d\ln T_c} = 8 > \frac{d\ln T_{\rm eff}^4}{d\ln T_c} \qquad [3.135]$$

and the disk is thermally unstable. This solution is represented by the middle branch with negative slope on the S-curve in Figure 3.8 (see also equation [3.177]).

The presence of this instability in the model is one of the unsolved problems of the accretion disk theory because it contradicts observations which do not show any unstable behavior in the range of luminosities where disks should be in the radiative pressure and electron-scattering opacity domination regime. This suggests that the disk description in this regime is incomplete.

For example, the radiation–pressure instability can be quenched if the disk vertical support is provided not by radiation but by a magnetic field. This field has to be brought into the disk from the surrounding medium, because the MRI dynamo is unable to produce a field with the required strength.

The heating rate can be rewritten as

$$Q^+ = \alpha P_{\rm rad} H \frac{{\rm d}\Omega}{{\rm d}R} = \alpha P_{\rm rad}^2 \frac{\Sigma}{\Omega_{\rm K}^2}\frac{{\rm d}\Omega}{{\rm d}R}, \qquad [3.136]$$

and from

$$\left.\frac{{\rm d}\log Q^+}{{\rm d}\log P_{\rm rad}}\right|_\Sigma = 2; \qquad \left.\frac{{\rm d}\log Q^-}{{\rm d}\log P_{\rm rad}}\right|_\Sigma = 1, \qquad [3.137]$$

we deduce, as before, that a radiation–pressure-dominated disk with electron-scattering opacity is thermally unstable.

However, with a magnetic-field pressure $P_{\rm mag}$, introducing

$$\beta' = \frac{P_{\rm mag}}{P_{\rm rad} + P_{\rm mag}} \qquad [3.138]$$

we can write the heating rate as

$$Q^+ = \frac{\alpha P_{\rm rad}^2}{(1-\beta')^2}\frac{\Sigma}{\Omega_{\rm K}^2}\frac{{\rm d}\Omega}{{\rm d}R}, \qquad [3.139]$$

from which we obtain

$$\left.\frac{{\rm d}\log Q^+}{{\rm d}\log P_{\rm rad}}\right|_{\Sigma, P_{\rm mag}} = 2\,(1-\beta')\,. \qquad [3.140]$$

Radiative cooling is independent of the magnetic pressure, so for $\beta' > 0.5$, the heating rate no longer increases more rapidly with the central radiation pressure, than the cooling rate and the disk is thermally stable. It is not clear, however, how and from where a stabilizing magnetic field will find itself in the right place and the right moment to quench the disk's thermal instability (see a related discussion in section 3.8).

3.5.2. *Thermal equilibria: the S-curve*

We will consider thermal equilibria of an accretion disk in which heating is due only to local turbulence, neglecting the effects of irradiation, because in the case of AGN disks (contrary to disks in X-ray binaries) irradiation does not make much difference to the instability's consequences. We put therefore $T_{\rm irr} = \widetilde{Q} = 0$.

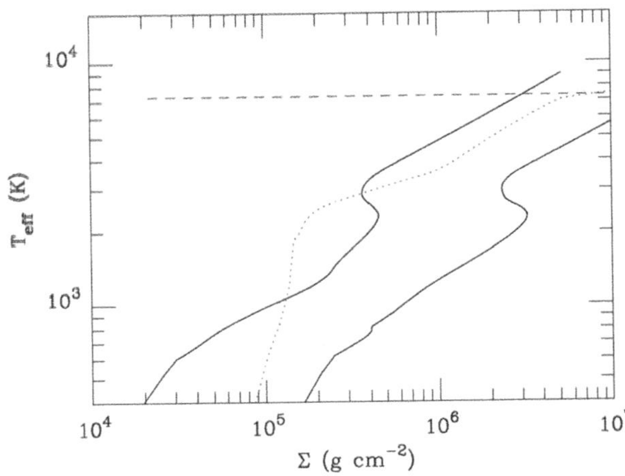

Figure 3.5. *S-curves in the $\Sigma - T_{\rm eff}$ plane, for $M = 10^8 {\rm M}_\odot$ and $R = 2 \times 10^{16}$ cm. S-curves for two values of the viscosity parameter are shown: $\alpha = 0.1$ (left) and $\alpha = 0.01$ (right). The dotted curve corresponds to $Q_T = 1$ (see equation [3.64]), and the dashed curve to $H/R = 0.1$. The model applies only to regions below the dashed curve and above the dotted one, that is, to a geometrically thin ($H/R \ll 1$), non-self-gravitating ($Q_T < 1$) accretion disks. In this case, the thin disk approximation applies to $r \lesssim 600$*

The thermal equilibrium in the disk is defined by the equation $Q^- = Q^+$ (see equation [3.56]), that is, by

$$\sigma T_{\rm eff}^4 = \frac{9}{8} \nu \Sigma \Omega_{\rm K}^2 \qquad [3.141]$$

(equation [3.30]). In general, ν is a function of density and temperature, and in the following, we will use the standard α-prescription equation [3.21]. The energy transfer equation provides a relation between the effective and the disk mid-plane temperatures so that thermal equilibria can be represented as a $T_{\text{eff}}(\Sigma)$ – relation (or equivalently a $\dot{M}(\Sigma)$–relation). In the relevant range of temperatures ($10^3 \lesssim T_{\text{eff}} \lesssim 10^5$), this relation forms an $\boldsymbol{S}$ on the (Σ, T_{eff}) plane, like in Figure 3.5. The upper, hot branch corresponds to the Shakura–Sunyaev solution presented in section 3.4.8. The two other branches correspond to solutions for cold (partially ionized) disks. Near the upper bend of the $\boldsymbol{S}$ and below it, convection can play an important role in the energy transfer.

Each point on the (Σ, T_{eff}) $\boldsymbol{S}$-curve represents an accretion disk's thermal equilibrium at a given radius, that is, a thermal equilibrium of a ring at radius R. In other words, each point of the $\boldsymbol{S}$-curve is a solution of the $Q^+ = Q^-$ equation. Points not on the $\boldsymbol{S}$-curve correspond to solutions of equation [3.56] *out of thermal equilibrium*: on the left of the equilibrium curve, cooling dominates overheating, $Q^+ < Q^-$; on the right, heating over cooling $Q^+ > Q^-$. Therefore, a positive slope of the $T_{\text{eff}}(\Sigma)$ curve corresponds to *stable solutions*, since a small increase of temperature of an equilibrium state (an upward perturbation) on the upper branch, say, will bring the ring to a state where $Q^+ < Q^-$ so it will cool down when getting back to equilibrium. In a similar way, a downward perturbation will provoke increased heating bringing back the system to equilibrium.

The opposite is happening along the $\boldsymbol{S}$-curve's segment with a negative slope as both a temperature increase and decrease lead to a runaway. The middle branch of the $\boldsymbol{S}$-curve corresponds therefore to *thermally unstable* equilibria.

A stable disk equilibrium can be represented only by a point on the lower, cold or the upper, hot branch of the $\boldsymbol{S}$-curve. This means that the surface density in a stable cold state must be *lower* than the maximal value on the cold branch: Σ_{max}, whereas the surface density in the hot stable state must be *larger* than the minimum value on this branch: Σ_{min}. Both these critical densities are functions of the viscosity parameter α, the mass of the accreting object, the distance from the center, and depend on the disk's chemical composition. In the case of solar composition, the critical surface densities are

$$\Sigma_{\text{min}} = 2.90 \times 10^3 \alpha^{-0.74} R_{15}^{1.04} m_8^{-0.35} \text{gcm}^{-2} \qquad [3.142]$$

$$\Sigma_{\text{max}} = 3.85 \times 10^3 \alpha^{-0.82} R_{15}^{0.99} m_8^{-0.33} \text{gcm}^{-2} \qquad [3.143]$$

and the corresponding effective temperatures

$$T_{\text{eff}}(\Sigma_{\text{min}}) = 4,300 R_{15}^{-0.12} \text{K} \qquad [3.144]$$

$$T_{\text{eff}}(\Sigma_{\text{max}}) = 3,300 R_{15}^{-0.12} \text{K}, \qquad [3.145]$$

where $R =: R_{15}10^{15}$cm. Using the equation (see equation [3.76])

$$\dot{M} = \sigma T_{\rm eff}^4 \frac{8\pi R^3}{3GM}, \qquad [3.146]$$

we obtain for the critical accretion rates

$$\dot{M}_{\rm crit}^+ := \dot{M}(\Sigma_{\rm min}) = 1.22 \times 10^{22} R_{15}^{2.52} {\rm g\,s^{-1}} \qquad [3.147]$$

$$\dot{M}_{\rm crit}^- := \dot{M}(\Sigma_{\rm max}) = 4.22 \times 10^{21} R_{15}^{2.52} {\rm g\,s^{-1}}. \qquad [3.148]$$

A stationary accretion disk in which there is a ring with effective temperature contained between the critical values of equations [3.144] and [3.145] cannot be stable. Since in a stationary disk, the effective temperature and the surface density both decrease with radius, the stability of a disk depends on the accretion rate and the disk size. For a given accretion rate, a stable disk cannot have an outer radius larger than the value corresponding to equation [3.147].

A disk is hot and stable if the rate at which mass is brought to its outer edge ($R \sim R_D$) is larger than the critical accretion rate at this radius $\dot{M}_{\rm crit}^+(R_D)$. On the other hand, a disk is cold and stable if the mass arrival rate to the disk is lower than the critical accretion rate at its inner radius $\dot{M}_{\rm crit}^-(R_{\rm in})$.

Taking $R_{\rm in} = 3R_S$ and $R_D = 2 \times 10^{16}$cm, for $10^8\,{\rm M}_\odot$ black hole, an accretion disk will be unstable if the mass arrival rate is contained in the range

$$9.6 \times 10^{20}\,{\rm g\,s^{-1}} < \dot{\rm M} < 2.3 \times 10^{25}\,{\rm g\,s^{-1}} \text{ or}, \qquad [3.149]$$

$$6.8 \times 10^{-7}\,{\rm M_\odot\,yr^{-1}} < \dot{\rm M} < 1.7 \times 10^{-2}\,{\rm M_\odot\,yr^{-1}} \qquad [3.150]$$

or

$$6.0 \times 10^{-6} < \dot{m} < 0.6. \qquad [3.151]$$

This means that a typical accretion disk in an AGN *is thermally unstable*.

3.5.3. *Variability of unstable AGN disks*

The thermal instability at a certain disk's radius produces a steep temperature gradient. This in turn creates a heating front that propagates radially bringing the cold disk regions to a hot state in which the heated up disk's rings are on the upper branch of the S-curve. The rise in temperature increases the viscosity (equation [3.23]). In the disks around stellar-mass objects, such as white dwarfs, neutron stars and black holes, this process redistributes the mass of the disk, creating a quasi-stationary disk configuration with an accretion rate $\dot{M} \approx$ const. $\approx$ few $\times\ \dot{M}_{\rm crit}^+(R_D)$, the hot critical accretion rate at the outer disk radius. This corresponds to the outburst maximum, from which the accretion rate and luminosity decay because, by construction, in an unstable disk, the mass-feeding rate is less than the hot critical accretion rate $\dot{M}^{+}_{\rm crit}(R_D)$ (see equation [3.149]). When the accretion rate drops below this value, a cooling-front forms in the outer disk regions and propagates towards the center, bringing the disk behind it to a cold state, thus switching-off the outburst. During the following quiescence, the disk, emptied during the outburst, fills up again until it reaches somewhere near the critical temperature, leading to the next outburst in the cycle. A typical dwarf-nova outburst or an X-ray binary transient event is characterized by a fast rise and slow decay which correspond to a viscous decay of a disk with a shrinking outer radius.

As can be seen in Figures 3.6 and 3.7, which show the light curve produced by an unstable disk around a supermassive black hole, the thermal disk instability also produces outbursts in AGNs, but a closer look shows that they are rather different from the eruptions observed in systems with less massive compact objects.

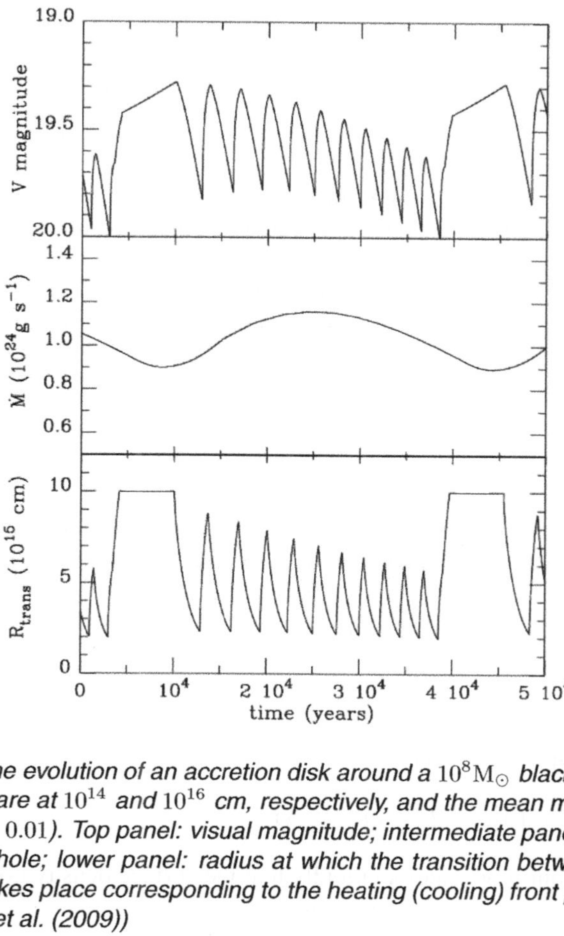

Figure 3.6. *Time evolution of an accretion disk around a $10^8 M_\odot$ black hole. The inner and outer radii are at 10^{14} and 10^{16} cm, respectively, and the mean mass feeding rate is 10^{24} g/s ($\dot{m} \approx 0.01$). Top panel: visual magnitude; intermediate panel: accretion rate onto the black hole; lower panel: radius at which the transition between the hot and cold regimes takes place corresponding to the heating (cooling) front position (source: from Hameury et al. (2009))*

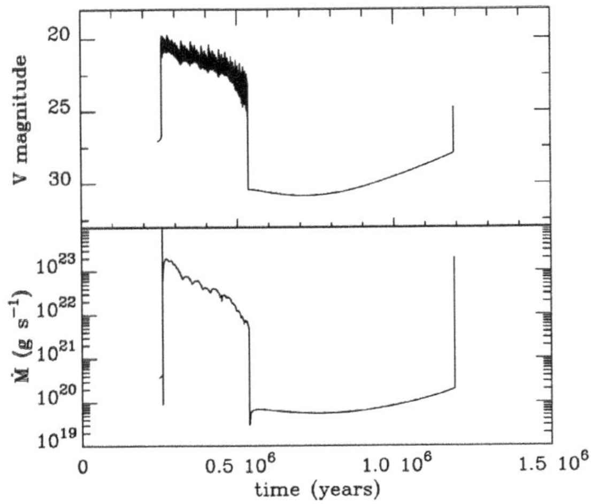

Figure 3.7. *Time evolution of an accretion disk with the same parameter as in Figure 3.6, except for the mean mass feeding rate which is here* 2×10^{22} *g/s (*$\dot{m} \approx 0.0001$*). Top panel: visual magnitude; lower panel: accretion rate onto the black hole (source: from Hameury et al. (2009))*

For relatively high mass-feeding rates ($\dot{m} \approx 0.01$ – Figure 3.6), the instability results in low-amplitude (~ 1 magnitude) outbursts corresponding to tiny modulations of the accretion rate. In fact, we can see that the timescale of brightness variations, corresponding to the temperature oscillations, is much shorter than the timescale of the accretion rate variations. The basic reason for this is that the front propagation speed is approximately α times the sound speed which means that the front propagation timescale is

$$t_{\rm front} \approx \frac{R}{\alpha c_{\rm s}} = \frac{R}{H} t_{\rm th}, \qquad [3.152]$$

where $t_{\rm th}$ is the thermal timescale. Hence, $t_{\rm front}$ is shorter than the viscous time $t_{\rm visc} = (R/H)^2 t_{\rm th}$ by a factor R/H, that is, by several orders of magnitude:

$$\frac{H}{R} \approx 3.2 \times 10^{-4} T_4^{1/2} m_8^{-1/2} R_{15}^{1/2} \qquad [3.153]$$

$$= 5.5 \times 10^{-5} T_4^{1/2} r^{1/2}. \qquad [3.154]$$

The viscous time is therefore

$$t_{\rm vis} \approx 1.5 \times 10^4 \alpha^{-1} T_4^{1/2} m_8 \, r^{1/2} \, {\rm yr}. \qquad [3.155]$$

Hence, the front propagates so rapidly that the surface density does not have the time to react to the changes of temperature.

In the case of stellar-mass compact objects, $t_{\rm front}$ is shorter than $t_{\rm visc}$, but not by such a large amount and strong gradients in the disk make the effective viscous time comparable to the front propagation time. The reason is that while in dwarf-nova stars and X-ray binaries, the unstable disk region is at $r \simeq 10^4$, in AGNs it is rather at $r < 300$, much deeper in the gravitational well of the accretor.

When we lower the accretion rate to $\dot{m} \approx 0.0001$, the disk manages to descend to quiescence (Figure 3.7), but it does this through a pure viscous decay, without a cooling front propagating down through the whole disk. The outburst phase lasts 3×10^5 years, which is roughly the disk viscous time. During this slow decay from maximum, heating/cooling fronts get reflected 400 times, creating as before low amplitude optical magnitude oscillations and tiny modulations of the accretion rate.

Neither disk irradiation, nor inner-disk truncation change the AGN outburst properties. This is not surprising since in stellar-mass systems, these effects play an important role, mainly by affecting the cooling from propagation, which in the AGN case are of no importance.

It remains to be seen if in AGNs, thermal-instability outbursts occur in an observable way or if they are even occurring at all.

3.6. Beyond thin disks

Until now, we have assumed that accretion disks are thin, that is, that $H/R \ll 1$. One of the consequences of this assumption was neglecting the advection terms in the energy and momentum equations for stationary accretion flows. The vertically averaged advective "cooling" term can be written as

$$Q^{\rm adv} = \frac{\Sigma T v_r}{R}\frac{ds}{d\ln R} = \frac{\dot{M}}{2\pi R^2} c_{\rm s}^2 \xi_a, \qquad [3.156]$$

where ξ_a is a slowly varying function related to the entropy gradient and is usually ~ 1. We then have

$$\frac{Q^{\rm adv}}{Q^+} \sim \frac{c_s^2}{\Omega_K^2 R^2} \approx \left(\frac{H}{R}\right)^2, \qquad [3.157]$$

where we used equations [3.57] and [3.75]. Therefore, neglecting the advective term in the energy equation

$$Q^+ = Q^- + Q^{\rm adv} \qquad [3.158]$$

is justified if we assume $H/R \ll 1$.

There are two regimes of parameters where this assumption is not valid, in both cases for the same reason: low radiative efficiency when the time for radial motion towards the black hole is shorter than the radiative cooling time. Low density (low accretion rate), hot, optically thin accretion flows are poor coolers and they are one of the two configurations where advection instead of radiation is the dominant evacuation-of-energy ("cooling") mechanism. Such optically thin flows are called ADAFs, for Advection Dominated Accretion Flows[3]. Also, advection-dominated flows are high-luminosity flows, accreting at high rates, but they are called "slim disks" to account for them not being thin, but still being described as if this were not of much importance.

We will begin with optically thin flows.

– ADAFs

Advection-dominated accretion flows (ADAFs) is a term that describes the accretion of matter with angular momentum, in which radiation efficiency is very low. In their applications, ADAFs are supposed to describe inflows onto compact bodies, such as black holes or neutron stars; but very hot, optically thin flows are bad radiators in general so, in principle, ADAFs are also possible in other contexts. Of course, in the vicinity of black holes or neutron stars, the virial (gravitational) temperature is $T_{\rm vir} \approx 5 \times 10^{12}(R_S/R)$ K, so that in optically thin plasmas, at such temperatures, both the coupling between ions and electrons and the efficiency of radiation processes are rather feeble. In such a situation, the thermal energy released in the flow by the viscosity, which drives accretion by removing angular momentum, is not going to be radiated away, but will be *advected* towards the compact body. If this compact body is a black hole, the heat will be lost forever, so that advection, in this case, acts as sort of a "global" cooling mechanism (we should keep in mind that, in general, advection may also be responsible for heating, depending on the sign of the temperature gradient – in some conditions, near the black hole, advection heats up electrons in a two-temperature ADAF).

In general, the role of advection in an accretion flow depends on the radiation efficiency, which in turns depends on the microscopic state of matter and on the absence or presence of a magnetic field. If, for a given accretion rate, radiative cooling is not efficient, advection is necessarily dominant, assuming that a stationary solution is possible.

– Slim disks

At high accretion rates, disks around black holes become dominated by radiation pressure in their inner regions, close to the black hole. At the same time, the opacity is dominated by electron scattering. In such disks, H/R is no longer $\ll 1$. However, this means that terms involving the radial velocity are no longer negligible since

3. Some people prefer to call such flows RIAFs, from *Radiatively Inefficient Accretion Flows*, but I prefer to stick to the name ADAF, which I had the pleasure to introduce.

$v_r \sim \alpha c_s(H/R)$. In particular, the advective term in the energy conservation equation $v_r \partial S/\partial R$ (see equation [3.53]) becomes important and finally, at super-Eddington rates, dominant. When $Q^+ = Q^{\mathrm{adv}}$, the accretion flow is advection-dominated and called a slim disk.

– Windy disks

In a super-Eddington accreting disk, the radiative pressure can blow out matter creating an outflow that will limit the luminosity to its local Eddington value. In such a case, $\dot{m} \sim r$.

3.6.1. *Advection-dominated accretion flow toy models*

We can illustrate fundamental properties of ADAFs and slim disks with a simple toy model. The advection "cooling" (per unit surface) term in the energy equation can be written as

$$Q^{\mathrm{adv}} = \frac{\dot{M}}{2\pi R^2} c_{\mathrm{s}}^2 \xi_a \qquad [3.159]$$

(equation [3.156]).

Using the (non-relativistic) hydrostatic equilibrium equation

$$\frac{H}{R} \approx \frac{c_{\mathrm{s}}}{v_{\mathrm{K}}} \qquad [3.160]$$

we can write the advection term as

$$Q^{\mathrm{adv}} = \Upsilon \frac{\kappa_{\mathrm{es}} c}{2R} \left(\frac{\dot{M}}{\eta}\right) \xi_a \left(\frac{H}{R}\right)^2 \qquad [3.161]$$

whereas the viscous heating term can be written as

$$Q^{+} = \Upsilon \frac{3}{8} \frac{\kappa_{\mathrm{es}} c}{R} \left(\frac{\dot{M}}{\eta}\right), \qquad [3.162]$$

where

$$\Upsilon = \left(\frac{c}{\kappa_{\mathrm{es}} r}\right)^2. \qquad [3.163]$$

Since $\xi_a \sim 1$,

$$Q^{\mathrm{adv}} \approx Q^{+} \left(\frac{H}{R}\right)^2 \qquad [3.164]$$

and, as said before, for geometrically thin disks ($H/R \ll 1$), the advective term $Q^{\rm adv}$ is negligible compared to the heating term Q^+ in thermal equilibrium, and viscous heating must be compensated by radiative cooling. Things are different at very high temperatures, when $(H/R) \sim 1$. Then, the advection term is comparable to the viscous term and cannot be neglected in the equation of thermal equilibrium. In some cases, this term is larger than the radiative cooling term Q^- and (most of) the heat released by viscosity is *advected* toward the accreting body instead of being locally radiated away, as happens in geometrically thin disks.

From equation [3.75], we can obtain a useful expression for the square of the relative disk height (or aspect ratio):

$$\left(\frac{H}{R}\right)^2 = \frac{\sqrt{2}}{\kappa_{\rm es}}\left(\frac{\dot{m}}{\eta}\right)(\alpha\Sigma)^{-1}r^{-1/2}. \quad [3.165]$$

Deriving equation [3.165], we used the viscosity prescription $\nu = (2/3)\alpha c_{\rm s}^2/\Omega_K$.

Using this equation, we can write for the advective cooling

$$Q^{\rm adv} = \Upsilon\Omega_K\xi_a(\alpha\Sigma)^{-1}\left(\frac{\dot{m}}{\eta}\right)^2. \quad [3.166]$$

The thermal equilibrium (energy) equation is

$$Q^+ = Q^{\rm adv} + Q^-. \quad [3.167]$$

The form of the radiative cooling term depends on the state of the accreting matter, that is, on its temperature, density and chemical composition. Let us consider two cases of accretion flows: those that are **optically thick**, and those that are **optically thin**.

For the optically thick case, we will use the diffusion approximation formula

$$Q^- = \frac{8}{3}\frac{\sigma T_c^4}{\kappa_{\rm R}\Sigma}, \quad [3.168]$$

and assume $\kappa_{\rm R} = \kappa_{\rm es}$. With the help of equation [3.165], this can be brought to the form

$$Q^-_{\rm thick} = 8\Upsilon\left(\frac{\kappa_{\rm es}R_S}{c}\right)^{1/2} r^2\Omega_K^{3/2}(\alpha\Sigma)^{-1/2}\left(\frac{\dot{M}}{\eta}\right)^{1/2}. \quad [3.169]$$

For the optical thin case of bremsstrahlung radiation, we have

$$Q^{-} = 1.24 \times 10^{21} H \rho^2 T^{1/2} \qquad [3.170]$$

which, using equation [3.154], can be written as

$$Q^{-}_{\text{thin}} = 3.4 \times 10^{-6} \Upsilon r^2 \Omega_K \alpha^{-2} (\alpha\Sigma)^2 . \qquad [3.171]$$

– In the **optically thick** case, we have therefore

$$\xi_a \left(\frac{\dot{M}}{\eta}\right)^2 + 0.18 r^{1/2} (\alpha\Sigma) \left(\frac{\dot{M}}{\eta}\right) + \\ +2.3 r^{5/4} (\alpha\Sigma)^{1/2} \left(\frac{\dot{M}}{\eta}\right)^{1/2} = 0 \qquad [3.172]$$

– In the **optically thin** case, the energy equation has the form

$$\xi_a \left(\frac{\dot{M}}{\eta}\right)^2 + 0.18 r^{1/2} (\alpha\Sigma) \left(\frac{\dot{M}}{\eta}\right) + \\ +3 \times 10^{-6} \alpha^{-2} r^2 (\alpha\Sigma)^3 = 0 \qquad [3.173]$$

There are two distinct types of advection-dominated accretion flows also: optically thin and optically thick. We will first deal with optically thin flows, which are the true ADAFs.

3.6.1.1. *Optically thin flows: ADAFs*

For prescribed values α and ξ_a, equation [3.173] is a quadratic equation in $(\dot{m}/\eta)$ whose solutions in the form of $\dot{m}(\Sigma)$ describe thermal equilibria at a given value of R. Obviously, for a given Σ, this equation has at most two solutions. The solutions form two branches on the $\dot{m}(\alpha\Sigma)$ – plane:

– the ADAF branch

$$\dot{m} = 0.53 \kappa_{\text{es}} \eta r^{1/2} \xi_a^{-1} \alpha\Sigma. \qquad [3.174]$$

and

– the radiatively cooled branch

$$\dot{m} = 1.9 \times 10^{-5} \eta r^{3/2} \xi_a^{-1} \alpha^{-2} (\alpha\Sigma)^2 . \qquad [3.175]$$

From equations [3.174] and [3.175], it is clear that there exists a maximum accretion rate for which only one solution of equation [3.173] exists. This implies the existence of a maximum accretion rate at

$$\dot{m}_{\max} \approx 1.7 \times 10^3 \eta \, \alpha^2 r^{1/2} . \qquad [3.176]$$

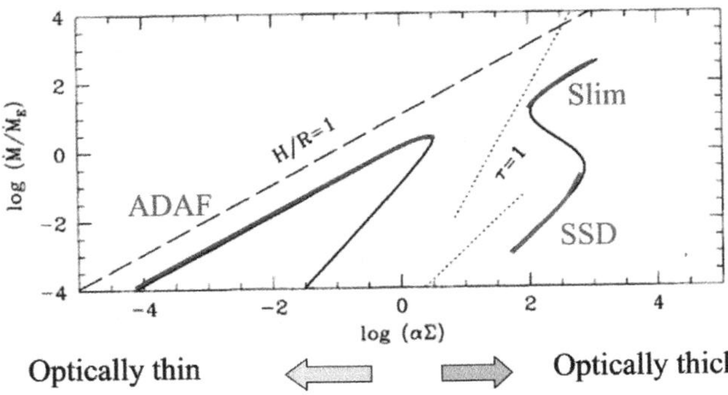

Figure 3.8. *Thermal equilibria for optically thick (the right solid S-shaped line) and optically thin (the left solid line) accretion flows. The upper branches correspond to an advection-dominated solution (ADAF and slim). SSD correspond to the Shakura–Sunyaev solution. Flows above the dotted lines $\tau = 1$ are optically thin, with τ being the effective optical depth calculated for radiation pressure-dominated (upper line) or gas-dominated (lower line) configurations. This figure corresponds to $M_{\rm BH} = 10{\rm M}_\odot$, $r = 5$, $\alpha = 0.1$ and $\xi_a = 1$ (source: adapted from Abramowicz et al. (1995)). For a color version of this figure, see www.iste.co.uk/combes/nuclei.zip*

This is where the two branches formed by the thermal equilibrium solutions on the $\dot{M}(\alpha\Sigma)-$ plane meet as shown in Figure 3.8.

The value of $\dot{m}_{\rm max}$ depends on the cooling mechanism in the accretion flow and the non-relativistic free-free cooling is not a realistic description of the emission in the vicinity ($(R/R_S) \lesssim 10^3$) of a black hole. The flow there will most probably form a two-temperature plasma. In such a case $\dot{m}_{\rm max} \approx 10\alpha^2$, with almost no dependence on radius. For larger radii, $\dot{m}_{\rm max}$ decreases with radius.

3.6.1.2. *Optically thick flows: slim disks*

Since the first two terms in equation [3.172] are the same as in equation [3.173], the high $\dot{m}$, advection-dominated solution is the same as in the optically thin case but now represents the *slim disk branch*:

$$\dot{m} = 0.53\, \kappa_{\rm es}\, \eta r^{1/2} \xi_a^{-1} \alpha\Sigma. \qquad [3.174]$$

Now, the full equation [3.172] is a cubic equation in $\dot{m}^{1/2}$ and on the $\dot{m}(\alpha\Sigma)-$ plane, its solution forms the two upper branches of the $\boldsymbol{S}$-curve shown in Figure 3.8. The uppermost branch corresponds to slim disks while the branch with negative slope represents the Shakura–Sunayev solution in the regime *(a.)* (see section 3.4.8), that is *a radiatively cooled, radiation pressure-dominated accretion disk*:

$$\dot{m} = 160\, \kappa_{\rm es}^{-1}\, \eta r^{3/2}\, (\alpha\Sigma)^{-1} \qquad [3.177]$$

3.6.1.3. *Slim disks and super-Eddington accretion*

From equations [3.165] and [3.177], we obtain for the disk aspect ratio

$$\frac{H}{R} = 0.11\left(\frac{\dot{m}}{\eta}\right)r^{-1} \qquad [3.178]$$

which shows that the height of a radiation-dominated disk is constant with radius and proportional to the accretion rate.

However, this means that with increasing $\dot{m}$ advection becomes more and more important (see, for example, equation [3.164]) and for

$$\frac{\dot{m}}{\eta} \approx 9.2r \qquad [3.179]$$

advection will take over radiation as the dominant cooling mechanism and the solution will represent a slim disk. Equation [3.179] can also be interpreted as giving the *transition radius* between a radiatively and advectively cooled disk for a given accretion rate $\dot{m}$:

$$r_{\rm trans} \approx \frac{0.1}{\eta}\dot{m} \qquad [3.180]$$

Another radius of interest is the *trapping radius* at which the photon diffusion (escape) time $H\tau/c$ is equal to the viscous infall time R/v_r

$$R_{\rm trapp} = \frac{H\tau\, v_r}{c} = \frac{H\kappa\Sigma}{c}\frac{\dot{M}}{2\pi R\Sigma} = \frac{H}{R}\left(\frac{\dot{m}}{\eta}\right)R_S. \qquad [3.181]$$

Note that both $R_{\rm trans}$ and $R_{\rm trapp}$ are proportional to the accretion rate.

In an advection-dominated disk, the aspect ratio H/R is independent of the accretion rate:

$$\frac{H}{R} = 0.86\,\xi_a\, r^{1/4}, \qquad [3.182]$$

therefore contrary to radiatively cooled disks, slim disks do not puff up with increasing accretion rate.

Putting [3.182] into equation [3.181], we obtain

$$r_{\rm trapp} = 0.86\,\xi_a^{-1/2} r^{1/4}\left(\frac{\dot{m}}{\eta}\right). \qquad [3.183]$$

Radiation inside the trapping radius is unable to stop accretion and since $R_{\rm trapp} \sim \dot{m}$, there is no limit on the accretion rate onto a black hole.

The luminosity of the toy-model slim disk can be calculated from equations [3.169] and [3.174] giving

$$Q^{-} = \sigma T_{\rm eff}^4 = \frac{0.1}{\xi_a}\frac{L_{\rm Edd}}{R^2}, \qquad [3.184]$$

which implies $T_{\rm eff} \sim 1/R^{1/2}$. The luminosity of the slim disk part of the accretion flow is then

$$L_{\rm slim} = 2\int_{R_{\rm in}}^{R_{\rm trans}} \sigma T_{\rm eff}^4 2\pi R dR = \frac{0.8}{\xi_a} L_{\rm Edd} \cdot \ln\frac{R_{\rm trans}}{R_{\rm in}} \approx L_{\rm Edd} \ln \dot{m}, \qquad [3.185]$$

where we used equation [3.180].

Therefore, the total disk luminosity

$$L_{\rm total} = L_{\rm thin} + L_{\rm slim} = \qquad [3.186]$$

$$4\pi\left(\int_{R_{\rm in}}^{R_{\rm trans}} \sigma T_{\rm eff}^4 R dR + \int_{R_{\rm trans}}^{R_\infty} \sigma T_{\rm eff}^4 R dR\right) \approx L_{\rm Edd}(1 + \ln \dot{m}),$$

where $L_{\rm thin}$ is the luminosity of the radiation-cooled disk for which equation [3.76] applies.

Such logarithmic luminosity–accretion-rate dependence is observed in the inner regions of simulated accretion flows.

It is easy to see that the same luminosity formula $L \approx L_{\rm Edd}(1 + \ln \dot{m})$ is obtained when we assume mass loss from the disk, resulting in a radially variable accretion rate: $\dot{M} \sim R$ (see section 3.8).

3.7. Disk coronae

The X-ray emission observed from AGN accretion flows require the presence of emitters other than the optically thick accretion disk that we have described up until this point. The highest temperature we can get from the inner regions of such a disk is $\lesssim 10^6$K (see equation [3.79]). In analogy to solar and stellar coronae, the optically thin hot structures, much hotter than the underlying photosphere, the X-ray radiating structure in AGNs and other accreting compact body systems, is called a *disk corona*. The main idea is that the hot electrons of such a corona, inverse-Compton upscatter to high energies the soft photons emitted by the underlying disk. There is not a single or even leading, or dominating disk corona model. Figure 3.9 shows various models or scenarios that have been proposed in the literature. Various models might correspond to various observed states of the accretion disk. The bottom configuration corresponds to the oldest proposed model whose idea was that differential rotation, together with convection and magnetic fields present in the disk, could produce loop-like structures, forming a magnetically confined hot corona, in analogy to what is observed in the Sun. As for the Sun, magnetic-field reconnection could play a role in accelerating electrons to high energies. More recently, for different physical reasons (without invoking convection), models of magnetically heated coronae have been constructed. In this case it is the MRI amplification mechanism that, from the poloidal field, produces a large amount of toroidal field in the upper layers of the disk. Other models invoke accretion-disk "evaporation", a mechanism that, for low accretion rates, is supposed to produce ADAFs in the inner part of accretion disks. In fact, an ADAF could play the role of a disk corona, as in the second panel from the top.

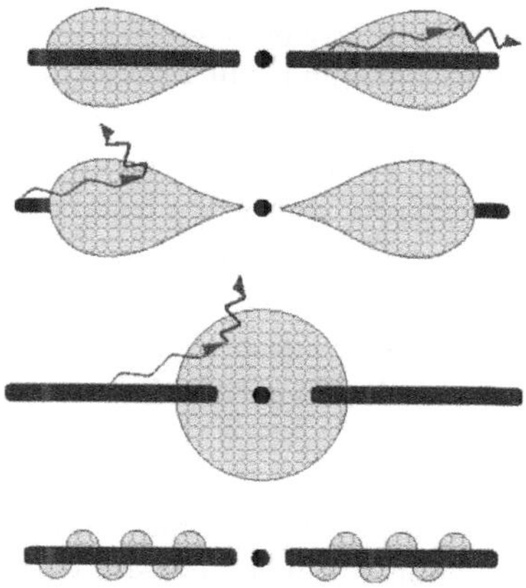

Figure 3.9. *Various possible (in principle) disk-coronal structures (courtesy of Chris Done). For a color version of this figure, see www.iste.co.uk/combes/nuclei.zip*

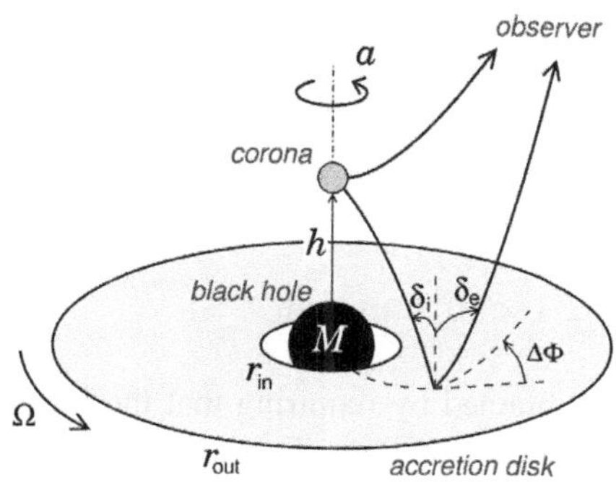

Figure 3.10. *A scheme representing the lamppost "coronal" model. The feature labeled "corona" is the lamppost in question (source: from Caballero-Garcia et al. (2019)). For a color version of this figure, see www.iste.co.uk/combes/nuclei.zip*

However, the most used "coronal" set-up is the so-called *lamppost* model, whose principle is shown in Figure 3.10. As its name demonstrate, this model does not have exaggerated ambitions to represent a physical reality, but is extremely useful when considering reverberation mapping. Of course, this lamppost is a corona only in name. Most often one says that it could correspond to the base of a jet.

3.8. Disks, winds and jets

Accretion in AGN is the source of the inverse phenomenon: outflows (ejections) that with radiation are the key elements of the feedback cycle linking the supermassive central black hole to its host galaxy. Radiation and collimated outflows in the form of jets interact with the interstellar medium leading to ejection or heating of the gas, but accretion disk winds apparently play the fundamental role in this interaction. Winds and collimated jets are produced by all systems containing a disk-accreting celestial body, from young stellar objects, through white dwarfs and neutron stars, to black holes of all masses, so we can expect that a common mechanism operates in accretion disks at all scales. The nature of this mechanism is still subject to controversy. In the case of winds, three possible mechanisms are invoked. Outer disk regions most probably eject thermal-driven winds that are accelerated by the thermal gas pressure. Their velocity is at most $\sim 1,000$ km/s.

The radiative pressure due to the intense radiation of most AGN can be a very effective way to drive an accretion disk wind. For sub-Eddington accretion rates, UV absorption lines are the main source of opacity (as for winds of massive stars), while near the Eddington luminosity, Compton scattering is the wind blowing driver.

3.8.1. *The "forgotten" Shakura–Sunyaev solution*

In their seminal 1973 paper, Shakura and Sunyaev also considered the case of super-Eddington accretion and found a solution, alternative to the slim disk solution (which they discarded). For mass-feeding rates $\dot{m} > 1$, they identified the *spherization* radius where the luminosity is close to the local Eddington value as

$$R_{\rm sph} \simeq \frac{27}{8}\dot{m}R_S \simeq 1 \times 10^{14}\dot{m}m_8\,\mathrm{cm} \qquad [3.187]$$

Then, the solution is obtained by requiring that the local emission within $R_{\rm sph}$ nowhere exceeds its local Eddington limit. This will be true if the outflow is such that the accretion rate through the disk decreases as

$$\dot{m} \simeq \dot{m}\frac{R}{R_{\rm sph}}. \qquad [3.188]$$

Then, the total luminosity is

$$L \simeq L_{\rm Edd}(1 + \ln \dot{m}) \qquad [3.189]$$

(see equation [3.186]). At very high accretion rates ($\dot{m} \gg 1$), the disk emission will also be strongly beamed by the flow geometry so that an observer situated in the beam of the emitting system will infer a luminosity

$$L_{\rm sph} = \frac{1}{b} L_{\rm Edd}(1 + \ln \dot{m}), \qquad [3.190]$$

where b is a beaming factor. Seen from the "side" such a source with very large apparent luminosities might appear to be rather dim. On the contrary, if, for example, $b \sim 1/\dot{m}^2$, a stellar mass ultra-luminous X-ray source (by definition $L > 10^{39}$ erg/s) accreting at $\dot{m} \simeq 10^4$, seen along the beam, could have an apparent luminosity $\sim 10^{45}$ erg/s and look like an AGN, but will be positioned off the galactic center.

3.8.2. *Relativistic jets*

Magnetic fields may play a fundamental role in the production of disk winds. In this case, the wind is accelerated by the centrifugal force of the magnetic field lines anchored on the disk and the magnetic pressure (Blandford–Payne mechanism, see Blandford and Payne (1982)). Magnetic fields are usually involved at the generation and acceleration phase and are always supposed to play a decisive role in keeping the jet collimated. Quite often in models, the jet is a collimated part of an outflow from the accretion disk, but in other cases, winds and jets can be launched by different mechanisms. Such a case is shown schematically in Figure 3.11 where the relativistic jet is powered by the black hole rotation while the source of the sub-relativistic wind energy is gravitational.

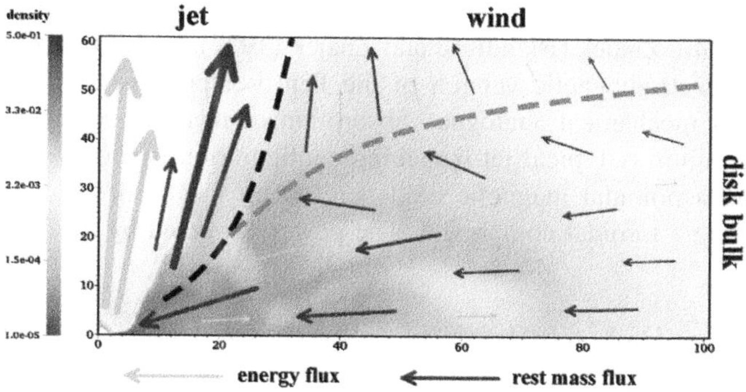

Figure 3.11. *Schematic picture of a disk-wind-jet structure near a rotating black hole (source: from Sądowski et al. (2013)). For a color version of this figure, see www.iste.co.uk/combes/nuclei.zip*

For launching relativistic jets observed in AGN, the best performing models involve large-scale magnetic fields anchored in the rapidly rotating matter of the inner parts of accretion disks. In the case of a rotating black hole, two sources of jet launching energy are possible: the gravitational energy of accretion and the black hole rotational energy that can be tapped through the electromagnetic Penrose process. This final mechanism is possible only when the accretor is a black hole because other rotating compact bodies do not have ergoregions.

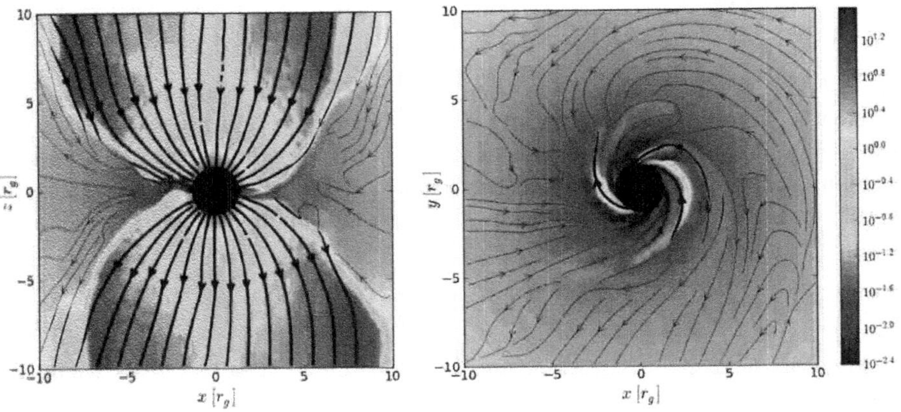

Figure 3.12. *Snapshot at $t \approx 7,806\, r/c$ from general-relativistic (GR) MHD simulation of the Blandford–Znajek process showing logarithm of rest-mass density in color (see the scale on the right-hand side) in both the $z - x$ plane at $y = 0$ (top left-hand panel) and the $y - x$ plane at $z = 0$ (top right-hand panel). The black lines trace field lines, where the thicker black lines show where field is lightly mass-loaded. In this simulation, $a = 0.9375$ and accreting matter is in the form of an ADAF (source: adapted from McKinney et al. (2012)). For a color version of this figure, see www.iste.co.uk/combes/nuclei.zip*

The Blandford–Znajek (Blandford and Znajek (1977), hereafter BZ) mechanism, which is the electromagnetic version of the Penrose, process works on the same principle as its mechanical analogue: absorption of negative energy and negative angular momentum. A typical jet-launching configuration is shown in Figure 3.12, where we see a poloidal magnetic field penetrating the black hole surface being twisted, forming a toroidal component. The power extracted by the BZ mechanism is equal to

$$P_{\mathrm{BZ}} = \frac{\kappa}{4\pi c} \Omega_{\mathrm{H}}^2 \Phi_{\mathrm{BH}}^2 f\left(\Omega_{\mathrm{H}}\right), \qquad [3.191]$$

where

$$\Phi_{\mathrm{BH}} = (1/2) \int_\theta \int_\varphi |B^r|\, \mathrm{d}A_{\theta\varphi}, \qquad [3.192]$$

is the magnetic flux threading one hemisphere of the black hole horizon. $dA_{\theta\varphi}$ is the area element in the $\theta - \varphi$ plane. κ is a numerical constant depending on the magnetic field geometry ($\kappa = 0.053$ for the so-called split-monopole geometry, such as in Figure 3.12). Quite often, equation [3.191] is written with $f(\Omega_H) = 1$, which is a good approximation for black hole spins up to $a \approx 0.95$, but for larger spins, of special interest when considering the BZ mechanism

$$f\left(\Omega_{\mathrm{H}}\right) \approx 1 + 0.345\left(\Omega_{\mathrm{H}} r_S/c\right)^2 - 0.575\left(\Omega_{\mathrm{H}} r_S/c\right)^4. \quad [3.193]$$

is a more suitable approximation.

We define the efficiency of the BZ mechanism as the ratio of the time-averaged electromagnetic flux extracting black hole rotational energy to the averaged rate at which the black hole absorbs rest-mass energy:

$$\eta_{\mathrm{BZ}} := \frac{\langle P_{\mathrm{BZ}}\rangle}{\langle \dot{M}\rangle c^2} \times 100\% \quad [3.194]$$

$$= \frac{\kappa}{16\pi}\left(\frac{\Omega_{\mathrm{H}} r_S}{c}\right)^2 \langle \phi_{\mathrm{BH}}^2\rangle f\left(\Omega_{\mathrm{H}}\right) \times 100\%, \quad [3.195]$$

where

$$\phi_{\mathrm{BH}} = \Phi_{\mathrm{BH}}\left(\langle \dot{M}\rangle r_S^2 c\right)^{1/2} \approx 10^4 \eta_{0.1}^{1/2} m_8^{1/2} \dot{m}^{1/2}\left(\frac{\Phi_{\mathrm{BH}}}{0.1\,\mathrm{pc^2 G}}\right) \quad [3.196]$$

is the dimensionless magnetic flux threading the black hole.

Therefore, the efficiency of the BZ mechanism depends strongly on the black hole spin and on the magnetic flux threading the black hole surface. In equation [3.196], we have normalized Φ_{BH} by its observed interstellar-medium value. In the most efficient configuration, the black hole receives as much large-scale magnetic flux as can be pushed into it by accretion. By supplying even more flux than this, some of it remains outside the horizon where it impedes the accreting gas, leading to a "magnetically arrested disk" (MAD). It is the MAD configuration that produces the most powerful relativistic jets.

This requires bringing this field to the innermost disk region where it is needed as an extractor of the black hole rotational energy. It is far from obvious how this can be done.

The simplest problem to address in this context is to consider a poloidal field threading a Keplerian accretion disk. The turbulent viscosity-driven accretion tends to drag the field-lines inwards but due to resistivity they diffuse outwards. Numerical simulation of this problem has shown that the condition for the magnetic field lines to be significantly dragged inwards is

$$\mathcal{B} = \frac{3H\,|v_R|}{2\eta} = \left(\frac{H}{R}\right)\left(\frac{\nu}{\eta}\right) = \left(\frac{H}{R}\right)\mathcal{P}_{\rm tm} \gtrsim 1, \qquad [3.197]$$

where ν, as before, is the kinematic viscosity coefficient, and

$$\eta = \frac{c^2}{4\pi\sigma} \qquad [3.198]$$

is the resistivity, with σ being the electric conductivity. $\mathcal{P}_{\rm tm}$ is known as the magnetic Prandtl number. The usual definition of the Prandtl number involves microphysical turbulence and viscosity, but here we are interested in transport coefficients of turbulent origin, hence the letter "t" in the index of the symbol. Therefore, in thin accretion disks, the magnetic field will be dragged inwards only for very large Prandtl numbers: $\mathcal{P}_{\rm tm} \gg 1$ or $\eta \ll \nu$. In MRI accretion disks, $\mathcal{P}_{\rm tm} \gtrsim 1$, so "thick" disks, such as slim disks or ADAFs, are needed if the magnetic flux is supposed to accumulate at the inner disk, near the black hole surface.

Another possibility is that the disk's angular momentum is not removed by turbulent viscosity, but by, for example, magnetic wind. Then, equation [3.197] does not apply and the condition for field dragging becomes

$$\mathcal{B} = \frac{3H\,|v_R|}{2\eta} \gtrsim 1. \qquad [3.199]$$

In some models $|v_R| \sim c_s$ and $\eta \sim Hc_s$, so field dragging is possible.

3.9. References

Abramowicz, M.A., Chen, X., Kato, S., Lasota, J.-P., Regev, O. (1995). *ApJ*, 438, L37.

Balbus, S.A. and Hawley, J.F. (1991). *ApJ*, 376, 214.

Bardeen, J.M. and Petterson, J.A. (1975). *ApJ*, 195, L65.

Blandford, R.D. and Payne, D.G. (1982). *MNRAS*, 199, 883.

Blandford, R.D. and Znajek, R.L. (1977). *MNRAS*, 179, 433.

Caballero-Garcia, M.D., Dovciak, M., Bursa, M., Papadakis, I., Karas, V. (2019). *Proceedings of Science*. arXiv:1901.04357.

Hameury, J.-M., Marck, J.-A., Pelat, D. (1994). *A&A*, 287, 795.

Hameury, J.-M., Viallet, M., Lasota, J.-P. (2009). *A&A*, 496, 413.

Horne, K. and Marsh, T.R. (1986). *MNRAS*, 218, 761.

McKinney, J.C., Tchekhovskoy, A., Blandford, R.D. (2012). *MNRAS*, 423, 3083.

Sadowski, A. (2011). Slim accretion disks around black holes. PhD Thesis, Nicolaus Copernicus Astronomical Center, Warsaw. arXiv:1108.0396.

Sadowski, A., Narayan, R., Penna, R., Zhu, Y. (2013). *MNRAS*, 436, 3856.

Shakura, N.I. and Sunyaev, R.A. (1973). *A&A*, 24, 337.

Toomre, A. (1964). *ApJ*, 139, 1217.

4

Relativistic Jets and Very High Energy Mechanisms

Hélène SOL
LUTH, Paris Observatory, CNRS, France

Research on active galactic nuclei (AGN) has recently taken decisive steps, with, in 2016, the almost irrefutable confirmation of the existence of black holes in our cosmos, thanks to gravitational wave detections from the coalescence of binary black hole systems, and direct proof of their presence in the heart of active galaxies, in 2019, thanks to the radio mapping of the shadow of the central black hole in the galaxy M87. The old "AGN paradigm" was then confirmed, supermassive black holes are indeed at the origin of the energies in these sources as well as gigantic jets that sometimes escape from them. Paradoxically, launched from the immediate vicinity of black holes, below the parsec, these AGN jets are likely to propagate and interact with the ambient environments on scales from kpc to Mpc, while maintaining stability and confinement. They represent the remarkable activity of supermassive black holes, which, unexpectedly, turn out to be just as effective as "propellants" as they are "absorbers". The influx of high-quality data across the entire electromagnetic spectrum and rapid developments in multi-messenger astronomy boosts ongoing work on these complex and still enigmatic objects. Current advances in very high energy gamma astronomy, in particular, shed new light on AGNs and their jets, which were among the first cosmic sources to be identified at the teraelectronvolt (TeV), and fully open the doors to multi-spectral exploration and a very promising radio-gamma synergy that this chapter aims to illustrate.

For a color version of the figures in this chapter, see www.iste.co.uk/combes/nuclei.zip.

Active Galactic Nuclei,
coordinated by Françoise COMBES.

4.1. The advent of ground-based gamma-ray astronomy

It was half a century after the discovery of gamma rays in the laboratory by Paul Villard, from samples of radium (Villard 1900) that various theoretical considerations led astrophysicists to confirm that such radiation could be emitted from the cosmos (Feenberg and Primakoff 1948; Morrison 1958). The 1950s also saw the first detections of Cherenkov flashes of light from extensive air showers induced by the interaction of cosmic rays and very high energy gamma rays with the upper atmosphere (see Hillas 2013; Degrange and Fontaine 2016; Fegan 2019). These interactions produce cascades of secondary particles, which move at a speed greater than that of light in the ambient air and, by the Cherenkov effect, radiate bluish flashes of a few nanoseconds, potentially observable from the ground. The following decade saw the rise of X-ray and gamma-ray astronomy (Figure 4.1). Space gamma-ray astronomy concerns the so-called "high energy" (HE) gamma rays in the range now extending beyond 0.5 MeV and up to a few hundred GeV. The detection techniques of ground-based gamma-ray astronomy, which now targets the "very high energy" (VHE) spectral range of a few tens of GeV to 300 TeV[1], took longer to set up, despite the investment of several teams and the development of various projects around the planet, in particular because of the difficulty in distinguishing extensive air showers, triggered by gamma rays, among the much more numerous hadronic showers, triggered by cosmic rays, which create considerable noise. The development of Imaging Atmospheric Cherenkov Telescopes (IACTs) overcame this problem and, in the late 1980s, led to the first solid detection of a TeV cosmic gamma-ray source by the Whipple collaboration, the Crab Nebula (Weekes *et al.* 1989), followed several years later by the TeV detection of two AGNs, Mrk 421 (Punch *et al.* 1992) and Mrk 501 (Quinn *et al.* 1996).

Shortly thereafter, an intense outburst from Mrk 501 in April 1997 was observed independently by the three experiments: Whipple, CAT and HEGRA. The light curves obtained showed significant and perfectly compatible variations over 9 days, providing the first precise limits on the physical processes involved in an AGN outburst at very high energies and definitively validating the success of the Cherenkov imaging techniques (Figure 4.2). Instrument performance has since gradually improved, in particular thanks to progress in instrumentation of the focal plane, detectors and very fast electronics, to the operation with telescope arrays allowing a stereoscopic analysis of the shower images, to the development of massive Monte Carlo simulations and to the implementation of new data analysis

1. 1 TeV = 10^{12} eV = 1.6×10^{-7} J = 1.6 ergs ≡ 2.42×10^{26} Hz ≡ 1.24×10^{-12} µm.

methods (de Naurois and Mazin 2015). The opening of the window to the VHE gamma sky took place in a remarkable way with the entry of the HESS experiment into service at the end of 2003. The so-called third-generation Cherenkov instruments currently in operation, in particular with the HESS, MAGIC and VERITAS[2] imaging telescopes, led, in early 2021, to a sample of around 229 sources[3] detected at VHE (Figure 4.1), for which they provide high-quality spectra, light curves and maps. The VHE catalog is also enriched by TeV data collected by particle samplers, which developed in parallel with the first detections by MILAGRO in the 2000s. These samplers, of which HAWC[4] is the most advanced current representative, do not compete with IACTs in terms of sensitivity and angular, spectral and temporal resolutions but are very complementary to them. In fact, they benefit from a much larger field of view and a better duty cycle, allowing them to survey the entire sky and to monitor bright transient phenomena.

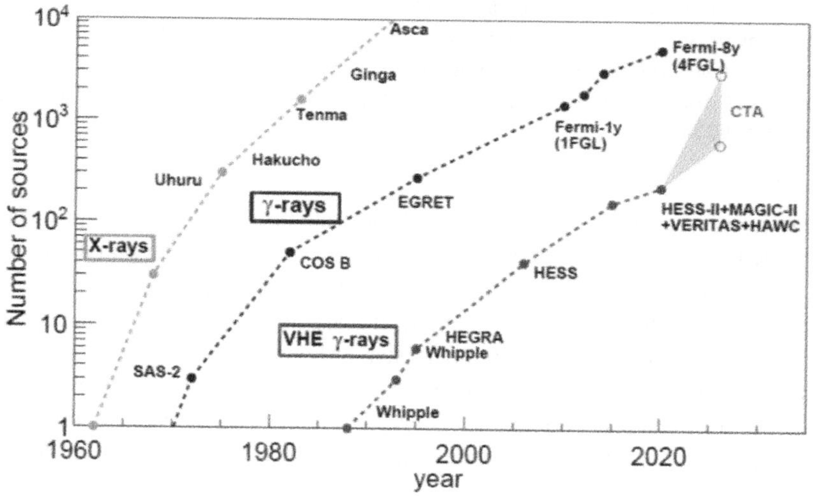

Figure 4.1. *Evolution of the number of cosmic sources detected in X-rays (green), HE gamma rays (blue) and VHE gamma rays (red) since 1960, and the plans for VHE with CTA to reach several thousand sources within a decade (source: M. de Naurois, adapted from de Naurois and Mazin 2015)*

2. https://www.mpi-hd.mpg.de/hfm/HESS/, https://wwwmagic.mpp.mpg.de/, https://veritas.sao.arizona.edu/.

3. See online catalog "TeVCat": http://tevcat.uchicago.edu/, "default catalog" and "newly announced" sources.

4. https://www.hawc-observatory.org/; see also LHAASO at: http://english.ihep.cas.cn/lhaaso/.

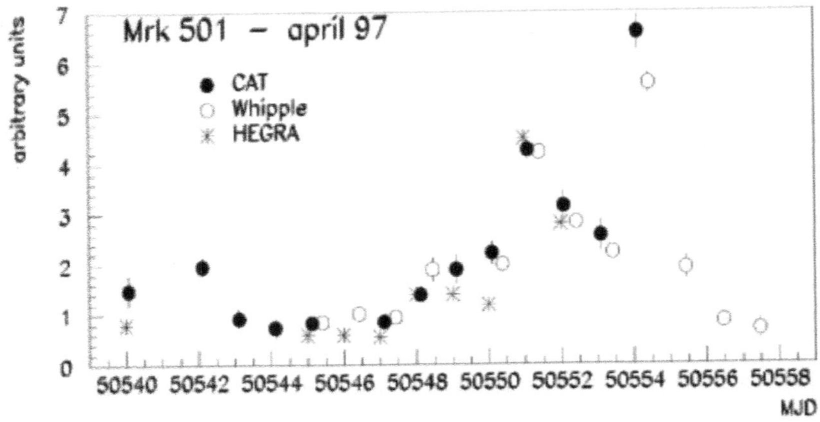

Figure 4.2. *First light curves of an AGN outburst obtained simultaneously at TeV energies by three independent experiments, CAT, HEGRA and Whipple, in 1997 (Protheroe et al. 1997). This is a remarkable outburst of the blazar Mrk 501, one of the brightest sources of VHE gamma rays in the extragalactic space located at redshift z = 0.034*

The diversity of the sky at VHE is well illustrated by the different types of point or extended sources already detected at TeV (Hinton and Hofmann 2009; Aharonian *et al.* 2013; Park 2018). In our galaxy and in the Large Magellanic Cloud, we can count, in addition to our Galactic Center, at least 34 pulsar wind nebulae and 26 supernovae remnants, 4 pulsars, 10 binary star systems, a few star clusters and a few diffuse TeV sources. In extragalactic space, at least 80 AGNs have been recorded, making them the largest population of cosmic sources among VHE emitters. Two starburst galaxies, and four gamma-ray bursts (GRB), captured at TeV in 2019 and 2020, complete the current picture. In addition, around 60 VHE sources still remain unidentified.

There are essentially only two possible pathways for the production of cosmic gamma rays: (1) bottom-up scenarios in which radiating particles are accelerated to very high energies, we then speak of leptonic or hadronic models, depending on the nature of the particles dominating the gamma-ray emission, and (2) top-down scenarios in which the decay or annihilation of exotic particles, in particular dark matter, induces high energy photons either directly or through secondary particles (De Angelis *et al.* 2018). In addition, there is the special case of gamma-ray outbursts that are potentially emitted during the final stage of evaporation of possible primordial low-mass black holes by the Hawking process. All gamma-ray emissions observed so far are *a priori* bottom-up scenarios and research into signatures of other types of radiation have so far only led to upper limits. Ground-based gamma-ray astronomy thus explores, as a priority, the real power plants of our universe, which house powerful particle accelerators mainly generated by large flows of matter inducing shocks and turbulences. It also studies the environments surrounding these accelerators, which the accelerated particles and their emitted gamma rays pass through and with which the two interact. It offers unique opportunities for the study of extreme particle acceleration mechanisms, in particular, around compact objects such as black holes or neutron stars, and for better understanding the precise origin of galactic and extragalactic cosmic rays, a century after their discovery.

4.2. Very high energy gamma-ray observations of active galactic nuclei

The TeV domain provides a renewed vision of AGNs by giving privileged access to the immediate environment of massive black holes, to the mechanisms of jet formation and extreme acceleration of particles, to non-thermal radiation and to their temporal evolution (Prandini 2017; Park 2018; Sol 2018). This section provides an overview from an observational point of view.

4.2.1. *The current sample: characteristics and observational biases*

The group of 80 AGN confirmed at VHE to date, typically between 100 GeV and 30 TeV, is dominated by a population of 74 blazars[5] with, in particular, (1) 51 BL Lac objects of the HBL type including Mrk 421, Mrk 501 and PKS 2155-304, (2) nine intermediate IBL types including 3C66A, W Com and BL Lac itself, (3) two LBL types including AP Lib, (4) eight FSRQs including 3C 279 and (5) OJ 287, a peculiar BL Lac object, a probable massive black hole binary system. Added to this population of blazars are six AGNs associated with four well-known Fanaroff–Riley I (FRI)-type radio galaxies – M87, Cen A, NGC 1275 and 3C 264 – and two more difficult to classify radio galaxies, 3C 310 and PKS 0625-35, possible sources of transition between BL Lac objects and FRIs.

---

5. BL Lac objects (BL Lacs) and Flat Spectrum Radio Quasars (FSRQs) constitute the class of blazars, one class of AGNs and quasars with a relativistic jet pointing towards the observer, orientated at a small angle θ from the line of sight. FSRQs show bright broad emission lines, whereas BL Lacs do not, or if they do, have very weak ones. HBLs, IBLs and LBLs are three subgroups of BL Lacs, respectively, High-, Intermediate- and Low-frequency peaked BL Lacs (see section 4.2.4).

We can also note the presence of four still unidentified VHE sources that are located outside the galactic plane and which are potentially AGNs. Three of them were detected by HAWC (including two with extended emission). The fourth, discovered in 2009 by MAGIC, could correspond to the FRI radio galaxy 3C66B, but this association requires confirmation. The two starburst galaxies NGC 253 and M82 detected at TeV could also harbor a weakly active nucleus, as part of the still-uncertain connection between "starbursts" and the activity of galactic nuclei. Also taking into account the Galactic Center, a low-luminosity galactic nucleus, the extended sample of sources related to the AGN phenomenon thus amounts to 87 sources, or about 38% of all known VHE cosmic emitters.

To date, all the AGNs detected at TeV are strong radio-loud AGN and more than 92% of them are blazars. The standard model of AGN unification (see Chapter 1) therefore indicates a very strong observational bias in favor of "blazar"-type AGNs whose jets are directed towards the Earth, and thus benefit from a significant boosting of their radiation by the relativistic Doppler effect, making them the most powerful persistent sources that we can observe in our universe (see section 4.3). This observation has several consequences. It reveals that the zones dominating the VHE emission are likely to be embedded in relativistic flows and are therefore located somewhere along the jets. It also demonstrates that we currently only see an extremely small fraction of TeV-emitting AGNs, whose jets are oriented at a very small angle from their line of sight. The improved sensitivity should thus allow a much larger population to be obtained, including more and more unbeamed sources. The new generation of instruments will be needed in order to access more in-depth statistical studies, which will show whether the VHE emission is a generic property of all radio-loud AGN or not, and vice versa.

The redshift distribution of AGNs at TeV is dominated by weak z values, with 50% of the sources whose redshift is known at $z < 0.15$, and extends to $z \lesssim 1$ with the two FSRQs, PKS 1441 + 25 ($z = 0.939$) and S3 0218 + 35 ($z = 0.954$) a gravitational lens whose deflector galaxy is located at $z = 0.684$. At least three observational biases impact this redshift distribution: (i) the flux limitation problem obviously favors the detection of the closest sources, (ii) this effect is further reinforced by the absorption phenomenon of VHE photons during their propagation and their interaction with the diffuse extragalactic background light (see section 4.4.4) and moreover (iii) the redshift of nearly a third of the AGNs detected at VHE remains undetermined, in particular for the most distant sources, the BL Lac objects dominating the sample characterized by the weakness or absence of spectral lines, which further reduces the number of AGNs detected at TeV and explicitly recognized as sources with large redshifts.

Another important bias that shapes the current sample is due to the high variability of some sources and to the "under alert" observation strategies, which often encourage observations of high-state AGNs when they are particularly bright.

4.2.2. *Variability of AGNs at very high energies*

Despite the limitations on the temporal coverage of their VHE light curves, several AGNs are already recognized as TeV-variable, and likely to show states of extreme variability (Figure 4.3). Flux variations have been spotted at all time scales, from several years down to a few minutes, which are the fastest variabilities ever detected in other spectral domains from AGNs. The origin of the variabilities is not yet clearly identified, due to the multitude of possibilities, and it is likely that several phenomena come into play, in fact. The study of the variability properties in HE and VHE gamma rays of some archetypal sources has, moreover, shown that the noise observed during *high* states of high activity was not the same as the noise observed during *low* states. A change in power spectral density ($\mathcal{P}(\nu_t) \propto \nu_t^{-\xi}$, where ν_t is the temporal frequency) was found, for example, in the light curves of the source PKS 2155-304, which presents a red or Brownian noise ($\xi{\sim}2$) during intense flares on time scales of less than three hours, and a flicker ($\xi{\sim}1$) for longer-term variations from a few days to several years (Abdalla *et al.* 2017), testifying to a different nature of the variability depending on the activity level.

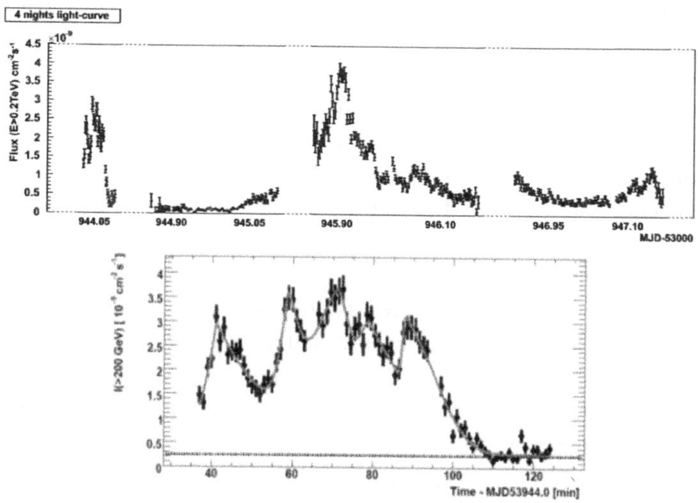

Figure 4.3. *Light curve above 0.2 TeV of the HBL PKS 2155-304, during an exceptional period of activity in July 2006.* ***(Top).*** *Two high-intensity flares separated by two days (Abramowski et al. 2010).* ***(Bottom).*** *Close-up of the first flare, showing very rapid variability detected down to the 4-minute scale (Aharonian et al. 2007)*

The extremely rapid variabilities sometimes detected indicate that the emission is coming from a very compact zone and could suggest that VHE emission is coming from the environment that is very close to the central black hole. Indeed, the causality argument applied to highly variable VHE phenomena imposes a significant constraint on the size R of their emitting zone,

$$R \leq ct_{var}\,\delta/(1+z),$$

with c the speed of light in a vacuum, t_{var} the typical time of flux variation observed at TeV and

$$\delta = 1\,/\,[\Gamma(1 - \beta_J\,cos\theta)],$$

the Doppler factor of the emitting zone, due to the change of relativistic reference frame between this zone and the observer, which tends to amplify the properties of variability detected in the blazars. Here, $\Gamma = (1-\beta_J^2)^{-1/2}$ is the bulk Lorentz factor of the emitting zone, $u = \beta_J c$ is its velocity and θ is its angle of motion with respect to the line of sight (see section 4.3). In several cases of extreme variability, for example, in the blazars PKS 2155-304, Mrk 501 and PKS 1222 + 216, the radio galaxy M87 and the transition source IC 310, this argument leads to very small emitting zones, comparable to or even less than the Schwarzschild radius R_s of the black hole, even for high values of Doppler factors δ possible in blazars. We therefore expect them to be located towards the base of the jets near the horizon, since the value of R_s provides *a priori* the minimal spatial scale of the global phenomena in action along the relativistic jets, formed in the immediate environment of the central black hole and which gradually widen with the radial distance (see section 4.3, and Chapter 3, section 3.8). The incredible compactness of some regions emitting variable phenomena, however, remains critical to explain. The inconsistency can be solved if the most rapidly varying flares correspond to events with particularly high Doppler factors, which makes it possible to relax the causal constraint. This explanation has not yet been proven, however, and remains partly problematic, in particular, in the case of radio galaxies such as M87, whose Doppler factors are, in principle, rather low, and also in the case of FSRQs like PKS 1222 + 216, where the VHE gamma-ray radiation from a region close to the central black hole should *a priori* be self-absorbed during its interaction with the strong ambient radiation field due to the accretion disk and the broad line region[6] (BLR). The very rapid variabilities pose a real challenge to the VHE radiation models currently debated (see sections 4.4 and 4.5) and are still a mystery.

6. See Chapter 1.

The development of global alert networks allows a greater multispectral monitoring of AGN active states, but the number of flares that are well described in time and frequency is still very limited. In general, the low-energy gamma-ray and X-ray emissions observed during flares appear to be highly correlated with VHE emissions, as illustrated in Figure 4.4, which makes it possible to strongly constrain the radiative models. There are, however, exceptions, with, in particular, the existence of "orphan" VHE flares, detected only at VHE, such as, for example, in the HBL 1ES 1959 + 650 (Krawczynski *et al.* 2004). More indirect links between emission at VHE and at lower optical and radio energies have been proven, increasing the probability of detecting a flare at VHE during high states spotted in optics, but they still remain quite elusive (Figure 4.5) and reflect a gap in the global understanding of the phenomenon of AGN activity.

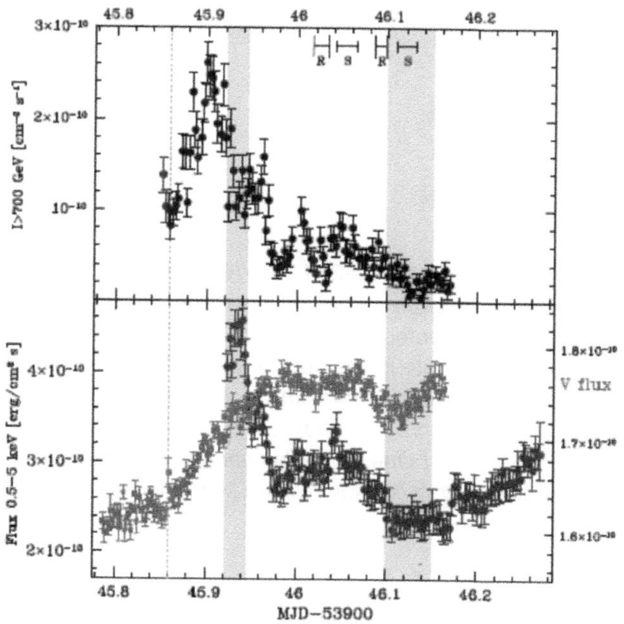

Figure 4.4. *VHE (black), X-ray (blue) and optical (red) light curves of the second flare of the remarkable eruption of PKS 2155-304 shown in Figure 4.3. An in-depth analysis shows that the VHE and X-ray emissions and spectra are strongly correlated, while the optical flux shows only a slight increase (~15%), which starts at the moment of the flare and reaches a plateau after two hours (Aharonian et al. 2009)*

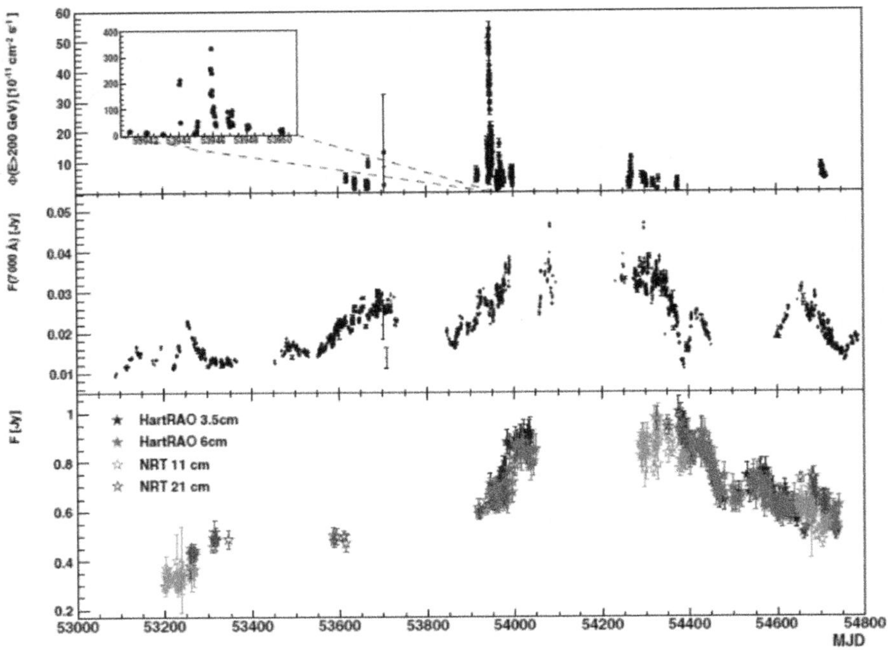

Figure 4.5. *Light curves of PKS2155-304 at VHE* ***(top)****, in the optical spectrum* ***(middle)*** *and in the radio spectrum* ***(bottom)*** *over approximately five years around the state of VHE activity presented in Figures 4.3 and 4.4. The VHE flare occurs during a high and increasing state in the optical and radio fluxes. The flux at 3.5 cm reaches a first maximum about a month and a half after the VHE flare, and the optical flux begins a long decrease after about 15 months. This multi-lambda approach suggested, for the first time, a link between intense activity at VHE and the long-term evolution of low-energy flows in blazars (Abramowski et al. 2012a)*

4.2.3. *Spectral energy distribution*

A peculiarity of the blazars and VHE emitting AGNs is their spectral energy distribution (SED) from radio to VHE, which shows two main bumps, one at "low energy" occurring in an interval between the millimeter range and the soft X-rays, and the other at "high energy", peaking in the gamma-ray domain. Two representative examples of SEDs of BL Lacs and FSRQs, shown in Figures 4.6 and 4.7, illustrate this characteristic.

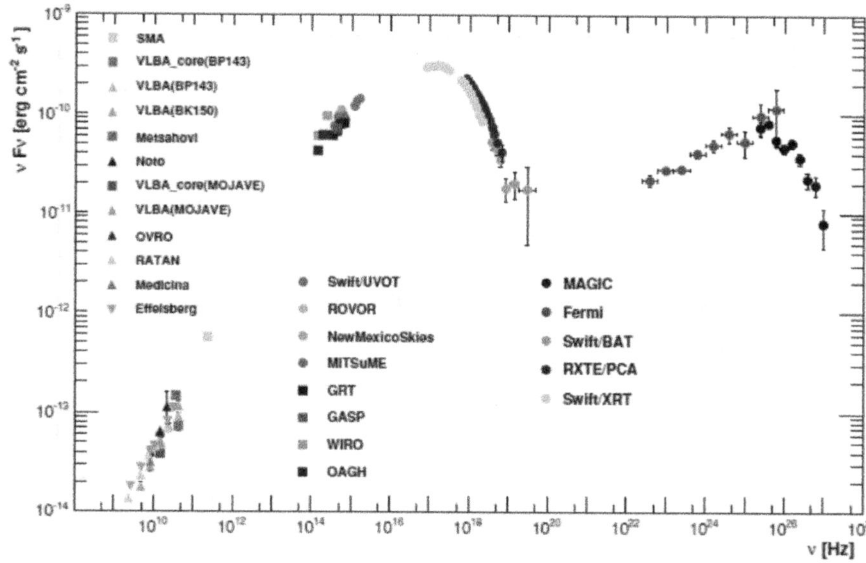

Figure 4.6. *Spectral energy distribution νF_ν in erg cm^{-2} s^{-1}, observed in the low and steady-state phase of HBL Mrk421 at z = 0.031, from the radio to the VHE range. The SED represented in this way makes it possible to immediately compare the relative significance of the energy production by the source according to the spectral range (Abdo et al. 2011)*

The SED of the steady state of Mrk 421 was obtained during a multispectral observational campaign involving more than 20 instruments, carried out in the first half of 2009, while this source was in a low-activity state. The series of quasi-simultaneous SEDs of 3C279 were carried out thanks to an in-depth monitoring of the source over 10 years, particularly involving the GASP-WEBT program, with 34 ground-based radio, near-infrared and optical observatories, and Fermi and Swift satellites. The emissions inducing the two bumps observed in the blazar SED come from the jet directed towards the observer. Amplified by relativistic effects, they generally dominate the other radiations expected from the accretion disk, the dust torus and the BLR of the AGN. There is consensus in attributing the low-energy bump to the synchrotron emission of relativistic particles in the magnetized jet, and the high energy bump either to their inverse Compton emission or to hadronic processes (see section 4.4).

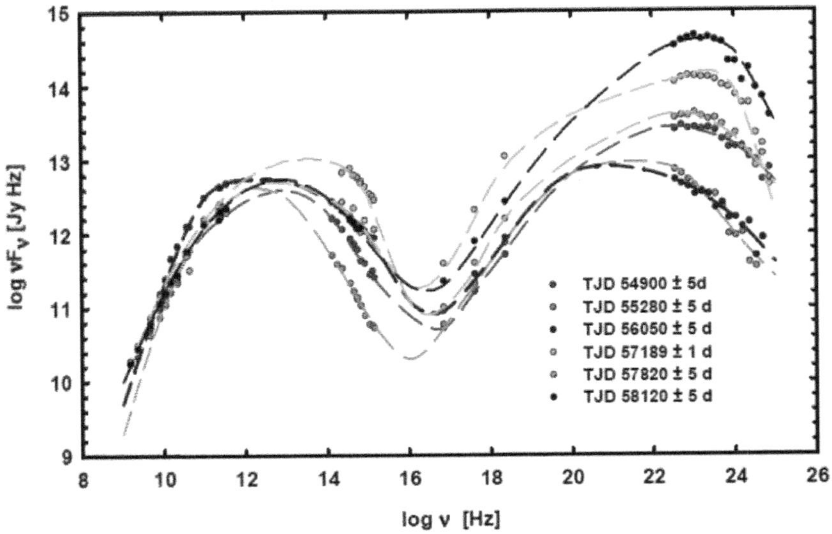

Figure 4.7. *Temporal evolution of the spectral energy distribution νF_ν in JyHz (1 Jy = 10^{-23} erg cm^{-2} s^{-1} Hz^{-1}) of the source 3C279, archetype of FSRQs at* $z = 0.536$, *monitored in multi-lambda observations over the decade 2008–2018. The colored dashed curves connect quasi-simultaneous datasets. The good correlation observed between the X and gamma fluxes suggests a common origin to these types of radiation (Larionov et al. 2020)*

4.2.4. Classification of sources

Despite their very similar characteristic two-humped shape, the SEDs of BL Lacs and FSRQs exhibit specific differences, illustrated in Figures 4.6 and 4.7, which led to a classification of the various types of blazars already mentioned (FSRQ, LBL, IBL, HBL), known as the blazar sequence (Figure 4.8). A first study based on a representative sample of 126 blazars detected in radio and X-rays (Fossatti *et al.* 1998) in fact revealed a shift in the "average" SED of the blazars as a function of their bolometric luminosity L_{bol}: their SED sits at increasingly higher energies, and the relative importance of the high energy peak decreases as their L_{bol} decreases, thus establishing a continuous sequence from FSRQs to extreme HBLs, through LBLs and IBLs. The validity and universality of this sequence have been the subject of many debates, in particular due to the existence of sources that do not easily fit into this classification, including the recent example of blazar 1ES 1215 + 303, which has been found to be able to change type by going from an IBL to an HBL during a flare (Valverde *et al.* 2020) or even Mrk 501, HBL archetype, which has been observed in a state of "extreme HBL" with a sharp gamma-ray peak above 0.5 TeV in 2012 (Ahnen *et al.* 2018). However, an updated version of the blazar sequence, illustrated in Figure 4.8, constructed with a sample enlarged to 747 blazars, grouped by gamma-ray luminosity level at GeV, specifies the behaviors and confirms the general trends already identified in the original study (Ghisellini *et al.* 2017).

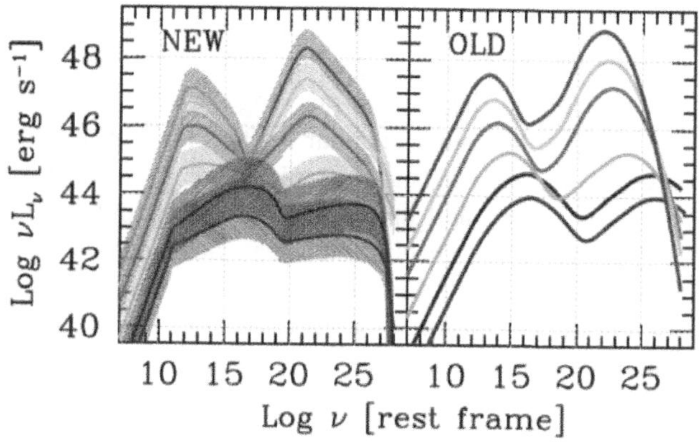

Figure 4.8. *Revisited blazar sequence. The average SEDs of the different types of blazars, grouped according to their gamma-ray luminosity level between 0.1 and 100 GeV* ***(left)*** *or radio luminosity at 5 GHz* ***(right)****, show a joint shift from high luminosities to low luminosities, and from low frequencies to high frequencies, when going from the most powerful FSRQs (in red) to extreme HBLs (in purple), through intermediate classes of FSRQ, LBL, IBL and HBL in yellow, green, light blue and dark blue. The new illustration of the blazar sequence (left) is based on a sample of 448 FSRQs and 299 BL Lacs from 3LAC, the third catalog of gamma-ray AGNs detected by the LAT instrument of the Fermi satellite (Ghisellini et al. 2017)*

The relative importance of the gamma-ray peak in SEDs of FSRQs compared to BL Lacs can be interpreted quite naturally in the context of "leptonic" radiative scenarios (see section 4.4). FSRQs have standard luminous accretion disks, of the "Shakura-Sunyaev" type and derivatives, whose ionizing radiation is re-emitted by the BLRs, as well as large dust tori. A strong gamma-ray emission can therefore be generated by the inverse Compton effect of the relativistic electrons of the jet in intense external radiation media surrounding the AGN. Conversely, the accretion regime of BL Lacs with Advection Dominated Accretion Flow[7] (ADAF) corresponds

7. See Chapter 3.

to sub-luminous accretion disks ($L_{disk} < 0.01\ L_{Eddington}$), inducing neither BLRs nor intense ambient radiation fields. The less luminous gamma-ray peak observed is then attributed mainly to the "synchrotron-self-Compton" (SSC) emission, due to the inverse Compton effect of relativistic electrons interacting with their own synchrotron radiation. The global radiative cooling of the electrons, weaker than in the FSRQs, allows them to be accelerated more efficiently and thus explains the shift of the SEDs of BL Lacs to the highest energies.

Although interesting, this unifying scheme has not yet been proven, and it is also possible that selection effects strongly impact or even induce the observed sequence. When it becomes possible in the gamma-ray range, a statistical approach of FRI radio galaxies, assumed to be sources similar to BL Lacs but oriented at a greater angle with respect to their line of sight, should make it possible to advance our understanding of classification and unification of blazars and AGNs (Rieger and Levinson 2018). To date, a few specific radio galaxies, such as 3C 310 and PKS 0625-35, make it possible to probe the intermediate zone between BL Lacs and FRI and to test the strength of the current unification scenarios. Great progress could also come from the multispectral approach. In particular, as all these AGNs detected at the VHE are strong radio-loud sources, the exploration of their compact cores at very high angular resolution is accessible thanks to radio astronomy techniques, which sheds great light on the properties of relativistic jets in the environment of the central black hole. Although still difficult to relate to VHE phenomena, a global vision of the activity phenomenon should emerge from the confrontation of these different perceptions of AGNs.

4.3. Extragalactic jets, structures observed from sub-parsec to mega-parsec scales

From the outset, radio astronomy played an essential role in the field of AGNs with the analysis of the first extragalactic radio sources, such as Cygnus A in the 1950s (Jennison *et al.* 1953), which led to the discovery of quasars, thanks to the identification of the sources 3C 48 and 3C 273. Several decades before the development of gamma-ray astronomy, radio astronomy offered a wide field of investigation of ejection phenomena, in parallel with studies on the physics of accretion, carried out mainly from infrared to X-rays. The study of extragalactic jets is now based on all spectral domains, which continuously provide very diverse results (Boettcher *et al.* 2012; Romero *et al.* 2017; Blandford *et al.* 2019; Davis and Tchekhovskoy 2020). It should be noted, however, that less than 10% of AGNs are strong radio emitters and produce jets, although the origin of the dichotomy between radio-loud and radio-quiet AGNs is not yet well understood. It is often attributed to the properties of the central core and, in particular, to the rapid rotation of the hole, which allows the production of jets. However, several Seyfert AGNs, without strong radio emissions or jets, turned out to host a rapidly rotating black hole, which seems contradictory. It is therefore likely that other phenomena interfere in the production of jets by AGNs, such as the various accretion regimes which can successfully become long-lasting or cannot, magnetic field properties of the disks which can allow the emergence of jets or not, or environmental effects which can induce various large-scale flows of matter towards the AGN that favors (or does not favor) the establishment of jets and which also impact the propagation, the collimation and the stability of the jets once formed.

4.3.1. *Nuclear jets studied by radio interferometry below the parsec scale*

The angular resolution of a few arcminutes attainable at VHE is remarkable for the gamma domain, but it is still very limited. This limitation is partly offset by the excellent temporal resolution obtained for gamma-ray flares from the jets. The very fast variabilities observed at TeV indicate that they come from very small emission zones, *a priori* located in the environment close to the black hole, at the base of the jet. These regions have been studied in depth in radio astronomy due to their synchrotron emission (see section 4.4), which has made it possible to understand their main properties and characteristics. Very long baseline interferometry (VLBI) makes it possible to map the base of the jets at very high angular resolution, in milliarcseconds and below, up to the immediate vicinity of the black hole (Figure 4.9). VHE and VLBI data therefore provide very complementary information and the coordinated exploration of the TeV emission zones and the compact cores and radio jets detected in VLBI, which sometimes seem possibly linked (Marscher 1999; Charlot *et al.* 2006; Marscher *et al.* 2008; Acciari *et al.* 2009; Akiyama *et al.* 2015; Rani *et al.* 2018), is particularly interesting. It simultaneously covers the intricate issues of jet formation and particle acceleration mechanisms.

Nuclear jets, or black hole jets, form in a zone of acceleration and first collimation that was observed in M87 (Figure 4.9) at the sub-parsec scale, typically below 1,000 R_S, and cross the sphere of gravitational influence of the black hole up to the Bondi radius of a few hundred parsecs (in 3D). They generally propagate beyond, in their host galaxy (galaxy jets), and then towards the extended lobes or halos (lobe jets). Nuclear jets generally appear as a rectilinear structure formed of bright knots, emanating from one side of a compact flat spectrum radio core (Figure 4.10).

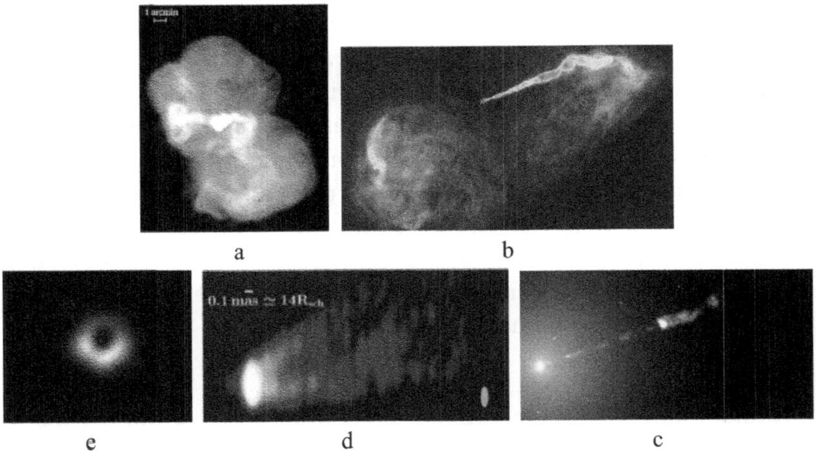

Figure 4.9. *Multi-scale images of the M87 radio galaxy, with clockwise successive zooms towards the active nucleus from* **(a)** *the extended halo observed by LOFAR at 50 MHz to* **(e)** *the shadow of the black hole that is revealed on the 1.3 mm emission from the plasma orbiting the black hole, detected by the intercontinental network of the Event Horizon Telescope (EHT) project (Akiyama et al. 2019). As the jet is emitted by the central core along the black hole's axis of rotation, analysis of the luminous crescent-shaped accretion flow detected in* **(e)** *suggests that it contains a non-stationary hotspot or anisotropic emissions, which could explain its apparent orientation with respect to the direction of the jet (Nalewajko et al. 2020). In* **(a)**, *the halo with a projected length in the sky plane of 74 kpc is fed by the jet and the counter jet (which is not very visible) and illustrates the complexity of the feedback effects of AGNs on the surrounding environment (Gasperin et al. 2020). The relativistic jet remains predominant from the kpc scale to the 7 Schwarzschild radius scale* **(b, c and d)**. *The nuclear jet emerges from the immediate vicinity of the black hole at less than 7* R_s, *and appears bright at the edges on the map obtained by the GMVA at 86 GHz* **(d)**, *suggesting a center-edge stratified jet structure emanating from the inner region of the accretion disk (Kim et al. 2018). It turns out to be oriented at ~17 degrees from the line of sight, so the projection effects are quite large (Walker et al. 2018). The collimation is maintained up to about 2 kpc (in the plane of the sky) as illustrated by the image obtained in optics by the HST* **(c)**, *then the jet begins to disperse in an inner lobe clearly visible on the map at 2 cm by the VLA* **(b)**. *The plasma flow continues, however, and fills the extended halo* **(a)**. *At a distance of M87 (z = 0.0043, 16.7 Mpc), 1 milliarcsecond is approximately 0.08 pc, therefore 140* R_s *for a black hole of* $6 \times 10^9 M_\odot$. *(sources: (a) Gasperin et al. (2020); (b) Image by the VLA at 2 cm (Owen et al. 1999), NRAO (https://www.spacetelescope.org/images/opo9943b/ or http://www.aoc.nrao.edu/%7Efowen/M87_inner.html); (c) HST, image in Kellerman et al.* Astrophysics and Space Science, *September 2007; (d) Kim et al. (2018) and image in* Effelsberg Newsletter, *9(2), 9, May 2018; (e) Akiyama et al. (2019))*

A remarkable phenomenon, predicted by modelers and highlighted by radio interferometry (see, for example, Cohen *et al.* 1971), is the frequent detection of apparent superluminal motion in these jets (Figure 4.10). This effect of relativistic origin comes from the interpretation of the maps obtained at a given time as snapshots of the source, when, in reality, taking into account the propagation time of the light from the different parts of the jet makes these maps spread out in space and time. It is easily explained in the context of special relativity, by considering, for example, a jet oriented at a low angle θ with respect to the line of sight, and a radio knot N, which outflows at the speed $\beta_N c$ of the radio core C, assumed to be at rest at $t = 0$ in the core reference frame. After time t, the knot N has traveled a distance $d = \beta_N c\,(\sin\theta)t$ in the sky plane, and as it approaches the observer of $l \sim \beta_N c\,(\cos\theta)t$, the time elapsed for the observer from the moment when N exits C is $t_{obs} \sim (1 - \beta_N\cos\theta)t$. The apparent speed of motion of N with respect to C in the sky plane measured by the observer is thus:

$$u_{obs} = d/t_{obs} = \frac{\beta_N \sin\theta}{1 - \beta_N\cos\theta}\,c$$

A simple derivation shows that this apparent velocity reaches its maximum $u_{obs} = \Gamma\beta_N c = c\sqrt{\Gamma^2 - 1}$ for $\cos\theta = \beta_N$ (or $\sin\theta = 1/\Gamma$). The jets at $\beta_N > 1/\sqrt{2}$ can therefore appear superluminal for sufficiently low angles of orientation towards the observer. More precisely, we see, by considering the limit values, that the detection of a superluminal velocity $u_{obs} > c$ indicates a Lorentz factor of the knot $\Gamma \geq \sqrt{1 + u_{obs}^2/c^2}$ and a jet orientation angle lower than θ_{max}, such that $\cos\theta_{max} = (u_{obs}^2 - c^2)/(u_{obs}^2 + c^2)$.

The measurement of superluminal velocities has been a key element in confirming the relativistic nature of the base of many jets and constraining their overall velocity and orientation properties. This is how the VLBI was able to demonstrate that the gamma-ray emitting FSRQs had very strongly relativistic jets pointed directly at the observer, for example, the source 3C 279, illustrated in Figure 4.10, which has superluminal velocities of more than 20c at the milliarcsecond scale and can be modeled by a jet with Lorentz factor $\Gamma > 22$ orientated at an angle $\theta < 2°.6$ with respect to the line of sight (Rani *et al.* 2018). A recent result of the EHT collaboration made it possible to probe the base of the jet with an unprecedented resolution of 20 microarcseconds at 230 GHz and highlights the presence of similar speeds at distances of around 0.13 pc from the active core, that is, 1,700 R_s for a black hole of $8 \times 10^8 M_\odot$. The discovery of a three-component nuclear structure, elongated almost transversely to the general direction of the jet, sheds new light on the zone where the jets were formed. The nature of this novel structure is still unknown. It could be explained by the presence of curvatures in the jet, perhaps helical, a phenomenon already suspected on a larger scale. Further in-depth analyses are underway (Kim *et al.* 2020a).

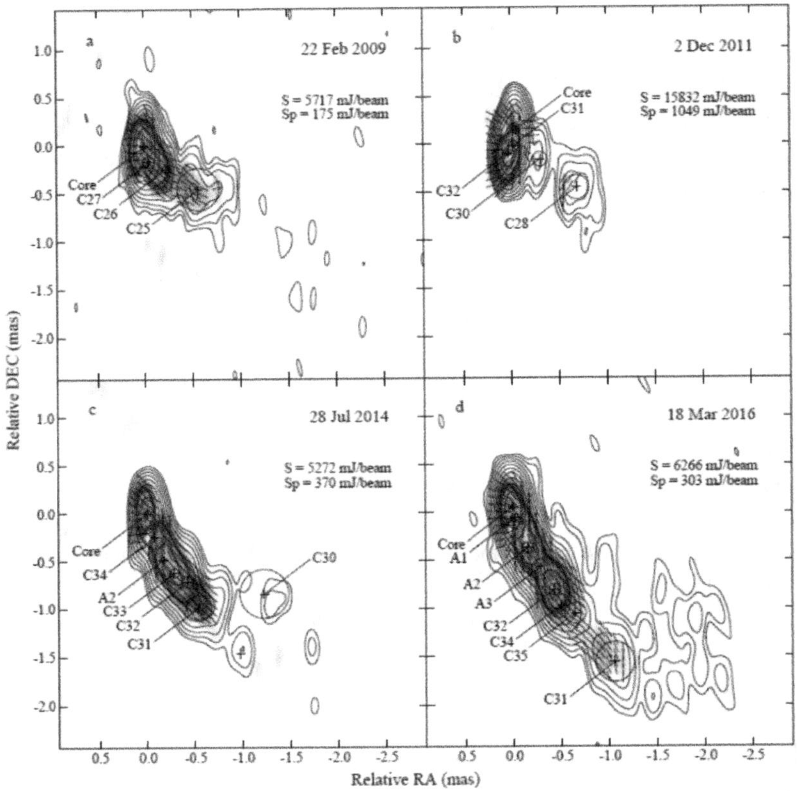

Figure 4.10. *A time series of 3C 279 maps at 43 GHz in total (contours) and polarized intensity (colors; the segments indicate the electric polarization vector). The comparison between the different periods observed makes it possible to identify the knots which escape from the core and shows strong changes in the structure of the jet on the milliarcsecond scale and below (1 mas = 6.5 pc for z = 0.536). The pattern of motion of radio knots is found to be very complex, with a range of observed velocities ranging from apparently stationary knots to superluminal velocities of 30 to 40c (source: Larionov et al. (2020))*

The quality of the radio maps now produced at very high angular resolution provides increasingly detailed information, with, in some cases, the demonstration of a complex flow with knots at different velocities (stationary, subluminal and superluminal), non-ballistic movements with nonlinear trajectories, sometimes acceleration or deceleration phenomena and signs of interactions that are still poorly understood between the knots. The simultaneous presence of superluminal and stationary knots in the same source also underlines an ambiguity that has hitherto been unavoidable in the interpretation made of the movements observed, some of which seem to reveal the effective velocity of the flow of the plasma constituting the jet, and others the speeds of waves or shocks that develop along the jet.

The relativistic speeds of the jets orientated towards the observer induce another effect that is essential for the understanding of radio-loud AGNs, known as Doppler enhancement. The radiation from a source carried at Lorentz factor Γ in the jet and radiating isotropically in the jet reference frame is, in fact, focused in a radiation cone with a half-opening angle $1/\Gamma$ by relativistic aberration ("beaming" effect) in the observer's frame of reference, and the observed flows can be amplified very significantly. This amplification follows directly from the Lorentz transformations between the two reference frames of the jet and the observer, according to which frequencies and intensities are related as follows: $\nu_{obs} = \delta\nu_{jet}$ and $I_{obs_{\nu_{obs}}}(\nu_{obs}) = \delta^3 I_{jet_{\nu_{jet}}}(\nu_{jet})$, where δ is the Doppler factor of the jet and the cosmological correction has been disregarded. The precise value of the flux amplification factor depends on the radiative properties of the emitting source but can be very high for blazars in particular, where it can easily exceed 10^4 for jets at $\Gamma > 20$ that are oriented a few degrees from the line of sight. This effect provides a direct explanation of several properties of the observed sources, such as the detection of a single unidirectional jet in the majority of the sources that have been mapped by radio VLBI, because the radiation received from the counter jet moving away from the observer remains much less, as well as the very important observational bias in favor of the priority detection of blazars by the first VHE detectors. It is, with the presence of obscuring dust tori around the accretion disks, the basis of any classification and unified scheme for AGNs according to their angle of orientation with respect to the observer.

Evidence of a relationship between the gamma-ray emission of AGNs and the properties of their VLBI jets has accumulated in recent years, although the origin of the links is not yet fully understood. Thus, out of 295 AGNs of the MOJAVE program, those detected by Fermi/LAT showed, on average, speeds observed in their VLBI jets that are greater than those of undetected AGNs, which indicates a strong correlation between the kinematics of the radio jets at the parsec scale and the Doppler amplification factor of the GeV gamma-ray emitting zone (Lister *et al.* 2016). The classification of blazars into FSRQ, LBL, IBL and HBL, based on their spectral properties (Figure 4.8), seems to correspond to an alternative classification based on purely kinematic criteria at the level of their VLBI jets. Jets dominated by superluminal knots appear to correspond to FSRQs and LBLs, those dominated by quasi-stationary knots to HBLs, and those with both types of knots to intermediate IBLs (Hervet *et al.* 2016). Moreover, a growing number of clues suggests that some of the flares observed at HE and VHE are correlated with various events in the VLBI structure, such as variations in the radio core, emergence of knots at the base of the jets and the formation of stationary knots (Marscher 1999; Charlot *et al.* 2006; Acciari *et al.* 2009; Akiyama *et al.* 2015; Jorstad and Marscher 2016; Rani *et al.* 2018; Larionov *et al.* 2020; Kim *et al.* 2020b). This research should soon make it possible to better constrain the current scenarios of AGN flares, which still have difficulty in precisely locating the gamma emission zone but mainly locate it in the formation and acceleration region of nuclear jets, or even in the magnetosphere of black holes (see sections 4.4 and 4.5). However, several types of flares with different properties have already been observed, possibly coming from different regions, including from the same source (see, for example, Abramowski *et al.* 2012b). Part of the VHE emissions may also come from more extended regions.

4.3.2. *Jets extended to the kpc scale and beyond*

Emerging from a very compact region that is sometimes below an astronomical unit, the jets can reach distances of several megaparsecs, for the largest of them. Two main families of extended jets have been identified: (i) those of the less bright FR I-type radio sources, which radiate strongly from their base then slow down, become subsonic and gradually disperse in the ambient medium, and (ii) those of the brightest FR II-type radio sources, which are very collimated and not very luminous, which end in very luminous hotspots, shocks zone with the ambient medium, and which feed two large radio emission lobes. The AGNs unification scenario by their angle of orientation associates BL Lacs with FR I sources and FSRQs with FR II sources. However, due to their specific orientation towards the observer, the extended jets of blazars detected at VHE do not appear well developed on maps carried out on Earth, due to projection effects. For now, only the extended jets of the radio galaxies detected at the VHE are accessible to us, less aligned than the blazars, relative to their line of sight, such as M87 (Figure 4.9) and Cen A (Figure 4.11), two nearby sources that are generally considered as special cases of the FR I-type. No FR II-type source showing a well-developed extended jet has yet been detected at TeV, but several of them have been detected at GeV by the Fermi satellite, such as Pictor A, a Seyfert 1 radio galaxy with broad emission lines (Figure 4.12), which provides an impressive example of a very fine and very collimated jet, detected in radio and X-rays; there is no interaction with the media crossed that seems to affect over a hundred kpc, until the final shock with the intergalactic medium at the hot spot.

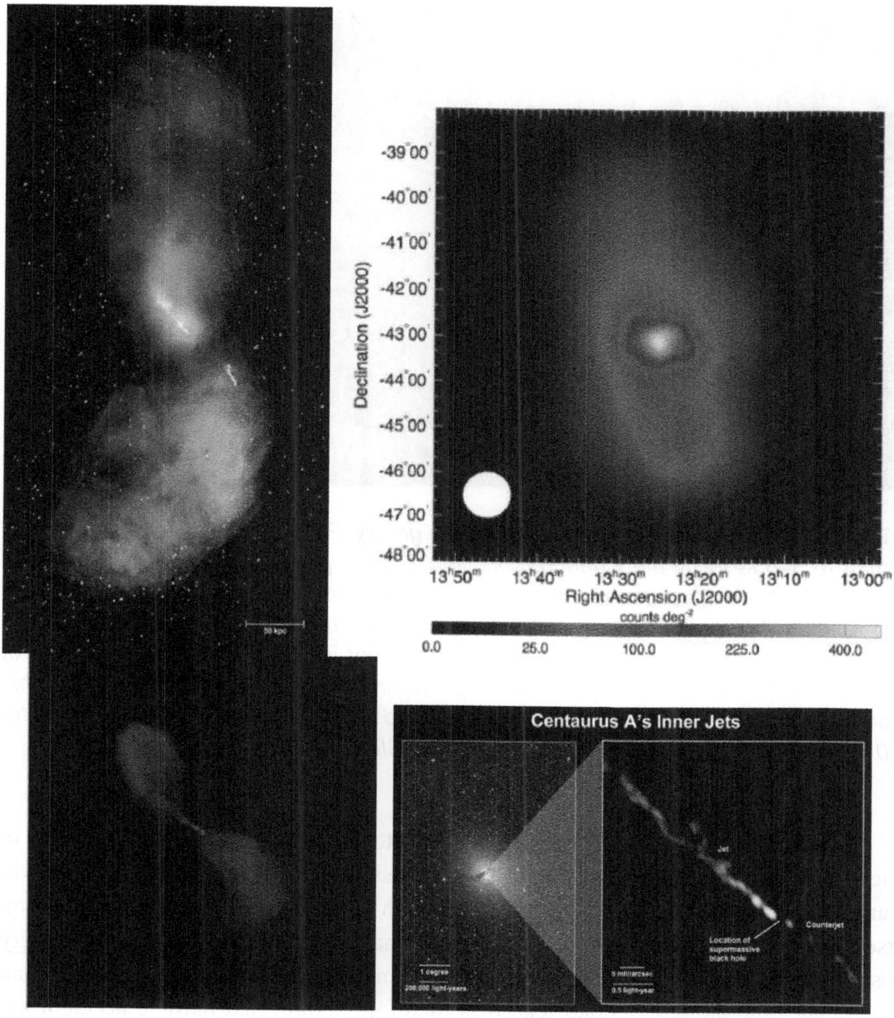

Figure 4.11. *The Centaurus A galaxy at different scales and wavelengths.* ***(Top left).*** *Giant radio lobes mapped at 1.4 GHz by the CSIRO's ATCA and Parkes radio telescopes in Australia (see https://www.scienceimage.csiro.au/image/10850/the-centaurus-a-galaxy/).* ***(Top right)****. The extended diffuse emission observed in GeV gamma rays from both sides of the galaxy (Abdo et al. 2010).* ***(Bottom left)****. The jets and inner lobes (Clarke, VLA at 1.4 GHz), in the center the galaxy seen in the visible range (HST).* ***(Bottom right)****. The nuclear jets mapped in VLBI at 8.4 GHz (source: project Tanami, Tracking Active Galactic Nuclei with Austral Milliarcsecond Interferometry, Muller et al. 2011)*

Figure 4.12. *Composite image of the radio galaxy Pictor A, X-ray view in the keV range (0.5–5 keV) by the Chandra satellite (blue), superimposed on the radio lobes (pink). The perfectly confined 95 kpc jet escapes from the active core and ends in a very bright hotspot where it collides with ambient gas. This jet is not directly orientated towards the observer and the counter jet is faintly visible in the center of the left lobe. The X-ray emission of the jet is probably of synchrotron origin in this source (source: Hardcastle et al. 2016 and https://chandra.harvard.edu/photo/2016/pictora/; radio lobes according to ATCA/CSIRO)*

While the very fast VHE flares seem likely to come from very compact zones in the vicinity of the black hole, it is quite possible that the continuous or slowly varying gamma-ray emissions come from much larger regions. The radio emission itself requires particle re-acceleration phenomena along the jets, because the energy loss time of the particles by synchrotron emission (see section 4.4) is often less than the transit time along extended jets. It is therefore certain that particle acceleration mechanisms are at work even at a great distance from the black hole. The Cen A radio galaxy also offers an example of an AGN where a diffuse gamma emission was detected by Fermi above 100 MeV in two large lobes extending on either side of the galactic core (Figure 4.11) and where an an extension in the direction of the jet was demonstrated in the image obtained at VHE (Abdalla *et al.* 2020). The flux of the diffuse emission detected by Fermi is greater than that coming from the Cen A core and represents a power comparable to the kinetic power expected in the jets. It can be interpreted as being due to the inverse Compton radiation of a very relativistic population of electrons interacting with the ambient photons of the 3K cosmic background and of the extragalactic background light (EBL).

At the moment, wide-field and low-frequency radio observations provide, thanks to the large LOFAR surveys, an impressive flow of data with, to date, a sample of 23,344 probable AGNs, for example, and a sample of 98 blazars including 30 gamma-ray emitters, highlighting the diversity of the populations observed, which will perhaps renew our current vision of extragalactic radio jets and sources (Hardcastle *et al.* 2019; Mooney *et al.* 2019). Modeling the radio jets and halos thus observed at very large scales makes it possible, among other things, to estimate the total power transported by the plasma stream ejected from the AGN. The comparison of the energy balances between these extended jets and their VLBI bases in the vicinity of the black hole and the gamma-ray emission zones could provide a key element to identify the origin, the detailed structure and the nature of the electron–positron and electron–proton plasma constituting the jets.

4.4. Non-thermal emissions and instantaneous radiative models

The full range of physical phenomena at play in extragalactic jets is not yet fully identified and understood. However, the most recent observations confirm the majority of the fundamental interpretations proposed so far regarding their mode of non-thermal emissions and help to refine and complete them. Here, we will give a quick overview of the radiation mechanisms (synchrotron, inverse Compton, hadronic processes) which have so far emerged as the most significant to account for the radiative properties of radio and gamma jets.

4.4.1. *Synchrotron emission*

The spectral, polarization and variability properties of radio radiation from nuclear jets at the parsec scale and jets extended to the kpc and Mpc scales quickly led to their interpretation in a very general way in the context of synchrotron emission of relativistic particles in motion in a magnetic field. This interpretation has so far proven to be universal, and its application has provided access to the determination of many physical parameters of sources. It combines the presence of very relativistic particles and a magnetic field, in perfect coherence with the main models of jet formation (see Chapter 3), by anticipating that a fraction of the plasma particles with a bulk Lorentz factor Γ, from a few units to a few dozen, can be accelerated to individual Lorentz factors γ typically reaching $\gamma \sim 10^7$.

Charged particles in accelerated motion emit electromagnetic radiation. A qualitative argument developed by Thomson (Longair 1992) makes it possible to visualize the origin of this radiation by considering a charge q at rest at the point O at the initial instant, with its radial electric field, and which is accelerated at the speed Δv in a time Δt. At a later time t, field lines are still radial from point O beyond a radius $R'' = ct$ because the information about the change in velocity is not yet known. On the other hand, within a radius $R' = c(t - \Delta t)$, the information has already arrived and the field lines point radially to the new position of the particle. This requires that there be a tangential component of the field in the shell with thickness $c\Delta t$ between the spheres with radii R' and R'', which propagates from the charge at the speed of light and corresponds to radiation. Note that this tangential component vanishes in the direction of the particle's acceleration. This radiation can be precisely deduced from Maxwell's equations in a vacuum and is described at all points $\boldsymbol{r}$ at instant t by the electric and magnetic radiation fields (which are added to the Coulomb field),

$$\boldsymbol{E}_{rad}(\boldsymbol{r},t) = \frac{q}{c}\left[\frac{\boldsymbol{n}}{\kappa^3 D} \times \{(\boldsymbol{n} - \boldsymbol{\beta}_{ret}) \times \dot{\boldsymbol{\beta}}_{ret}\}\right]$$

$$\boldsymbol{B}_{rad}(\boldsymbol{r},t) = \boldsymbol{n} \times \boldsymbol{E}_{rad},$$

with $\boldsymbol{n} = \boldsymbol{D}/D$ unit vector between the position of the particle at the retarded time and the point of interest ($\boldsymbol{r}$, t), $\boldsymbol{D}$ and D the vector and the distance between these two points[8], $\boldsymbol{w} = \boldsymbol{\beta}_{ret}c$ and $\dot{\boldsymbol{\beta}}_{ret}c$, particle velocity and acceleration at the retarded position and time.

The factor $\kappa = 1 - (\boldsymbol{n}.\boldsymbol{w}/c)$ plays an essential role for relativistic particles and describes the beaming of their individual radiation in a narrow cone around their velocity.

This radiation field can be calculated by passing through the retarded Liénard–Wiechert potentials for a point charge (Jackson 1975; Rybicki and Lightman 1979). It makes it possible to describe multiple radiative phenomena, associated with typical trajectories for emitting particles, and, in particular, the synchrotron emission of relativistic particles in motion in a magnetic field. Ignoring energy losses, the expression of the Lorentz force in special relativity indicates that these particles describe a helical path around the magnetic field, with a constant individual Lorentz factor γ, a constant velocity $v_{/\!/}$ along the magnetic field $\boldsymbol{B}$ and a velocity $\boldsymbol{v}_\perp$ in the plane perpendicular to $\boldsymbol{B}$ such that:

$$\frac{d\boldsymbol{v}_\perp}{dt} = \frac{q}{\gamma mc}\ \boldsymbol{v}_\perp \times \boldsymbol{B}$$

8. In this chapter, the equations are expressed in the CGS (Gaussian) system of units. Symbols in bold represent vectors, and those in thin characters represent scalars.

with $|\boldsymbol{v}_\perp|$ = constant, which corresponds to a uniform circular motion with an angular gyration frequency $\omega_B = |q|B/\gamma mc$ and a radius $R_L = \gamma mcv_\perp/|q|B$, called the relativistic Larmor radius, around magnetic lines.

The spectrum of radiation, expressed in erg.cm^{-2}.Hz^{-1} as energy per unit area and frequency ($\omega = 2\pi\nu$), is deduced from the temporal variation of the associated electric field $E(t)$ to which it is linked by Fourier transform $\hat{E}(\omega)$,

$$\frac{dW}{dA\, d\omega} = c|\hat{E}(\omega)|^2$$

A qualitative approach makes it possible to understand its nature here. The relativistic "beaming" effect concentrating the radiation in a cone with half opening angle $\sim 1/\gamma$ around the velocity of the particle, the observer receives a very short emission peak of duration Δt_{syn} at each gyration period, when the luminous beam sweeps its line of sight. The associated spectrum is thus formed from the fundamental frequency ω_B and a large number of harmonics. Fourier theory indicates that this spectrum is broad and can include frequencies up to $\omega \sim 1/\Delta t_{syn}$. The duration of a peak, $\Delta t_{syn} \sim 1/\gamma^3\omega_B$, is estimated from the time spent by the particle in the opening cone $2/\gamma$ towards the observer, relative to the gyration period $2\pi/\omega_B$ corrected by the factor $1 - \beta \sim 1/2\gamma^2$ for $\gamma \gg 1$ due to the motion of the particle towards the observer (here ignoring $v_{/\!/}$). This makes it possible to define a characteristic synchrotron frequency of the particle spectrum, beyond which the spectrum falls away, $\nu_c = \omega_c/2\pi \sim 3\gamma^3\,\omega_B/4\pi = 3\,\gamma^2|q|B/4\pi mc$. For electrons and positrons, we generally adopt $\nu_c \sim \gamma^2 B\ MHz$ with B expressed in Gauss. In the source frame of reference, a population of electrons with individual Lorentz factor $\gamma{\sim}10^4$ thus radiates in the sub-millimeter range around 100 GHz in a magnetic field $B{\sim}1$ mG and reaches the X-ray domain towards 10^{17} Hz at $\gamma{\sim}10^6$ for $B{\sim}0.1$G, which illustrates the possibilities of easily sweeping the spectral range of the low-energy bumps of the SEDs observed (Figures 4.6 and 4.7)[9].

The total power emitted by an individual particle is calculated from the flux of the Poynting vector $\boldsymbol{S} = c\boldsymbol{E} \times \boldsymbol{B}/4\pi$ integrated on the whole solid angle, $P_{ind} = 2|q|^4B^2\gamma^2v_\perp^2/3m^2c^5$. For a population of particles with an isotropic distribution of velocities at $\beta = v/c$, the average power radiated by the electron in the source

9. Note that an accurate comparison to SED blazars requires taking into account the overall motion of the source towards the observer, characterized by its bulk Lorentz factor Γ and its angle of orientation θ, introduced in section 4.2.2.

reference frame[10] is obtained by averaging over the pitch angles ψ between the velocity and magnetic field, $P_{synch} = 4\,\sigma_T c\beta^2\gamma^2 U_B/3$, where $\sigma_T = 8\pi\, q^4/3m^2c^4$ is the Thomson cross-section of the particle and $U_B = B^2/8\pi$ the magnetic energy density. The mean synchrotron radiative cooling time of very relativistic particles can be deduced from this,

$$t_{cool} \sim \frac{\gamma mc^2}{P_{synch}} \sim \frac{3mc}{4\sigma_T\gamma U_B}$$

which corresponds, for the electrons, to $t_{cool} \sim 5 \times 10^8\,\gamma^{-1}B^{-2} \sim 6 \times 10^{11}\,B^{-3/2}\,\nu^{-1/2}$ seconds. Thus, electrons reaching the sub-millimeter (for $B{\sim}1$ mG) or the X-ray domain (for $B{\sim}0.1$G) cool in ~ 1,500 years and ~ 14 hours, respectively. Applied, for example, to the source Pictor A (Figure 4.12), these values turn out to be much lower than the transit time of $\sim 3 \times 10^5$ years along 95 kpc of the jet at speeds close to the speed of light, illustrating the need for *in situ* particle re-acceleration mechanisms along the jet.

An interesting property to note is that t_{cool} is proportional to m^4 for particles with a given energy $E = \gamma mc^2$. The protons therefore cool down $(m_p/m_e)^4 \sim 10^{13}$ times slower than electrons[11], which explains the very clear predominance of synchrotron radiation from electrons observed in the universe, without, however, excluding specific cases where the synchrotron emission of protons could play a role.

Detailed analysis of the spectrum and polarization of synchrotron radiation emitted by an individual particle and populations of particles requires lengthy calculations. Initiated by Schott in 1912 before the first detections in 1947 obtained from the first "synchrotron" particle accelerators, it has been presented by several authors including Ginzburg and Syrovatskii (1964, 1965), Rybicki and Lightman (1979) and Longair (1994). The energy per unit of frequency and solid angle of the radiation field of an individual relativistic particle is expressed by the Fourier transform,

$$\frac{dW}{d\omega d\Omega} = \frac{c}{4\pi^2}\left|\int [D\,\boldsymbol{E}_{rad}(\boldsymbol{t})]\, e^{i\omega t} dt\right|^2$$

10. In section 4.4.1, no collective Doppler enhancement effect, due to the bulk motion of the source to the observer, is taken into account. It is later re-introduced in section 4.4.2 for application to blazars.

11. m_p and m_e are the proton and electron masses, respectively.

where $DE_{rad}(t)$ is calculated at the retarded time $t' = t - D(t')/c$ and is expressed in terms of the velocity and acceleration of the particle. The development of this expression in two directions, parallel and perpendicular to the direction of the magnetic field seen by the observer, makes it possible to describe the two states of radiation polarization. After a lot of algebra including some simplifications applicable to a very relativistic particle, integration on solid angles provides the energy emitted per unit of frequency during an orbital period $2\pi/\omega_B$ of the particle. We deduce the power emitted per unit of frequency (also called the emissivity of the particle) according to the two polarization states:

$$P_{\perp}(\omega) = \frac{\sqrt{3}\,|q|^3 B\,sin\psi}{4\pi mc^2}\,[F(y) + G(y)]$$

$$P_{/\!/}(\omega) = \frac{\sqrt{3}\,|q|^3 B\,sin\psi}{4\pi mc^2}\,[F(y) - G(y)]$$

with $y = \omega/\omega_{crit}$, where $\omega_{crit} = \omega_c\,sin\,\psi = 3\gamma^2|q|B\,sin\,\psi/2mc$, $F(y) = y\int_y^{\infty} K_{5/3}(\zeta)d\zeta$ and $G(y) = y\,K_{2/3}(y)$, where the modified Bessel functions K_n of $5/3$ and $2/3$ order appear. The total power emitted by frequency $\nu = \omega/2\pi$ is deduced from this by adding the two polarization states,

$$P_{tot}(\nu) = \frac{\sqrt{3}\,|q|^3 B\,sin\psi}{mc^2}\,F(y)$$

The function $F(y)$ varies as $y^{1/3}$ for $y \ll 1$, reaches its maximum close to 1 for $\omega \sim 0.29\omega_{crit}$ then decreases as $e^{-y}\sqrt{y}$ at a large y. Most of the power is emitted around the critical frequency ω_{crit}.

The Stokes parameters can be expressed in terms of $P_{\perp}$ and $P_{/\!/}$. The degree of linear polarization $\Pi(\omega) = (P_{\perp} - P_{/\!/})/(P_{\perp} + P_{/\!/}) = G(y)/F(y)$. Integrated in frequency, the linear polarization can be very high and reach a maximum around 70% for very ordered magnetic fields, a maximum that could occasionally be observed in certain sources. The generally observed polarization, however, is rather only a few %, which can be explained by the presence of a more disordered magnetic field, which causes the degree of linear polarization typically to drop by a factor of ${B_u}^2/({B_u}^2 + {B_{iso}}^2)$, where B_u is the uniform component of the magnetic field on which an isotropic random component is superimposed B_{iso}. In general, the polarization properties of radio sources have been found to be very compatible with the characteristics of the synchrotron emission, which has supported the interpretation of their radiation in terms of synchrotron emission. The polarization data remain complex to analyze because their interpretation depends on the global models adopted to describe the source, but they can provide remarkable information, in particular on the configuration of the magnetic fields and on the thermal electron–proton plasma density by the observation of Faraday rotation and depolarization.

For optically thin plasma, the projection of the magnetic field in the sky plane is perpendicular to the electric polarization vector. The total spectrum is deduced by integration over the entire population of emitting particles. Various approximations that greatly simplify the calculations of the total emissivity of a population of particles can be applied depending on the situations studied (Böttcher and Reimer 2012). The simplest of them is to describe $P_{tot}(\nu)$ using a Dirac function at $\nu = \nu_c$ for an isotropic distribution of the pitch angles.

The case of a population of particles distributed in energy in a power law such that $n(\gamma)d\gamma = K\gamma^{-p}d\gamma$ is particularly interesting. Its spectrum is then also a power law, with the total power per unit of volume and frequency (or synchrotron emissivity) expressed in erg s^{-1} cm^{-3} Hz^{-1}

$$P_{pop}(\nu) = \frac{\sqrt{3}}{p+1}\left(\frac{3}{2\pi}\right)^{\alpha}\Gamma\left(\frac{p}{4}+\frac{19}{12}\right)\Gamma\left(\frac{p}{4}-\frac{1}{12}\right)KB^{1+\alpha}\frac{|q|^{3+\alpha}}{c^{2+\alpha}m^{1+\alpha}}(\sin\psi)^{1+\alpha}\,\nu^{-\alpha}$$

where the spectral index $\alpha = (p-1)/2$ is linearly related to the power law index of the particle distribution (Rybicki and Lightman 1979). Here, $\Gamma(Y)$ is the gamma function of argument Y. This property explains the universality of the spectra of extended extragalactic radio sources, generally in power law over large spectral domains. Indeed, the main particle acceleration mechanisms (see section 4.5) precisely generate power law populations, as remarkably observed in the spectrum of cosmic rays. In practice, the power law spectral domain still remains limited due to the existence of low- and high energy cut-offs, at individual Lorentz factors γ_{min} and γ_{max}, in the energy distribution of radiating particles.

In compact sources such as the radio core of AGNs, the phenomenon of synchrotron self-absorption also changes this power law towards low frequencies when the absorption of photons by relativistic particles is large. For a brief description of the phenomenon, it is useful to introduce the brightness temperature of the source defined by radio astronomers, such that:

$$T_B = \frac{c^2 I_\nu}{2k_B\nu^2} \sim \frac{c^2 F_\nu}{2k_B\nu^2\,\Delta\Omega}$$

where k_B is the Boltzmann constant, $F_\nu = \int I_\nu \, d\Omega \sim I_\nu \, \Delta\Omega$ is the flux density of the source observed in erg s^{-1} cm^{-2} Hz^{-1}, I_ν is its intensity (which is preserved along a light ray in the absence of diffusion and absorption) and $\Delta\Omega$ is the solid angle subtended by the source. This brightness temperature corresponds to the temperature of a black body radiating the same intensity I_ν as the source within the Rayleigh–Jeans limit of the Planck spectrum. From a thermodynamic point of view, the brightness temperature of an incoherent radio source cannot exceed the energy contained in the particles radiating at a given frequency, so that the self-absorption and the optically thick regime are established at low frequencies below the self-absorption frequency ν_{aa} according to the constraint $k_B T_B \leq \gamma mc^2$, which corresponds to:

$$F_\nu \propto B^{-1/2} \, \Delta\Omega \, \nu^{5/2}$$

The spectrum in the optically thick regime is therefore in power law $\nu^{5/2}$ and is clearly distinguished from the spectrum in $\nu^{-(p-1)/2}$ expected at higher frequencies in the optically thin regime. This spectral peculiarity makes it possible to identify compact radio cores in VLBI maps often showing several components at the base of the jets. The above formula provides an estimate of the magnetic field in the compact source if the synchrotron emission is attributed to electrons, if the data allows the precise determination of the frequency ν_{aa} at which the self-absorption occurs, as well as the measurements of the flux density and the apparent size of the source at this pivotal frequency.

Another approach to evaluate the magnetic field is possible thanks to the "synchrotron scenario", under the energy minimization hypothesis of the observed source, for a given extension and luminosity. Total energy can be expressed as the sum of the energies contained in the magnetic field and in relativistic electrons and protons,

$$U_T = U_B + U_e + U_p = \frac{B^2}{8\pi} f\mathcal{V} + (1+k)CL_{syn}B^{-3/2}$$

where $\mathcal{V}$ is the source volume and f is the filling factor by the magnetic field, L_{syn} is the synchrotron luminosity of the source due to electrons in the frequency band of ν_1 to ν_2 for an assumed power law spectrum and spectral index α and $C(\alpha, \nu_1, \nu_2)$ is a simple function of the spectral index and frequencies. The factor $k = U_p/U_e$ depends on the acceleration and is not yet well constrained in the AGN jets[12]. The minimum value of the total energy is obtained by canceling its derivative with

12. It is around 30 in the interstellar medium, but the conditions in the AGN jets can be very different.

respect to the magnetic field, which almost corresponds to an equipartition of energy between the magnetic field and the particles, and leads to $U_{Tmin} = 7(1+k)U_e/4$ and the equipartition magnetic field $B_{eq} \sim \left[6\pi(1+k)CL_{syn}f^{-1}V^{-1}\right]^{2/7}$, which is expressed entirely as a function of direct observables (apart from the filling factor) and can therefore be measured. The expected values for the magnetic field decrease throughout the jets, typically of the order of 0.1 G in the first VLBI knots until 10^{-5} G in extended radio lobes. We can also highlight that the minimum-energy hypothesis is equivalent in practice to a minimum pressure inside the source, which makes it possible, for example, to quantitatively constrain the models of jet confinement by the pressure exerted by the external environments. The physical conditions and the evolution of the emitting plasmas, however, do not guarantee that minimum energy and equipartition are systematically reached at all, but these estimates illustrate the power of the application of the "synchrotron scenario", which allows access to several parameters and fundamental phenomena for the understanding of sources.

However, although very useful for the modeling of radio-loud and gamma-loud AGNs, the classic description of synchrotron radiation presented here and widely used in the scientific literature becomes more complicated when it comes to rigorously describing the complex and sometimes very particular physical conditions of relativistic space plasmas. It also includes various questionable hypotheses depending on the applications. In particular, the above reasoning ceases to be valid for very energetic particles, when they lose a significant fraction of their energy in one gyration period, which will be the case if $t_{cool} < 2\pi/\omega_B$, that is, for electrons at $\gamma \gg 10^7 B^{-1/2}$, a situation which can be found in pulsars[13] or at the base of jets at the level of the black hole magnetosphere. In addition to possible quantum effects, several situations can also induce significant modifications of the observed radiation, such as, for example, velocities directed mainly along magnetic lines, significant irregularities in the plasma and the magnetic field or rapidly varying context. In practice, the jitter-type radiation regime must be considered when the correlation length of the magnetic field becomes smaller than the non-relativistic Larmor radius $R_{nrL} \sim m\,c^2/|q|B \sim 2 \times 10^3\,(m/m_e)\,B^{-1}\,cm$ (Medvedev 2000; Nishikawa *et al.* 2003; Kelner *et al.* 2013). It can potentially become significant in highly turbulent and magnetized plasmas where it then contributes to extending the spectra towards high energies, and could occur in the context of relativistic AGN jets, gamma-ray bursts, pulsar winds and supernovae remnants.

13. This effect seems, for example, to explain the abrupt cut observed just below 100 GeV in the spectrum of the Crab Nebula (Hillas 2013).

4.4.2. *Inverse Compton emission and leptonic models*

The Thomson, Compton and inverse Compton scattering effects describing the scattering of a photon on an electron (and vice versa) are inevitable in AGN jets due to the simultaneous presence of electrons and radiation. Inverse Compton scattering plays a particularly important role by inducing an energy transfer from the population of relativistic electrons to ambient photons, some of which are thus boosted in the high and very high energy ranges. This phenomenon offers a natural explanation for the origin of the second bump observed in gamma rays in the spectral energy distributions (Figures 4.6 and 4.7), in the context of so-called "leptonic" scenarios where non-thermal radiations are emitted mainly by a population of electrons or electrons–positrons.

The different regimes of this electron–photon interaction depend on the energies involved. In the electron reference frame, when the energy $\varepsilon = h\nu$ of the incident photon is weak compared to the rest mass energy of the electron $m_e c^2$ ($x = h\nu/m_e c^2 \ll 1$), this is Thomson scattering and the characteristics of the scattered radiation are calculated from the radiated electromagnetic field ($\boldsymbol{E_{rad}}, \boldsymbol{B_{rad}}$) presented in section 4.4.1, emitted by the oscillating electron (assumed to be non-relativistic) in the incident electromagnetic wave, which provides the value of the total cross-section of this interaction, $\sigma_T = \frac{8\pi}{3} r_0^2 = 6.65 \times 10^{-25}\, cm^2$, the Thomson cross-section for the electron, where $r_0 \equiv e^2/m_e c^2 = 2.82 \times 10^{-13}\, cm$ is the classical electron radius. When the energy of the incident photon becomes comparable to or greater than the rest mass energy of the electron ($x = h\nu/m_e c^2 \geq 1$), the interaction enters into the general Klein–Nishina regime, where it is necessary to take into account quantum effects. The photon scattering is no longer elastic due to the electron recoil, which gains energy during the collision, and the cross-section becomes dependent on the energy of the incident photon. The conservation of the energy-momentum quadrivector makes it possible to express the energy of the scattered photon $\varepsilon_d = \varepsilon/[1 + \varepsilon\,(1 - cos\varphi)/mc^2]$ where φ is the scattering angle of the photon, and the total cross-section of the Klein–Nishina process σ_{KN} is calculated within the framework of quantum electrodynamics,

$$\sigma_{KN} = \frac{3}{4}\sigma_T \left[\frac{1+x}{x^3}\left\{\frac{2x(1+x)}{1+2x} - \ln(1+2x)\right\} + \frac{1}{2x}\ln(1+2x) - \frac{1+3x}{(1+2x)^2}\right]$$

This expression tends towards σ_T in the non-relativistic regime for $x \ll 1$ and is simplified in $\sigma_{KN} = (3\sigma_T/8x)\,(ln2x + 1/2)$ in the ultra-relativistic regime for $x \gg 1$. The effective cross-section σ_{KN} decreases with increasing photon energy, and the Compton scattering becoming less efficient at high energies (Rybicki and Lightman 1979).

In contrast, inverse Compton scattering occurs when the electrons are not at rest but have sufficient kinetic energy compared to the energy of ambient photons and are thus able to transfer energy to them during the interaction. Its description in the laboratory reference frame (≡ of the source) is deduced from that of the direct process described in the electron reference frame by a simple change of frame, by applying the formula of the relativistic Doppler effect twice (once for the incident photon, and once for the scattered photon). In the Thomson regime and for relativistic electrons with large individual Lorentz factors $\gamma \gg 1$, it is thus shown quite easily that head-on collisions, which are the most efficient, boost the photons by a factor of $4\gamma^2$ in energy in the laboratory reference frame. For an isotropic distribution of the incident photons, the average energy reached after scattering by the photons is $\varepsilon_{d_{lab}} \sim 4\gamma^2\, \varepsilon_{lab}/3$. Very high energies can thus be achieved while remaining in the Thomson regime such that $\gamma\varepsilon_{lab} \ll m_e c^2$, beyond which quantum effects reduce the effectiveness of the phenomenon.

Describing the inverse Compton radiation emanating from astrophysical plasmas requires the analysis, not of the individual photon-electron interaction, but of the interaction of a population of photons with a population of electrons. The inverse Compton power and spectrum radiated by an isotropic distribution of photons interacting with an isotropic population of electrons were calculated in the Thomson (Blumenthal and Gould 1970) and Klein–Nishina (Jones 1968) regimes for single scattering. A characteristic of the Thomson regime is the total power radiated by an electron interacting with an isotropic distribution of photons, $P_{Compt} = 4\sigma_T c\beta^2\gamma^2 U_{ph}/3$, where U_{ph} is the initial photon energy density. Its expression is very similar to the average synchrotron power radiated by an electron of a population with isotropic velocity distribution, such that:

$$\frac{P_{synch}}{P_{Compt}} = \frac{U_B}{U_{ph}}$$

The total power radiated per unit volume by a population of relativistic electrons (with $\beta \sim 1$) distributed in power law $N(\gamma) = K\,\gamma^{-p}$ between γ_{min} and γ_{max} interacting with an isotropic distribution of photons is then expressed quite simply in erg s^{-1} cm^{-3},

$$P_{Compt_{tot}} = \frac{4}{3}\,\sigma_T\, cU_{ph}K\,(3-p)^{-1}(\gamma_{max}^{3-p} - \gamma_{min}^{3-p})$$

The spectrum resulting from the interaction in the Thomson regime of a wide power law distribution of relativistic electrons and an isotropic photon population of initial photon density $\rho(\varepsilon)$ with a peak at a certain value is expressed as the total power scattered by volume and by energy and can be approximated as follows over a certain domain in ε' (Rybicki and Lightman 1979):

$$\frac{dP}{dV\,d\varepsilon'} = \pi c r_0^2\, K\, A(p)\, \varepsilon'^{-(p-1)/2} \int \varepsilon^{(p-1)/2}\, \rho(\varepsilon)\, d\varepsilon$$

with

$$A(p) = 2^{p+3}\,(p^2 + 4p + 11)/[(p+3)^2\,(p+5)(p+1)].$$

This spectrum representative of the inverse Compton emission is then in power law with spectral index $\alpha = (p-1)/2$, like that of the synchrotron emission. However, in blazar jets, the Klein–Nishina regime is generally reached and quantum description (Jones 1968) becomes necessary, especially at very high energies. The multiple inverse Compton scattering, which further complicates the description, however, often seems negligible (see an example in Figure 4.14). In general, the formulas describing the inverse Compton scattering phenomenon in detail are costly in computing time, and, depending on the goal, various very useful approximations can be found in the literature (Boettcher and Reimer 2012; Aharonian 2013).

There are two categories of leptonic scenarios, depending on the nature of the scattered ambient photons. In synchrotron-self-Compton (SSC) scenarios, it is the same population of relativistic electrons which is scattered and which boosts the photons of the background synchrotron radiation that it has itself generated by the inverse Compton effect. Alternatively, in external inverse Compton (EIC) scenarios, the population of relativistic electrons at the origin of the synchrotron radiation mainly boosts ambient photons coming from other sources of radiation, such as accretion disk emissions (re-emitted by clouds of the BLR), emission from the dust torus or synchrotron emission from larger regions of the jet. The radiation emitted is calculated in the source reference frame (here the jet plasma), but the fluxes observed by our telescopes are boosted by Doppler amplification (see section 4.3), and it is therefore necessary to introduce this effect to compare emission models and observed data.

Leptonic scenarios were the first to be proposed to account for X-ray then gamma emissions detected from AGNs (Königl 1981; Ghisellini *et al.* 1985; Sikora *et al.* 1994; Inoue and Takahara 1996) and have been widely developed in parallel with advances in observational gamma astronomy. They are now well constrained by the available data, especially the simplest one-zone SSC models with a limited number of free parameters, typically with three parameters to describe the emission zone towards the base of the VLBI jet (Doppler factor, radius, magnetic field) and some parameters that characterize the spectrum of radiating particles (density, power law index, low- and high energy cut-offs). Despite these constraints, SSC models generally account for the SEDs obtained from HBL sources in their stationary state (Figure 4.13a), or even at a given time during eruptive states, very well. Completed by EIC contributions when they become significant, they can also reproduce SEDs from intermediate BL Lacs (Figure 4.14). In the case of FSRQs which have very bright accretion disks, it is likely that the EIC components dominate the SEDs compared with the SSC component (see section 4.5).

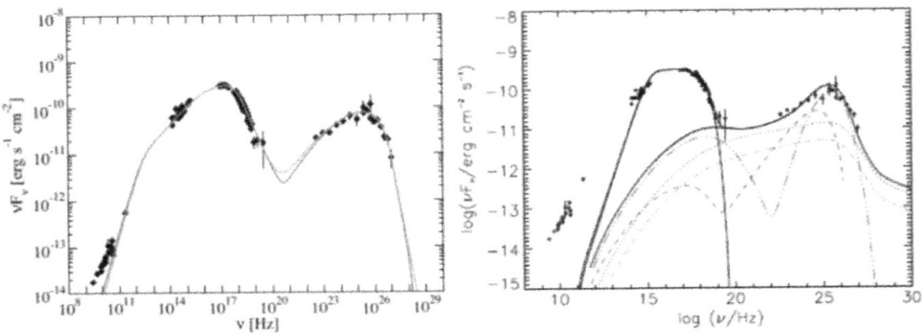

Figure 4.13. *Modeling the spectral energy distribution of the low quasi-stationary state of the HBL Mrk 421 presented in Figure 4.6, according to Abdo et al. 2011.* ***(Left).*** *Homogeneous one-zone SSC model, established for two minimum variability time scales (red: 1 day; green: 1 hour), corresponding, respectively, to radius of 5 x 10^{16} to 5 x 10^{15} cm, Doppler factors of 20 to 50, and magnetic fields of 0.04 to 0.08 G. In this scenario, the electron population is assumed to follow a broken power law, with two breaks and three spectral indices of 2.2, 2.7 and 4.7.*

(Right). *Hadronic model in which relativistic electrons and protons are injected into a strong magnetic field, B ~ 50 G, in a zone with a radius of 4 x 10^{14} cm and Doppler factor of 12. In this scenario, the protons interact (photomeson reactions and pair cascades) with the primary electron synchrotron radiation, which completely dominates the first soft X-ray bump. The second gamma bump is by synchrotron radiation from protons and muons above GeV, and by pion cascades from MeV to GeV (blue: muon synchrotron and associated cascades; red: proton synchrotron and associated cascades; black: neutral pion cascades; green: charged pion cascades) The attenuation of the flux at very high energies due to the interaction of VHE photons with the diffuse extragalactic background radiation (EBL) was taken into account in both models (see section 4.4.4). The excess radio emission observed at very low frequencies, particularly large in the hadronic model, can be attributed to the synchrotron radiation of the extended jet*

The popularity and established success of leptonic scenarios are also due to the fact that they directly explain the correlations often observed between the X-ray and gamma-ray fluxes detected (see Figure 4.7), since they attribute the origin of synchrotron and inverse Compton emissions, detected in these two spectral bands, to the same population of electrons (and positrons), respectively.

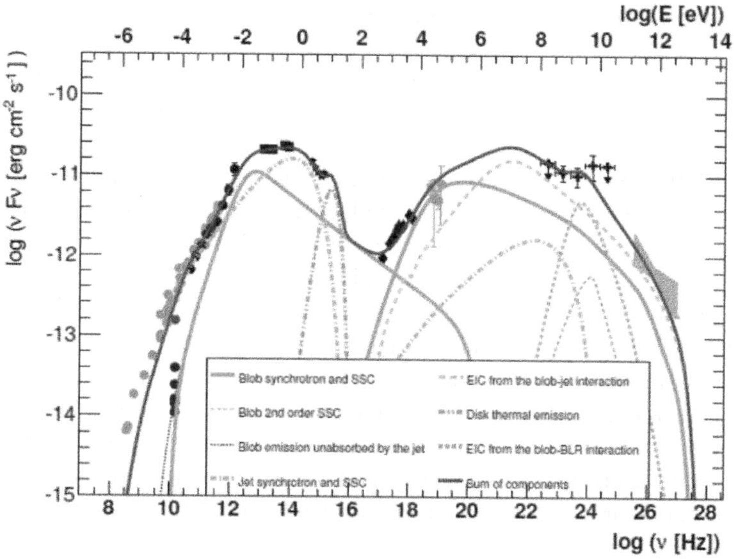

Figure 4.14. *Energy spectral distribution of the intermediate BL Lac AP Librae, reproduced by an SSC model in a blob (in blue) immersed in an extended jet, also taking into account the presence of a faint accretion disk and the external inverse Compton (EIC) components induced by the interaction of the relativistic electrons in the blob with the radiations by the BLR and the extended jet, as well as second-order inverse Compton SSC scattering (negligible here). In the radio domain, the blue dots correspond to the VLBI core and the gray dots to the extended emission (Hervet et al. 2015)*

Difficulties arise, however, for one-zone SSC models in particular, when it comes to describing the overall evolution over time of a multispectral flare observed across the entire electromagnetic spectrum. Datasets of this kind that are complete with respect to spectral domains and with good temporal coverage are still very rare, but some are starting to strongly constrain the models and highlight their limitations. The detection of large components of hard-spectrum emissions at multi-TeV energies could also challenge leptonic models because the quantum effects of the Klein–Nishina regime reduce the efficiency of the inverse Compton emission in this domain (Zech *et al.* 2017; Abdalla *et al.* 2018). Solutions could then be provided by taking into account radiation of hadronic origin.

4.4.3. *Hadronic and lepto-hadronic models*

Basic hadronic models describe the gamma emission zone as a compact, highly magnetized region that is carried away in the jet and contains a population of relativistic electrons and protons (Mannheim 1993; Aharonian 2000; Mücke and Protheroe 2001). So far they have been less developed than leptonic models for AGN physics because they require very high energy protons whose presence has not yet been clearly proven in jets, and which seem less able to induce the sometimes observed highly variable flares, although this is debatable (Mastichiadis *et al.* 2020). They also require magnetic fields of a few Gauss to several tens of Gauss, more intense than those of leptonic models, and a potentially high global energy budget, with several variants and free parameters that are difficult to constrain with the currently available data. However, they have the advantage of being more complete than leptonic models, because they typically also integrate lepton physics, which develops with the creation of secondary particle populations in particular. We then speak of lepto-hadronic models when the gamma-ray emissions by the inverse Compton effect from the electrons–positrons are no longer negligible within an initially hadronic scenario, which may be the case for moderate magnetic fields. They are especially interesting in the context of multi-messenger astronomy because they allow exploration of the possible correlations between the detection of very high energy gamma rays from AGN and "neutrino" or "cosmic ray" events identified by IceCube- or AUGER-type infrastructures.

Hadronic emission models that aim to account for the gamma-ray spectra of AGN jets essentially take into account (i) the interaction of energetic protons with low-energy photons, producing pions during photomeson production such as

$$p + \gamma \rightarrow p' + n^0\pi^0 + n^+\pi^+ + n^-\pi^- + \cdots$$

and producing electrons and positrons during the Bethe–Heitler process of pair production,

$$p + \gamma \rightarrow p' + e^+ + e^-$$

as well as (ii) the interaction of energetic protons (and the secondary particles produced) with the magnetic field and their synchrotron emission up to very high energies. Neutral and charged pions decay rapidly according to different decay chains, and in particular according to the main channels,

$$\pi^0 \rightarrow \gamma + \gamma$$

$$\pi^+ \rightarrow \mu^+ + \nu_\mu$$

$$\pi^- \rightarrow \mu^- + \bar{\nu}_\mu$$

$$\mu^+ \rightarrow e^+ + \nu_e + \bar{\nu}_\mu$$

$$\mu^- \rightarrow e^- + \bar{\nu}_e + \nu_\mu$$

thus generating cascades of secondary particles, inevitably including gamma-ray photons, muons and neutrinos. Neutrons can also be produced during the proton–photon interaction and take some of the initial power with them in the jet. In general, the interactions of ultra-relativistic protons with photons have a small cross-section and are relatively slow, so the process requires compact sources with a fairly intense radiation field. However, this radiation field is constrained if the gamma source must remain optically thin (with an optical depth of $\tau_{\gamma\gamma} \leq 1$) for gamma rays relative to the phenomenon of pair production by photon–photon interactions (see section 4.4.4),

$$\gamma + \gamma \rightarrow e^+ + e^-$$

whose cross-section is much greater, which limits the efficiency of proton–photon interactions. Conversely, if these interactions are very effective, the optical depth increases and the electromagnetic cascade phenomenon is intensified in the source. The radiated spectrum then depends mainly on the development of the cascades and no longer on the spectrum of the primary ultra-relativistic protons and drifts towards lower energies.

Proton–proton interactions can also generate gamma rays and neutrinos. However, they seem less frequent in AGN jets, except in specific models of heavy jets, for example, the collision of an electron–proton jet with a star or a dense cloud, or at the base of the jets in interaction with the accretion flow. Various public versions of Monte-Carlo-type hadronic codes are available, in particular, SOPHIA for photomeson interactions and SYBILL for proton–proton interactions, which make it possible to calculate the spectral characteristics of the radiation emitted. Figure 4.13b presents the modeling of Mrk 421 in a low state via a hadronic scenario, compared to an SSC leptonic scenario. Both types of scenarios succeed in reproducing the observed SED, while maintaining a comparable total power in the jet of about 10^{44} erg s^{-1}. Note, however, that the physical parameters strongly depend on the model adopted and that the SSC and hadronic scenarios lead to very different properties of the VHE emitting zones, which should ultimately make it possible to distinguish between them.

Advances in multi-messenger astronomy will bring a new dimension to this field of study. The origin of ultra-high energy cosmic rays is still unknown, but there are indications of correlations between their directions of arrival and the flux distribution of different populations of nearby extragalactic gamma-ray sources, including AGNs (Aab *et al.* 2018). The 3σ detection of a high energy muon neutrino by IceCube in 2017, in spatial and temporal coincidence with the gamma-ray blazar TXS 0506 + 056 during an eruptive phase, and a 3.5σ emission of neutrinos previously detected in this direction of the sky (Aartsen *et al.* 2018a, 2018b) also suggest that blazars are sources of neutrinos. Any accurate confirmation of the significance of such types of correlation would identify AGNs as cosmic-ray accelerators. This would bring new constraints on gamma-ray emission models and their hadronic components (Cerruti *et al.* 2019; Zhang *et al.* 2020) and help to clarify the various feedback mechanisms of AGNs on their cosmic environment.

4.4.4. *Propagation of VHE photons: pair creation, absorption and electromagnetic cascades*

The phenomenon of electron–positron pair creation by photon–photon collision already mentioned, $\gamma\gamma \rightarrow e^+e^-$, can greatly affect the VHE gamma-ray photons emitted by AGNs. This reaction requires photons with energies ε_1 and ε_2 (in the interaction reference frame) high enough to be above the pair creation threshold, such that:

$$\varepsilon_1\varepsilon_2 \geq 2\ (m_e c^2)^2/(1 - \cos\chi)\ ,$$

so $\varepsilon_1\varepsilon_2 \geq (m_e c^2)^2$ for a head-on collision (χ being the collision angle). The effective cross-section $\sigma_{\gamma\gamma}(\varepsilon_1, \varepsilon_2, \chi)$ of this interaction involving photons and electrons is close to the effective Thomson cross-section at high energies, but due to the threshold effect, it sharply peaks around $\varepsilon_1\varepsilon_2 = 4\ (m_e c^2)^2/(1 - cos\chi)$, where it reaches $\sigma_{\gamma\gamma}^{peak} \sim 0.2\ \sigma_T$. Gamma photons with the energy ε_{VHE} therefore interact mainly with background radiation with energy ε_{bkgd} such that:

$$\varepsilon_{bkgd} \sim 2\ (m_e c^2)^2\ /\ \varepsilon_{VHE}$$

This pair-production phenomenon induces a very significant absorption of gamma rays by ambient radiation during their propagation, whether in extragalactic space or in the source itself.

On their way to Earth, gamma rays above the threshold of 0.25 TeV interact primarily with photons (around ~ 1 eV) from the diffuse extragalactic background radiation (EBL), and extreme gamma rays above the threshold of 250 TeV interact with photons (around ~ 0.001 eV) from the cosmological microwave background radiation (CMB). We can estimate the distance $\mathcal{D}$ from which the universe becomes opaque to gamma-ray photons of energy above the threshold, reached when the optical depth $\tau_{\gamma\gamma} \sim 0.2\, \sigma_T < n_{bkgd} > D$ is equal to 1, which defines a gamma-ray horizon as a function of the energy of VHE photons considered and the background photon density $n_{bkgd}\,(\varepsilon_{bkgd}, z)$, reducing the possibilities of exploring the cosmos at very high redshifts in gamma rays.

The absorption induced by the interaction $\gamma_{TeV} + \gamma_{EBL} \rightarrow e^+ + e^-$ must be taken into account for the reproduction of spectra currently observed in the TeV domain. For this, EBL models from 0.1 to 1,000 μm, such as that of Dominguez *et al.* (2011), allow precise calculation of optical depths and provide tabulated values as a function of gamma photon energy and source redshift. The EBL due to the light emitted by all stars, galaxies and cosmic dust is, however, complex to model, and its direct measurements are very difficult due to the presence of bright foreground emissions. The gamma observations of distant AGNs and the modeling of their intrinsic spectra can therefore inversely contribute to constraining the EBL (Abeysekara *et al.* 2019) and to exploring certain open questions in cosmology and fundamental physics (see section 4.6.4).

In extragalactic space, the electrons–positrons generated by pair production interact via the inverse Compton effect mainly with the (more numerous) photons of the CMB, $e + \gamma_{CMB} \rightarrow \gamma' + e$, which triggers the generation of electromagnetic cascades with the production of gamma photons and secondary particles. These cascades develop as long as the photons emitted by inverse Compton scattering retain an energy above the pair creation threshold and contribute to the diffuse extragalactic gamma-ray background detected. Under the hypothesis of being initiated by an extremely high energy photon (or electron) and entirely developed at large distances on the two CMB and EBL radiation backgrounds, these cascades present a fairly universal photon spectrum due to being independent of the spectrum of primary particles, which follows the following trend in total number: $n(\varepsilon) \propto \varepsilon^{-3/2}$ at low energies $\varepsilon \leq \mathcal{E}_X$ and $n(\varepsilon) \propto \varepsilon^{-2}$ at intermediate energies $\mathcal{E}_X \leq \varepsilon \leq \mathcal{E}_\gamma$, with a cut-off at high energies around the minimum energy $\mathcal{E}_\gamma$ (~ 0.4 TeV) of absorbed photons. The broken power law appears around the transition energy $\mathcal{E}_X$ (~ 0.1 GeV), energy of the photon radiated by the inverse Compton effect by the minimum energy electron produced by the energy photons $\mathcal{E}_\gamma$ (Berezinsky and Kalashev 2016; Aloisio *et al.* 2018). The presence of a non-zero intergalactic magnetic field (IGMF) modifies the development of such electromagnetic cascades because the electrons–positrons deviate from the initial cascade direction, which can induce an extension or a "pair halo" around the gamma-ray image of the primary source, as well as temporal delays in the photon arrival times according to their energy. Some of the energy from the cascades is also removed by synchrotron radiation if the IGMF is high enough. All these cascade effects, however, still remain essentially inaccessible to current gamma-ray instruments but could be studied by those of the next generation (Abdalla *et al.* 2021).

Pair creation, absorption and cascade development also play an important "*in situ*" role in the gamma-ray emitting region of AGNs and impose a lower limit on the radius R_γ of the zone from which gamma-ray photons of energy ε_γ can escape, corresponding to a "gamma-sphere", where the optical depth $\tau_{\gamma\gamma}$ of the pair production on ambient photons with energy $\varepsilon_{bkgd} \gtrsim m_e^2c^4/\varepsilon_\gamma$ is equal to 1 (Levinson and Blandford 1995). This lower limit must be compared with the upper limits on the size of the emitting zone deduced by the causality argument for the variable sources (see section 4.2.2). The constraint is reduced due to relativistic Doppler factors and does not seem to pose a big problem for BL Lacs in which, in the absence of a bright accretion disk, the ambient radiation at the level of the jets is *a priori* weak and internal $\gamma\gamma$ absorption is relatively negligible. On the other hand, the situation is more critical for FSRQs like 3C279 and PKS 1222 + 216, whose bright accretion disks and BLRs make the large "gamma-sphere" difficult to reconcile with the maximum size of their emission VHE zone deduced from their rapid variabilities (Prandini 2017; Blandford *et al.* 2019). Solutions can be found by considering more elaborate geometries, as well as the properties of the gamma region and ambient radiation, which sometimes constrain the location of the emission zone. However, it is still difficult to explain the rapid variabilities (Meyer *et al.* 2019; Abdalla *et al.* 2020b).

4.5. Particle acceleration and time-dependent radiative models

The non-thermal emissions presented in section 4.4 all come from populations of very energetic particles, the presence of which, in the central AGN cores and jets, seems logical but whose exact origin is not yet fully understood. Various particle acceleration processes can be established and induce these very relativistic populations, thus allowing the complete description of phenomena observed at very high energies.

4.5.1. *Radiative models and temporal evolution*

An overall description of the observed gamma-ray sources requires, in addition to the identification of the emission mechanisms presented in section 4.4, the energy distribution function of the relativistic radiating particles $f_i(E,\boldsymbol{r},t)$ and its spatio-temporal evolution to be modeled. A general kinetic equation that governs the evolution of this distribution function was initially established for the purposes of studying galactic cosmic rays (Leverett 1956; Kardashev 1962; Ginzburg and Syrovatskii 1964). This is a "Fokker–Planck" equation, which can be expressed as follows, for a population of particles of species i assumed to be isotropic following multiple deflections on the ambient plasma (diffusive approximation),

$$\frac{\partial f_i}{\partial t} = \nabla.(D_i\,\nabla f_i) - \nabla.(f_i\boldsymbol{u}) - \frac{\partial}{\partial E}(b_i f_i) + \frac{1}{2}\frac{\partial^2}{\partial E^2}(d_i f_i) + Q_i(E,\boldsymbol{r},t) - p_i f_i + \sum_k \int P_i^k\,(E',E)\,f_k(E',\boldsymbol{r},t)\,dE'$$

where the terms on the right-hand side of the equation represent, respectively, (1) the spatial diffusion of the particles, with the diffusion coefficient $D_i(E,\boldsymbol{r},t)$; (2) the effect of expansion or compression if there is a large-scale bulk motion of velocity $\boldsymbol{u}$; (3) the mean systematic energy change, where $b_i(E) = dE/dt$ represents the average energy increment (positive or negative) of particles per unit of time; (4) fluctuations in this continuous variation in energy, where the coefficient $d_i(E) \equiv d < (\Delta E)^2 >/dt$ can be expressed generally as the average square of the increment in energy per unit of time[14]; (5) the number of particles of energy E injected per unit of time by external sources into a volume element of the emitting zone considered; (6) the number of particles which escape or suddenly disappear from the energy domain per unit of time, p_i being the probability of such a process per unit of time; and (7) the influx of type i particles induced by "catastrophic" collisions with particles of type k, with $P_i^k(E',E)$ being the probability per unit of time and energy of the appearance/disappearance of a type i particle of energy E during a collision with a type k particle of energy E'.

All the complexity of the physics underlying this description resides in the different coefficients which, in principle, depend on the general properties (macro- and

14. The expressions adopted for terms 3 and 4 assume that the energy variations take place continuously by small successive changes and that the energy increment remains low compared to the initial energy of the particle, which is well justified for the mechanisms of statistical accelerations and radiations generally considered. The strong variations are described by terms 5, 6 and 7.

micro-physics) of the population of energetic particles and of the ambient medium with which it interacts (nature of the particles, velocity field, turbulence, magnetic field, background radiation, etc.), boundary conditions (geometry, injection, escape, etc.) and various mechanisms which develop and induce energy losses and gains through cooling and acceleration of particles. The knowledge already acquired on the general description of gamma-ray emitting AGN jets, however, makes it possible to set a certain study framework. Indeed, the phenomena of continuous energy losses will, in this case, be dominated by the radiative losses due to the radiations presented in section 4.4, to which can be added the phenomenon of particle escape, as well as a loss due to an adiabatic expansion of the emission zone with $\boldsymbol{u_r} \neq 0$, for example, in the event of non-confinement of the emission zone in the jet (compression will conversely induce a gain in energy). In addition, the effects of "catastrophic" collisions between different populations of particles and ambient plasmas can be expressed in terms of radiation, injection or escape, which significantly simplifies the general system of equations, both for hadronic and leptonic scenarios (Romero *et al.* 2017).

In the end, it is the energy gain phenomena that represent the most open question currently. Indeed, several particle acceleration mechanisms can develop. Each of them represents a complex field of study that continues for decades, and we still do not know which ones are dominant. In addition, the detection of VHE emissions and bright flares occurring on very short time scales place strong constraints on the strength, efficiency and speed of acceleration required. Due to the low density of the plasmas concerned, the various interactions allowing the particles to be efficiently accelerated ultimately pass through electromagnetic fields and the Lorentz force in $\boldsymbol{F} = q(\boldsymbol{E} + \boldsymbol{v} \times \boldsymbol{B}/c)$. These acceleration mechanisms therefore only relate to charged particles; any very energetic neutral particle can only be secondary. However, the strong electrical conductivity of astrophysical plasmas generally prevents the maintenance of a static electric field (except in the immediate environment of compact objects, see section 4.5.4), and no work can be transmitted to a particle in a static magnetic field. The acceleration processes must therefore mainly rely either on transient electric fields or on time-varying magnetic fields (which induce an electric field such as $\nabla \times \boldsymbol{E} = -\partial \boldsymbol{B}/c\partial t$). An electric field also appears from the magnetic field in a moving plasma, as expressed by the transformation laws of the electromagnetic tensor $\boldsymbol{E}' = \Gamma \boldsymbol{V} \times \boldsymbol{B}/c$ during a change of reference frame. A very general criterion (Hillas 1984) makes it possible to express the maximum energy E_{max} reachable by a particle of charge q in an accelerator of size Λ_{acc} by neglecting energy losses and specifying that the particle remains confined in the accelerator with a Larmor radius $R_L < \Lambda_{acc}$ and so

$E_{max} \sim |q|B\Lambda_{acc}$. In a jet with a bulk Lorentz factor Γ, the relativistic effect modifies this formula by $E_{max} \sim \Gamma|q|B\Lambda_{acc}$, which indicates the orders of magnitude required for the magnetic field and the dimensions of the acceleration zone to reach the domains beyond TeV. We recall below the main particle acceleration processes suspected of playing an important role in the context of gamma-ray jets. The scientific literature is abundant on this subject, and the reader can refer to various books and journals (Blandford and Eichler 1987; Eilek and Hughes 1991; Kirk 1994; Longair 1994; Parizot 2003; Balogh and Treumann 2013; Ohsawa 2014; Treumann and Baumjohann 2015; Gonzalez and Parker 2016; Anchordoqui 2019; Marcowith *et al.* 2020; Pohl *et al.* 2020; Nishikawa *et al.* 2021).

4.5.2. *Fermi acceleration mechanisms*

Fermi mechanisms (Fermi 1949, 1954) are universal and are the most widely regarded. They can generate and maintain populations of highly energetic particles with a power law energy spectrum of index $p \geq 2$, which are indeed observed in many cosmic sources and primarily in cosmic rays whose almost power law spectrum extends over nearly 12 orders of magnitude in energy. The fundamental process is based on the interaction of particles and their exchange of energy with moving scattering centers, "magnetic clouds" in initial studies, then, more generally, fluctuations of the magnetic field due to the turbulent superposition of waves in plasmas, mainly Afven waves. A resonant interaction can indeed take place between wave and particle when the Larmor radius is close to the wavelength of the perturbation, $R_L \sim \lambda$, which allows diffusion and transfer of energy. Within the framework of the quasi-linear theory, Fermi acceleration acts only on the particles that are energetic enough to be able to resonate with the Alfven waves, that is, when their speed and that of the wave correspond, which requires the initial injection of already pre-accelerated particles. The notion of wave-particle resonance evolves, however, within the framework of anisotropic turbulence and strong nonlinear turbulence theories where non-resonant interactions can occur and the particles interact with specific plasma structures (solitons, vortices, etc.) rather than with the waves.

We generally distinguish the type I Fermi mechanisms due to diffusive shock acceleration (DSA) and type II Fermi mechanisms due to stochastic acceleration by turbulence. Both are likely to be able to establish themselves in AGN jets where different types of shocks and turbulence develop. Following Fermi's initial model (1949), which illustrates a type II process, we can show by expressing the conservation of energy and total momentum that, during the elastic collision[15] of a

15. This is here a "non-collisional shock", in which the "particle–cloud" interaction passes through electromagnetic fields ("direct" collisions are negligible).

relativistic particle with a non-relativistic magnetic cloud, energy E of a particle varies by $\Delta E \sim -2E\boldsymbol{v_p}.\boldsymbol{u_N}/c^2$. Here, $\boldsymbol{v_p}$ and $\boldsymbol{u_N}$ are the initial velocities of the particle and the cloud ($v_p \sim c$). The particle gains energy in head-on collisions ($\boldsymbol{v_p}.\boldsymbol{u_N} < 0$) and loses some in rear collisions ($\boldsymbol{v_p}.\boldsymbol{u_N} > 0$). Assuming a distribution of random moving clouds and the population of particles isotropized, the probability of head-on collisions, in $v_p + u_N$, is higher than the probability of rear collisions, in $v_p - u_N$. The particles therefore gain energy overall, with an average gain per collision of:

$$< \Delta E > \propto \frac{u_N^2}{c^2} E$$

This average gain is proportional to the energy of the particle, which is characteristic of Fermi mechanisms, and also proportional to u_N^2/c^2 (or u_A^2/c^2 for interaction with Afven waves, with velocity u_A), hence the name type II or second-order Fermi process. The characteristic acceleration time is independent of energy, $\tau_{acc} = E/(dE/dt) \sim E/(< \Delta E >/\tau_{coll}) \propto \tau_{coll}c^2/u_N^2$, where τ_{coll} is the typical time between two collisions. There is therefore a bijection between the time t spent by the particle in the acceleration zone and the energy reached E, $t \sim \tau_{acc}\, Ln\,(E/E_0)$. If the characteristic time of escape of particles from the acceleration zone $\tau_{esc} = N(t)/(dN/dt)$ is also independent of energy, the particle density can be written as $N(t) = N_0 \exp(-t/\tau_{esc})$, which is expressed as a function of energy and provides the energy distribution of the population of accelerated particles $N[E(t)] \propto N_0\, E^{-\tau_{acc}/\tau_{esc}}$, hence the differential distribution $dN(E) \propto E^{-p} dE$ with a power law index $p = 1 + (\tau_{acc}/\tau_{esc})$.

The situation can become very different depending on the macroscopic configuration of the plasma, especially if the velocity fields are ordered in such a way that they favor collisions that transmit energy to the particles. This is what happens in the environment of strong non-collisional shocks, where the plasma upstream (respectively, downstream) of the shock front "sees" the plasma on the other side of the shock moving in its direction from its own reference frame. The fast particles are partly trapped in the vicinity of the shock by their diffusion on the magnetic turbulences, which they themselves have excited, which allows some of them to pass through the shock several times, each time undergoing a head-on collision with an energy gain (see Figure 4.15). The Fermi acceleration thus becomes first order (Bell 1978), with an average gain in energy per round trip through the shock of:

$$< \Delta E > \propto \frac{\Delta u}{c} E \propto \frac{r-1}{r} \frac{u_{shock}}{c}$$

where $\Delta u = u_{am} - u_{av} > 0$ is the relative speed between the two media upstream and downstream of the shock (u_{am} and u_{av} being their respective speeds in the front shock reference frame), u_{shock} ($= -u_{am}$) is the speed of the shock in the medium upstream reference frame (stationary) and r is the compression ratio of the shock. For a strong very supersonic shock, $r = (\Gamma_a + 1)/(\Gamma_a - 1)$, where Γ_a is the adiabatic index, equal to 5/3 for a monoatomic ideal gas which corresponds to $r = 4$. The acceleration time depends on the diffusion properties in both upstream and downstream media, but it is the same diffusion mechanisms that are responsible for both the confinement and escape of particles from the acceleration zone. This circumstance means that acceleration time and escape time have a similar dependence on energy with a ratio τ_{acc}/τ_{esc} ~ constant, hence a population of particles accelerated in power law. Calculating the escape probability effectively expresses the differential distribution of particles,

$$dN(E) \propto E^{-p}\, dE \text{ , with } p = \frac{r+2}{r-1}$$

The energy spectrum of a population of particles accelerated by this type I Fermi mechanism is in power law with index p, which, remarkably, is independent of the peculiarities of the underlying diffusion processes, and depends only on the structure of the shock, with a compression ratio $r \gtrsim 4$ for (very) strong shocks. The type I Fermi mechanism thus presents a universal character, which leads to power law populations with index $p \sim 2$, which is actually commonly observed in the cosmos.

Fermi mechanisms are still the subject of numerous theoretical and numerical works which attempt to describe the complexity of the nonlinear coupling between the accelerated particles and the magnetic turbulence which they themselves generate by the triggering of various instabilities which preside over their own acceleration, in order to achieve a fully self-consistent description of the whole phenomenon, for different configurations and types of plasmas, and to account for the multiple effects that enrich the first models mentioned above (shock precursor, instabilities, temporal evolution of the magnetic field, turbulence and shock, feedback of accelerated particles on the structure of the shock and turbulence, relativistic regime, radiative effects, boundary effects, etc.). In particular, "*ab initio*" kinetic simulations, which are very costly in computing time, should make it possible to understand the details of the long-term evolution of shocks and microturbulences, and associated accelerations, thanks to the current exponential growth in the power of supercomputers.

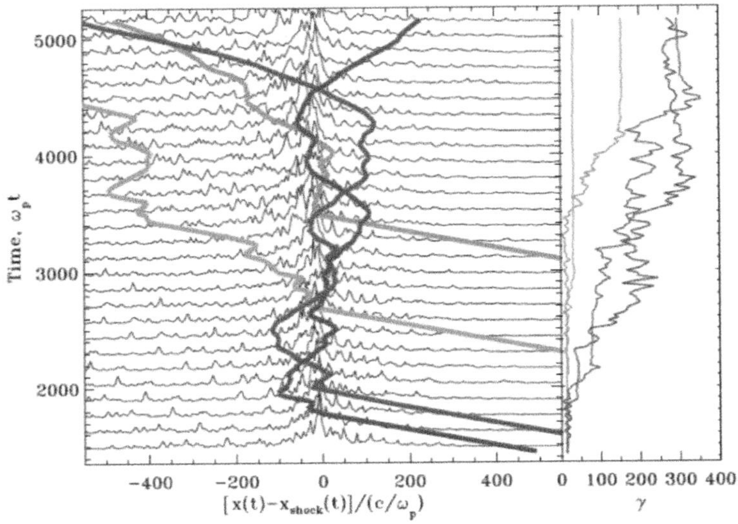

Figure 4.15. *Results of PIC simulations*[16] *obtaining, for the first time in a self-consistent manner, the onset of Fermi acceleration in a non-collisional relativistic shock* ($\Gamma_{shock} \sim 21$ *with respect to the upstream medium). The shock propagates through an initially unmagnetized electron–positron plasma, and the acceleration mechanism engages on its own as the shock and the generated magnetic fields and turbulence progress. At the end of the simulation, a non-thermal tail in power law with index ~2.4 (with exponential cut-off at high energies) has grown, containing ~1% of the particles and ~10% of the energy of the downstream medium. In the figure, all the variables are measured in the downstream reference frame.* ***(Left)****. The average profile of the magnetic energy (in black) is shown at different times (on the y axis), depending on the distance to the shock located at zero (on the x axis). The thickness of the shock is around several Larmor radius of thermal particles in the self-generated magnetic field. The trajectories of four representative particles are superimposed in color. Most of the particles in the flow pass through the shock once and do not come back upstream, like the orange particle. But some, like the green, blue and red particles, go through the shock several times and are significantly accelerated before escaping downstream or upstream, thus populating the suprathermal tail of the particle population.* ***(Right)****. The energy of the four representative particles (on the x axis) is given as a function of time (on the y axis). The particles that gain the most energy (blue and red) undergo several reflections between downstream and upstream, moving almost parallel to the surface of the shock. Their greatest energy gains appear during upstream reflections, the upstream magnetic fluctuations being advected towards the shock and carrying an electric field due to the motion, seen from the downstream reference frame (Spitkovsky 2008)*

16. PIC = particle-in-cell.

Relativistic shocks expected in AGN jets must be specially considered, in particular, ultrarelativistic shocks with Lorentz factor Γ_{shock} in the upstream medium much greater than 1 (Begelman and Kirk 1990; Ellison *et al.* 1990; Achterberg *et al.* 2001; Lemoine *et al.* 2006; Spitkovsky 2008; Sironi *et al.* 2015; Lemoine *et al.* 2019; Marcowith *et al.* 2020). Around such a fast shock, the particles in the downstream medium have difficulty catching up with the shock and crossing it upstream, only the particles pre-accelerated to Lorentz factors $\gamma \gg \Gamma_{shock}$ can achieve it. The amplification of a magnetic microturbulence excited from the shock precursor by Weibel instability (or filamentation), however, induces the launch of the Fermi I acceleration process, which then generates a population of particles accelerated in power law with index p typically of the order of 2.2. However, if the shock is magnetized, the passage of particles from downstream to upstream will only be possible in the case of almost parallel shocks, where the flow is almost aligned with the magnetic field, with a low angle Θ between the magnetic field and the shock normal, $\Theta \leq \Theta_{crit}$ ($\Theta_{crit} \sim 34°$ for very relativistic and weakly magnetized shocks). In fact, beyond this critical angle, the charged particles should progress at a speed greater than that of light along the magnetic lines in order to cross the shock. Type I Fermi mechanisms then lose their efficiency. Perpendicular or quasi-perpendicular shocks (the configuration generally expected for external shocks) become "superluminal", the self-generated turbulence remains weak and the type I Fermi acceleration appears to be inhibited, both for electron–positron plasmas and electron–ion plasmas (Sironi *et al.* 2015).

However, two particular types of acceleration, shock surfing (SSA) and shock drift (SDA) acceleration, can develop in quasi-perpendicular shocks because of the non-zero electric field in $\boldsymbol{V} \times \boldsymbol{B}/c$, which develops due to the motion of the magnetized plasma towards the shock. The particles can then increase their energy by more than an order of magnitude during a single encounter with the shock by a process sometimes referred to as a fast Fermi process. In a surfing situation, the particles are trapped at the level of the front shock, between the electrostatic potential of the shock and the Lorentz force upstream. The electric field accelerates these particles essentially along the shock front, until they escape from the shock potential. In a drift situation, a particle can acquire significant energy as its guiding center crosses the shock once, drifting in the direction parallel (or antiparallel) to the electric field[17]. By slightly moving away from the shock, it can then be forced back into it by the convective flow. This drift acceleration is enhanced in relativistic quasi-perpendicular shocks and has been proposed as an alternative to Fermi

17. See https://svs.gsfc.nasa.gov/4513 for a visualization of shock drift acceleration.

mechanisms to explain the intense radio emission from hot spots of FRII-type radio sources (Begelman and Kirk 1990). When the turbulence is low, drift acceleration can also be dominant in oblique shocks and lead to hard-spectrum population of particles, with power law index $p \sim 1$ (Summerlin and Baring 2012).

In addition to shocks, shearing flows with a strong transverse velocity gradient also offer a configuration that is conducive to an acceleration process by shearing, which qualitatively becomes similar to a stochastic Fermi process but in which the motion of the scattering centers is no longer random but organized according to the shear profile of the overall flow velocity. Several effects then combine to reinforce the phenomenon of turbulent acceleration (Rieger 2019). In a regular non-relativistic flow along z with transverse shear in the x direction, that is, for a velocity field $u_z(x)$ such that $\partial u_z/\partial x \neq 0$, the speed of the scattering centers changes from an effective speed $\bar{u} = l\,(\partial u_z/\partial x)$ for a particle that crosses the flow along the x axis on its mean free path $l = c\tau_{coll}$. The energy gain during a collision becomes:

$$\Delta E \propto \frac{\bar{u}^2}{c^2} E \propto \frac{l^2}{c^2}\left(\frac{\partial u_z}{\partial x}\right)^2 E$$

and the resulting acceleration time is inversely proportional to the mean free path, $\tau_{acc} \sim E/(<\Delta E>/\tau_{coll}) \propto c\,l^{-1}\,(\partial u_z/\partial x)^{-2}$, with the effective velocity $\bar{u}$ increasing with l. The shear acceleration is thus found to be faster for particles already pre-accelerated at high energies. Protons with a large mean free path are more easily accelerated than electrons. The detailed description of the microphysics of this process makes it possible to express its diffusion coefficients and shows that the distribution of the accelerated particles tends towards a power law whose index depends on the underlying turbulence, which can precisely be self-generated by the growth of instabilities excited by shearing, in particular the kinetic Kelvin–Helmholtz instability (Rieger and Duffy 2006).

The shear acceleration could very well be at work in the AGN jets, which for decades have shown increasing signs of the presence of a transverse gradient in their flow velocities, and in which a shear necessarily develops at the interface between the jet and the ambient medium in which it propagates. A first study (Rieger and Duffy 2021) illustrates how instabilities and turbulences induced by shear in a layer of thickness $\sim 0.1\,R_{jet}$ on the boundary of a jet with radius R_{jet} may be sufficient to efficiently accelerate electrons up to PeV (10^{15} eV) and protons up to EeV (10^{18} eV) by shear acceleration, without, however, these instabilities leading to the destruction of the jet.

4.5.3. *Magnetic reconnection*

Magnetic reconnection is an event in which magnetic lines "break" and "reconnect", leading to a change in the topology of the magnetic field and a release of magnetic energy, which is converted and transmitted to the ambient plasma in various forms. The phenomenon can be relatively slow, or extremely fast, according to the admissible framework to describe it. In the context of MHD, admissible to describe, for example, "collisional" reconnections in AGN disks, the temporal evolution of the magnetic field in a conductive fluid can be deduced directly from Maxwell's equations and Ohm's law,

$$\frac{\partial \boldsymbol{B}}{\partial t} = \nabla \times (\boldsymbol{u} \times \boldsymbol{B}) - \nabla \times \eta\, (\nabla \times \boldsymbol{B})$$

where $\boldsymbol{u}$ is the fluid velocity, $\eta = c^2/4\pi\sigma$ is the magnetic diffusivity (≡ resistivity) and σ is the conductivity. Since astrophysical plasmas are excellent conductors, the first term on the right-hand side of the equation is usually dominant and the magnetic field is said to be "frozen" in the plasma, carried along with the velocity field. However, there are situations where the second term becomes dominant when the plasma is resistive (i.e. not perfectly conductive), particularly in singular zones of the plasma flow where the term $\nabla \times (\boldsymbol{u} \times \boldsymbol{B}) = 0$, for example, around stagnation points where $\boldsymbol{u} = 0$. This is the case, for example, when the velocity field forces the convergence of plasmas with magnetic lines of opposite directions, or in vortices of magnetic turbulence. The magnetic field is then no longer "frozen" and diffuses with a characteristic time $\tau_{diff} \sim \Lambda_B^2/\eta$, where Λ_B is the spatial scale characteristic of variations in $\boldsymbol{B}$. In these singular zones, the magnetic configuration changes and the dissipation of the magnetic field can induce very strong currents, which form a current sheet and a non-zero electric field. The plasma is heated by the Joule effect and particles can be accelerated. Local strengthening of the gradients of $\boldsymbol{B}$ further reduces the diffusion time τ_{diff} and results in faster dissipation. The development of instabilities induced by the currents and the resulting turbulences can also lead to the appearance of anomalous resistivity, which further accentuates the phenomenon.

The MHD approach to the phenomenon is no longer suitable for small spatial scales where ions and electrons decouple and where plasma microphysics must be described. The conditions are then "non-collisional", and very fast or even explosive reconnections can occur. This reconnection regime is for diluted and hot plasmas and may apply to AGN jet plasmas. New terms that are difficult to model must be made explicit, in particular in the detailed expression of Ohm's law, such as the Hall current (Zweibel and Yamada 2009; Treumann and Baumjohann 2015). Numerical simulations make it possible to approach the problem by returning to the fundamental laws of kinetic physics described by the Vlasov–Maxwell equations. They show that reconnections in a non-collisional relativistic regime can generate power law populations of particles accelerated up to very high energies, with indices $p \sim 1.5$ to 1 when the magnetization is sufficient, which may potentially be harder than those provided by Fermi I and II processes (Melzani 2014a, 2014b; Sironi and Spitkovsky 2014; Marcovith *et al.* 2020).

The reconnection phenomenon is studied in the laboratory and is actually observed in solar flares and the Earth's magnetosphere. However, it is still not fundamentally understood and many questions remain open (how is it triggered, why is it so fast, what is its link with turbulence, what processes determine acceleration at the highest energies, etc.). It is probable that this phenomenon develops in the magnetized plasmas of AGNs and jets, where intense, sheared, turbulent velocity fields are likely to create singular zones conducive to the triggering of very fast reconnections. Some consensus has also been established on the idea that reconnection can contribute to the pre-acceleration of particles. The potentially explosive nature of the phenomenon also seems attractive to account for the fastest flares that are observed at the scale of a few minutes (Giannos *et al.* 2009; Nalewajko *et al.* 2011; Cerutti *et al.* 2012; Morris *et al.* 2019; Asano and Hayashida 2020), and, despite the uncertainties in the detailed description of the mechanism, increasing attention is being paid to scenarios involving reconnection.

4.5.4. *Acceleration in the magnetosphere of massive black holes*

Black holes rotating in a magnetic field behave like a kind of induction machine (Thorne *et al.* 1986) and their magnetosphere, like that of pulsars, can be a particular place of very efficient acceleration of particles, either by centrifugal force, which reigns in the vicinity of the black hole (Gangadhara and Lesch 1997; Osmanov *et al.* 2007), or by electric fields appearing in plasma void zones or "gaps" that are likely to develop on singular surfaces of the magnetosphere (Beskin *et al.* 1992; Hirotani and Okamoto 1998).

The first IACT detections of very variable outbursts in several AGNs such as M87, PKS 2155-304 and Mrk 501, having shown that, according to the causal argument, even with very high bulk Lorentz factors in jets, the size of gamma emission zones at VHE should not exceed a few Schwarzschild radii in these sources, the magnetospheric acceleration mechanisms appeared to be very promising since they are naturally concentrated in the environment that is very close to the black hole (Neronov and Aharonian 2007; Rieger and Aharonian 2008; Istomin and Sol 2009; Osmanov 2010; Broderick and Tchekhovskoy 2015; Katsoulakos and Rieger 2018). The gamma absorption issue seems to be overcome in the absence of bright accretion disks in the sources concerned, and for the magnetic field values of a few tens of Gauss, which is expected in the acceleration zones. The particles injected into the magnetosphere, by creation of e^+e^- or pre-accelerated particles coming from the accreting plasma, from a turbulent disk or from gaps, undergo centrifugal action and are directly accelerated *a priori* up to very high maximum energies, typically up to 10 or 100 TeV for electrons at the light cylinder and up to 10^{18}–10^{20} eV for protons which do not undergo high synchrotron losses. They can thus be at the origin of the variable emissions detected at VHE, by inverse Compton radiation for electrons or by p–p interactions for protons.

Interest in the radiative mechanisms expected in the immediate vicinity of black holes has recently been renewed with the advent of new observational tools such as the Event Horizon Telescope and GRAVITY, capable of resolving structures of a few Schwarzschild radii in nearby AGNs (Akiyama *et al.* 2019; Jimenez-Rosales *et al.* 2020). Recent numerical simulations of non-collisional relativistic electron–positron plasmas in general relativity carried out in 2D (GRPIC code) make it possible to explore the properties of the magnetosphere of Kerr black holes (Parfrey *et al.* 2019; Crinquand *et al.* 2020). Starting from solutions in a vacuum, they reach stationary states and illustrate the microphysics underlying the formation of jets around black holes.

A first study carried out in the "quasi-force-free" approximation, with injection of e^+e^- pairs but without radiation transfer, shows the activation of the Blandford–Znajek process with extraction of the black hole rotational energy and launch of a relativistic jet, as well as the formation of an important current sheet in the equatorial plane due to particles which fall towards the black hole along the field lines and accumulate on the lines of the ergosphere[18] not crossing the horizon (Figure 4.16a). In this current sheet, magnetic reconnection events generate plasmoids, which fall towards the black hole and cross the horizon. The jet includes the lines that enter the ergosphere and can carry strong currents. In the low e^+e^- injection regime, particles start to be ejected along the jet, but far from the horizon, the power of the jet is mainly transmitted as Poynting flux.

18. The ergosphere of a Kerr black hole is the zone around the horizon in which the black hole drags space and matter in its rotational motion.

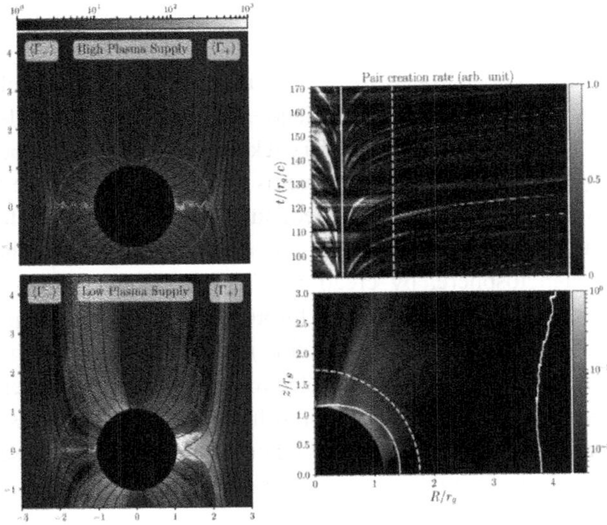

Figure 4.16. *Numerical simulations (GRPIC) of the launching of a jet and the creation and acceleration of particles in the magnetosphere of a black hole.* **Left (a)**. *The simulation evolves from the stationary Wald solution of a rotating black hole in vacuum, immersed in an asymptotically uniform magnetic field aligned with the angular momentum of the black hole, assumed to be in nearly maximum rotation (a = 0.999). This solution includes the electric field generated by the rotation of space–time, which rapidly fills the magnetosphere as electron-positron pairs are injected at each time step, according to two plasma supply regimes, high* **(top)** *and low* **(bottom)**. *The colors encode the average Lorentz factors measured in the FIDO reference frame of the local fiducial observer, on the left for electrons (Γ_-), and right for positrons (Γ_+). The ergosphere boundary is shown in green and the magnetic flux surfaces in black. Two stationary states are reached, with the jets launched by the Blandford–Znajek mechanism and the formation of equatorial current sheets. The most energetic particles are found in the current sheet and along the jet and reach higher Lorentz factors in the low plasma supply regime. The maximum Lorentz factor reached (10^3) corresponds here to the total potential drop generated by the rotation of the black hole (Parfrey et al. 2019).* **Right (b)**. *In this scenario, the magnetic field is assumed to be monopolar and the magnetosphere is initially free of particles but filled with monoenergetic photons above the pair creation threshold, which triggers the mechanism (a = 0.99). The* **bottom insert** *is the 2D map of the time-averaged pair creation rate. The solid white lines are the two light surfaces, inner and outer, averaged over time. The* **top insert** *shows the space-time diagram of the pair creation rate in the high-opacity regime along the 45-degree angle. The solid white line indicates the position of the inner light surface, where the intermittent gap develops. Although pair creation is continuous in the simulations, the trajectories appear discretized due to low sampling (Crinquand et al. 2020)*

Another series of simulations focuses on the development of gaps and pair production that self-consistently feed the magnetosphere (Figure 4.16b). Three species are introduced in these simulations, electrons, positrons and photons, as well as radiation transfer with inverse Compton scattering and the e^+e^- pair creation by photon–photon interaction. When the plasma density decreases, the electric field induced by the rotation is no longer screened, which creates an electrostatic gap that accelerates the particles. Various fiducial optical depths are explored. At small optical depths, a large gap forms where the particles are essentially accelerated and pair creation only takes place far from the black hole. At higher optical depths, an intermittent gap develops at the base of the jet that is very close to the internal light surface (i.e. where the rotational velocity of the magnetic lines reaches the speed of light). It ejects fragments of e^+e^- plasma and could be the source of the gamma-ray emissions and outbursts observed. This result could answer a long debate on the location of gaps in the magnetospheres of black holes. Coupling these GRPIC codes with a light ray tracing algorithm by the same authors allowed them to produce synthetic light curves expected from simulated magnetospheres (Crinquand *et al.* 2021). The amplitude of the flares directly generated by the intrinsic variability due to the pair production mechanism seems to be lower than that observed in the AGNs concerned. If the VHE emission is indeed of magnetospheric origin, the observed variabilities could be due to a fast response of the magnetosphere to sudden changes in its environment (variation in the accretion rate, accretion of a plasmoid). Alternatively, the flares could be induced by the disk–magnetosphere interaction.

4.5.5. *Variability and dynamic spectra*

No consensus has yet been reached on the mechanism(s) dominating the acceleration of particles in jets, nor on the location of the gamma emission zone, and it seems likely that several processes come into play according to the different types of sources. Thus, certain very fast flares observed in BL Lacs or FRI radio galaxies could be of magnetospheric origin, but this scenario seems to be excluded for flares originating from FSRQ sources due to gamma absorption. It is also likely that different types of outbursts and different gamma emission zones can be activated in the same source, as has perhaps been observed in M87, which shows flares with very different multi-wavelength properties (Abramowski *et al.* 2012b), or in the blazar 1ES 1959 + 650, where a fraction of the flares observed in VHE turn out to be orphans, with no counterpart detected in X-rays (Acciari *et al.* 2020). Several acceleration mechanisms can also be linked together, thus providing populations of pre-accelerated particles, capable of being subsequently accelerated to VHE. The main challenge therefore currently relates to the development of an overall scenario allowing the self-consistent insertion of the required processes in an adapted geometry.

Here, we illustrate the most studied scenarios, which describe the gamma emission zone as a fairly complex region swept away in the jet, in or around which particle acceleration mechanisms are engaged, often Fermi type I or II, whose effects expressed within the framework of the quasi-linear theory make it possible to complete the kinetic equation in section 4.5.1. This kinetic equation has no general solution, but several analytical solutions are available in the literature under various simplifying assumptions, including homogeneity of the emitting zone.

A first approach involves assuming the existence of a non-radiating particle acceleration zone and directly injecting into the emission zone the power-law population that it generates (with a cut-off at high energies). The kinetic equation in this emission zone is reduced to three terms on the right-hand side of the equation (radiative losses, escape and injection), that is

$$\frac{\partial N(\gamma,t)}{\partial t} = \frac{\partial}{\partial \gamma}\left[-\left(\frac{\partial \gamma}{\partial t}\right)_{rad} N(\gamma,t)\right] - \frac{N(\gamma,t)}{t_{esc}(\gamma)} + Q(\gamma,t)$$

which significantly simplifies the description while providing a stationary or time-dependent model, mainly based on the choice of the injection term $Q(\gamma,t)$. The term $(-\partial\gamma/\partial t)_{rad}$ is the cooling rate of a particle including all the radiative particle loss rates (see section 4.4) expected in the emitting zone, that is, in leptonic models, the synchrotron and inverse Compton cooling rates. This modeling makes it possible to focus on the radiation properties, regardless of the nature of the particle accelerator, but does not explain the origin of the energetic particles injected. The next step in complexity is to also describe the Fermi I and II accelerations and activate them as needed in the emission zone, or in its immediate environment, according to the typical kinetic equation:

$$\frac{\partial N(\gamma,t)}{\partial t} = \frac{\partial}{\partial \gamma}\{-[S(\gamma,t) + D_A(\gamma,t)]\, N(\gamma,t)\} +$$

$$+\frac{\partial}{\partial \gamma}\left\{D_p(\gamma,t)\frac{\partial N(\gamma,t)}{\partial \gamma}\right\} - \frac{N(\gamma,t)}{t_{esc}(\gamma)} + Q(\gamma,t)$$

where the term $S(\gamma,t)$ includes radiative losses and gains due to first-order acceleration, $D_p(\gamma,t)$ is the momentum diffusion coefficient governing the second-order stochastic acceleration terms and the term $D_A(\gamma,t) = 2D_p(\gamma,t)/\gamma$ represents the average energy change due to the momentum diffusion process. In a magnetized turbulent plasma, the diffusion coefficient can be expressed as a function of the ordered magnetic field B_0 and its fluctuations δB, $D_p \sim (u_A/c)^2 (\delta B/B_0)^2 (R_L/\lambda_{max})^{q-1} (cp^2/R_L)$, with u_A the Alfven velocity, R_L the Larmor radius and λ_{max} the maximum wavelength of the Alfven wave spectrum (Tramacere *et al.* 2011). The exponent q depends on the turbulence spectrum, with $q = 2$ (respectively, 5/3 or 3/2) for a "hard sphere" spectrum (respectively, for a Kolmogorov or Kraichnan spectrum).

The application of these descriptions to sources and their flares still comes up against the problem of collecting detailed data sets, with good temporal and multispectral coverage (from radio to TeV) of the activity phases, which are often not yet available because of the suddenness of flares and the instrumental and organizational constraints. The development of alert networks and multi-wavelength observations, as well as machine learning techniques (Sadeh 2020), should soon allow these difficulties to be overcome and eruptive phases of gamma-ray emitting AGNs to be studied in depth. However, some flares have already been well monitored during multispectral campaigns. This is the case of the highly variable blazar 3C 279 that we take here as an example, an archetype of the class of VHE gamma-loud FSRQs, of which several relatively well-sampled flares are now available and have been the subject of various studies (Aleksic *et al.* 2014; Diltz *et al.* 2015; Hayashida *et al.* 2015; Paliya *et al.* 2016; Rani *et al.* 2018; Abdalla *et al.* 2019; Böttcher and Baring 2019; Lewis *et al.* 2019; Asano and Hayashida 2020; Larionov *et al.* 2020).

A consensus seems to be emerging to locate the emission zone of 3C 279 flares beyond the BLR and therefore beyond several hundred Schwarzschild radii from the black hole to meet the constraints on the absorption of gamma rays. In this configuration, the detection of fast variabilities lasting a few minutes at HE (Ackermann *et al.* 2016) raises questions, since these variabilities must come from a more compact region. One possibility is to assign them to active turbulence cells within the main radiative zone. Models of protection by a plasma layer of the gamma emission zone that is internal to the jet, against its exposure to external ambient radiations, could offer an alternative (Meyer *et al.* 2019). Single-zone leptonic and hadronic models manage to reproduce the general evolution of a flare but struggle to explain in detail all the SEDs and multispectral light curves now available, suggesting a greater complexity of the emission zone (Figure 4.17). Time-dependent two-zone models may be more promising for successfully reproducing flares that are well sampled in time and frequency (Figure 4.18).

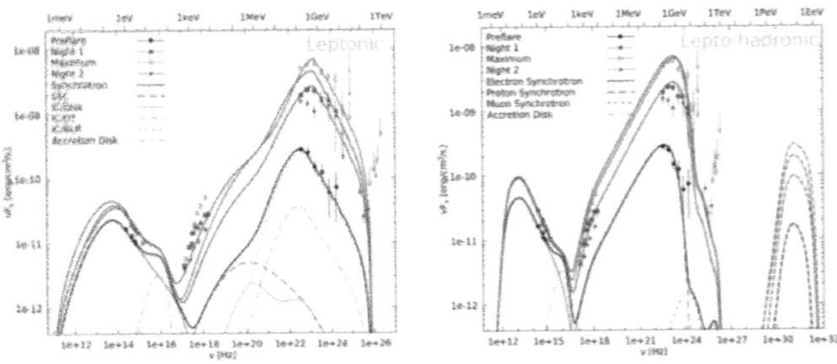

Figure 4.17. *The June 2015 flare of blazar 3C 279. Two one-zone models (in solid lines) are considered to account for four levels of activity (in black, red, green and blue) during the flare monitored in optical, X-rays and gamma-rays (data from SMARTS, ATOM, Swift-XRT, Fermi-LAT and HESS). The emission zone is placed more than 1.7 x 10^{17} cm from the black hole, with a radius of 1.8 x 10^{16} cm and a Doppler factor of 30, and is associated with a stationary shock such as a recollimation shock. The exploration of a large domain of parameters makes it possible to identify solutions that globally account for the observed SEDs, but neither of the two models manages to precisely reproduce all the available data, in particular during the decrease of the flare.* ***(Left).*** *One-zone leptonic scenario, with a magnetic field of 0.65 G, an electron injection power of 8 x 10^{41} erg/s and an acceleration time to escape time ratio of 1.* ***(Right).*** *One-zone lepto-hadronic scenario, with a magnetic field of 50 G, an electron injection power of 3.3 x 10^{41} erg/s and protons injection power of 7 x 10^{43} erg/s and an acceleration time to escape time ratio of 30. The model also provides the expected neutrino spectrum (dotted lines), which remains beyond the reach of current detectors like IceCube (Abdalla et al. 2019)*

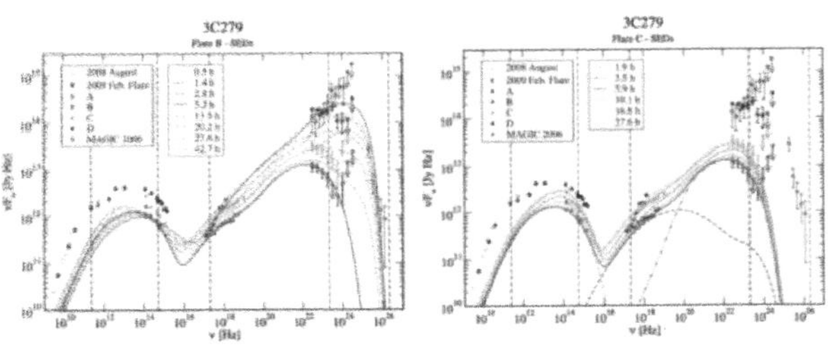

Figure 4.18. *SED of the blazar 3C279 for different levels of activity during two flares in December 2013: Flare B of December 20 (in red) on the* ***left****, and Flare C of December 31 (in yellow) on the* ***right****. The data points (Hayaskida et al. 2015) are reproduced by an entirely time-dependent leptonic model with two zones, including a zone with a weakly relativistic oblique shock where the Fermi I acceleration and drift mechanisms are established, surrounded by a larger emission zone into which are injected the accelerated particles which radiate via the synchrotron, SSC and external inverse Compton effect. In the figure on the right, the three emissions constituting the low state are indicated (synchrotron – dotted line; SSC – dashed line; EIC – dashed-dotted line). The light-colored full lines represent the rise then the fall of the flares (Böttcher and Baring 2019)*

4.6. Perspectives on active galactic nuclei at very high energies

4.6.1. *A contrasting current situation*

Ground-based gamma astronomy has now established itself as a new branch of astronomy following the number of discoveries over the past 15 years. It revolutionized the view of our universe, which has proven to be particularly rich in very high energy phenomena, which had, until then, been largely underestimated. VHE gamma emission emanates, in particular, from the most significant celestial power plants, sites of gigantic flows of matter that impact their environment and are part of cosmic history. Several new guidelines suggested in this chapter emerge for our understanding of AGNs and the extreme energy phenomena they conceal, with reasonable explanations of particle acceleration mechanisms and HE and VHE gamma-ray radiation, which are available and give an overall account of the electromagnetic properties observed in terms of spectra and light curves.

However, it remains difficult to identify the right scenario among the various plausible options with certainty, due to the lack of sufficient constraints, and equally, to overcome the challenges posed by particular cases of sources or events that turn out to be incompatible with available models. The data collected on gamma radiation has not yet made it possible to identify the general framework for specifying the macroscopic, geometric and dynamic properties of the plasmas which emit them. No consensus has yet been established on fundamental questions related to the initial production of jets; the acceleration of particles to extreme energies; the location of gamma emission zones; the ultimate origin of variability or even the nature of radiating particles, leptons or hadrons. This lack of a general framework limits the possibilities of data interpretations as they are often dependent on the models applied and favors the proliferation of alternatives, none of which is fully satisfactory or capable of providing a unified model.

The main challenge for the next few years is probably to put together all the available parts in a coherent way and thus to succeed in developing a general reference scheme for the field of AGNs, including their multispectral properties up to high and very high energies. This will require in-depth study of emblematic sources and a statistical approach of the various populations of AGNs and jets at different redshifts, as well as the inevitable pursuit of theoretical and numerical developments in relativistic plasma physics.

4.6.2. *A very promising future for instruments*

The international very high energy astronomical community is now engaged in the new-generation "Cherenkov Telescope Array" (CTA) project, a gamma-ray observatory made up of several dozen IACTs spread over two sites, in Chile and the Canary Islands, to ensure good coverage of the entire sky. The CTA is expected to improve sensitivity 10-fold, to widen the spectral range of IACTs from 20 GeV to 300 TeV and tp significantly improve all current instrumental performance in terms of angular, spectral and temporal resolution (Acharya *et al.* 2013). The construction of CTA is underway and will take several years. Open to the entire scientific community, CTA will offer great flexibility in terms of observational modes, in a complete array or by sub-arrays. The high sensitivity combined with fields of view of several degrees make CTA an instrument that is very suitable for wide sky surveys, and the pointing speed will also offer the possibility of operating under alert, favoring the study of transient phenomena down to time scales of a minute and below. The angular resolution of the order of arcminute (comparable to that of the human eye in the visible range) represents the best instrumental resolution achievable beyond the X-ray domain and will allow detailed mapping of extended gamma-ray sources.

The observational perspectives opened up by CTA are very rich for the field of AGNs (Sol *et al.* 2013; Inoue *et al.* 2019; Mazin *et al.* 2019; Zech *et al.* 2019). They should help us to understand the missing link between the physics of accretion, the magnetosphere of black holes and the jet formation mechanisms, and will shed new light on the still open questions of classification and unification of the various types of AGNs. Extrapolations of spectra to TeV from Fermi catalogs indicate that several hundred AGNs will be detectable with CTA, reaching redshifts $z \sim 2$. It will be necessary to gather representative samples to explore the phenomenon of activity beyond blazars, for jets oriented at a greater angle with respect to the line of sight, such as FRI and FRII radio galaxies, including certain Seyfert galaxies (Romano *et al.* 2020), and steep-spectrum radio quasars. The systematic survey of a quarter of the sky (outside the galactic plane) planned as a future key CTA program will make it possible to avoid the numerous observational biases affecting the current study of AGNs at VHE and to collect unbiased samples from 1,000 hours of observation of several tens or even hundreds of AGNs, making it possible to determine the luminosity functions and log N-log S diagrams[19], while promoting the serendipitous discovery of rare or still unknown phenomena. Confronted with measurements of the extragalactic gamma-ray diffuse background, log N-log S diagrams of BL Lacs and FSRQs will test the overall consistency of the models of gamma-ray emission and unification of AGNs according to their viewing angle.

19. Diagrams indicating the number of objects as a function of flux.

Dedicated observations of different types of AGNs over a wide range of redshifts will provide high-quality spectra as well as their temporal evolution over timescales of several years, down to a minute during very bright events. This program, coordinated with multi-lambda monitoring, will form a homogeneous database that is essential for the detailed study of the specific spectral characteristics of various effects and for the comparative analysis of the high state or low state variabilities, depending on the type of AGN and redshift. It should make it possible to distinguish the relative importance of leptonic or hadronic scenarios according to the types of sources and their states. Sophisticated alert systems will optimize the monitoring of eruptive states or bright outbursts, important in several ways because their strong signal-to-noise ratio makes it possible to obtain precise spectra, to explore the smallest time scales, to probe extreme phenomena and to extend the accessible domain to large redshifts. Real-time data analysis will make it possible, in less than 30 seconds, to identify important transients in the field of view of all operating IACTs and to launch internal alerts to the entire array to elicit more in-depth observations. Atmospheric shower detection arrays such as HAWC in Mexico and LHASSO in China, operating from 100 GeV to 100 TeV and beyond, will be very complementary to CTA on these projects, thanks to their large field and extended monitoring capacity. The quality of future CTA data could make it possible to better spatially resolve nearby sources such as M87 and Cen A and to identify diffuse emissions or bright regions along the jets, which, combined with the multi-spectral maps already available from radio up to X-rays, would offer a synthetic view of the acceleration and radiation zones, requiring that extreme particle acceleration mechanisms be effective even at a great distance from the black hole.

The complexity of AGNs makes them emitters across the whole electromagnetic spectrum and potential emitters of non-photonic messengers, at the crossroads of different fields of physics and different observational techniques. Studies will therefore fully benefit from the influx of multiple data expected from many research infrastructures in the field of multispectral astronomy and astroparticle physics for non-photonic messengers, which are driving the rapid development of our knowledge on AGNs. Without being exhaustive, we can cite, in the field of space gamma rays, Fermi and the SVOM project; in X-rays, Chandra, XMM, Swift, NuStar, ASTROSAT, HXMT, eROSITA and the future ATHENA satellite; in optics, the Vera Rubin observatory, JWST, ELT, GMT, TMT; in radio, LOFAR, ALMA, the VLBI networks including EHT, SKA and its precursors, as well as IceCube and the future KM3Net for the detection of high energy neutrinos; the Pierre Auger observatory for cosmic rays and the future LISA space interferometer for gravitational waves from massive black holes. All of these experiences obviously contribute to the rapid evolution of global alert networks for the exploration of the transient universe, which will be of great benefit to the study of blazars and AGNs.

X-ray and space gamma-ray monitoring data will naturally remain essential insofar as they generally probe the same populations of emitting particles as the VHE gamma rays and will make it possible to better test and improve existing models. The optical and infrared regions, where many phenomena are superimposed, some related to the physics of disks and host galaxies, and, for others, to that of jets, represent a complex spectral domain for understanding gamma-ray emitting AGNs, which may provide a missing link between accretion and ejection. Photometric and polarimetric monitoring in optics, such as the RoboPol project, will be important in the perspective of better understanding the correlations that sometimes appear between the optical properties and VHE of AGNs (Valverde *et al.* 2020). Obtaining spectroscopic redshifts also remains a key element for the study of BL Lacs (Pita *et al.* 2014; Goldoni *et al.* 2021). In addition, the developments of very high angular resolution aim to resolve all the structures of AGNs on the scale of a few parsecs, thus addressing the questions of interaction of the jets with the environment of the galactic core, and already make it possible to directly probe dust tori (Jaffe *et al.* 2004) and BLRs (Sturm *et al.* 2018), whose radiation governs VHE emissions by external inverse Compton scattering. As for the potential correlations of data collected at VHE with those of non-photonic detectors, the field is so new that any positive result would in itself constitute a remarkable breakthrough. We can think, for example, of the enigmatic origin of ultra-high energy cosmic rays, which could be associated with transient events such as AGN flares or with faint still unidentified phenomena.

For current research on AGNs at VHE, particularly promising links are also emerging in the radio domain and VLBI arrays on multiple subjects. For example, the advances of the mm VLBI with the EHT array will explore the formation zone of the jets in depth following the entry into service of new radio telescopes such as NOEMA on the Plateau de Bure, France. They should be able to follow the temporal evolution of the emitting plasma around the black hole, identify radio emissions from disks or jets and test the relevance of different models of jet formation, such as the Blandford–Znajek mechanism, which extracts rotational energy from the black hole and can produce ultra-relativistic jets (McKinney 2005), or the Blandford–Payne mechanism that launches the jet from the accretion disk, which EHT aims to distinguish (Jeter *et al.* 2020). The demonstration of concrete correlations between the evolution of the structure of this radio-emitting plasma and the outbursts or the fast variations in the VHE light curves would be a decisive step for the assembly of the current "multispectral puzzle" of AGNs. Another example is provided by SKA, which will be able, among other things, to characterize, by measurements of Faraday rotation and depolarization, the magnetized plasmas around the AGNs, in the jets and in the surrounding environments in which the emitting particles and the gamma rays propagate, which will detail the ambient conditions in which the VHE phenomena develop and evolve and will provide a more solid framework for their modeling than at present.

4.6.3. *The specific issue of a general reference model*

The study of AGNs is at a radical turn, with the confirmation of the presence of massive black holes at the center of their activity and the influx of various discoveries from different fields of astrophysics. It is now necessary to gather enough elements to identify the logic underlying these various manifestations and to succeed in developing and validating a "universal" scenario of multi-lambda and multi-messenger AGNs, on which to rely to access a detailed interpretation of all the data. Several paths can be taken for this and we illustrate here, in five questions, that of the VHE field.

Where is the gamma emission zone located? The question of the location of the variable emission zones at VHE should be clarified thanks to a better temporal and multispectral coverage of the flares and to the constitution of significant samples for BL Lacs, FSRQs and radio galaxies. Better data quality will make it possible to differentiate the different categories of flares and to constrain parameters more, which will justify the application of more detailed models, to ultimately confirm the existence of a dominant emission zone, or the need to multi-zone scenarios, depending on the case. Exploring the variability properties down to smaller and smaller timescales will drastically constrain the size of the emission zones. It will make it possible to test whether a lower limit for the variability is observed, which could then be related to the Schwarzschild radius of the central black hole (after correction by the Doppler factor) and to validate the idea of locating the origin of the most variable flares at the ultimate base of the jets, in the black hole magnetosphere. The detection of periodicities or quasi-periodicities could also be interpreted in this context. Finally, the localization of the VHE zone can be directly pinpointed if it is confirmed that certain flares are correlated over time with events spotted in high resolution at low energies, in particular, with changes in the VLBI structures near the radio core, or possibly more distant knots along the jets. Any quasi-periodicities could then be due to helical motion in the jets.

Observational data will *a priori* remain less restrictive in locating stationary emission zones. However, with precise VHE spectra revealing, for example, photon–photon absorption signatures and a better description of the ambient radiation backgrounds due to the disks, tori, BLRs and the jets themselves, taking into account the absorption effects will allow the VHE zone to be located in relation to the other components of AGNs. VHE astrometry of a few bright sources could also reveal whether the stationary emission is from the core of the AGN, from knots along the jets or possibly from hot spots on the FRII lobes, and the mapping could reveal the presence of extended emissions.

If it turned out to be impossible to meet the various constraints obtained on where the VHE emission zones are located, more unusual issues could be discussed, such as the existence of axion-like particles, which would allow radiation to come from the vicinity of the black hole or below the BLR by drastically reducing the $\gamma - \gamma$ absorption effects (see section 4.6.4).

What are VHE radiating particles and how are they accelerated? Leptonic and hadronic scenarios are distinguished by the precise forms of SEDs that they can produce, in particular, at energies of multi-TeV and beyond, and by their capacities to reproduce variable phenomena. Obtaining and modeling the temporal evolution of the detailed SEDs of AGN populations should therefore make it possible to discern which leptonic or hadronic scenario dominates the observed radiations and which are the main emission mechanisms at play at VHE, which may depend on the source type and, especially, its state of activity, quiet or eruptive. The detection of very fast variability coming from blazars favors leptonic scenarios, at least during very intense outbursts. However, the observation of new cases of spatial and temporal correlation between high energy cosmic neutrino events and AGN outbursts at VHE, following the first detection obtained in 2017, would also underline the importance of hadronic physics in jets. The observation of hard spectra in VHE, beyond the energies characteristic of gamma absorption by the diffuse extragalactic background, would be a signature of the production of ultra-high energy cosmic rays (UHECRs) by AGNs. Such a result would indicate that hadronic scenarios can be at work in AGNs. It would constitute one of the elements of AGN feedback mechanisms on the environment and would provide an answer as to the origin of these UHECRs.

Better knowledge of the nature of the radiating particles and the location of the emission zones would make it possible to identify their overall kinematics, their magnetic field and their environment. Combined with the study of the temporal profiles of flares directly linked to the processes of acceleration and cooling of particles, this should make it possible to distinguish the acceleration mechanisms that are actually at work – centrifugal or electrical forces in the magnetosphere of black holes, Fermi processes in various types of shocks and turbulence along the jets or magnetic reconnection effects – and in turn increase knowledge of the physical parameters describing the regions involved.

What are the origins of the variability and can we predict AGN flares at VHE? At very large temporal scales, the spatial extension of the jets indicates that a relatively stable activity cycle can last over 10^7 years and beyond for an AGN. The

VHE detection of halos of electron–positron pairs around AGNs could make it possible to independently estimate the duration of the periods of VHE activity, which would be interesting to compare with the long-term evolution of the active nucleus and its galactic environment. At shorter scales that are directly accessible with our measurements, some AGNs show periodicities over a few years or tens of years. This is the case with the BL Lac OJ 287, a probable massive black hole binary system that exhibits predictable multi-lambda flares approximately every 12 years around the periastral phase, which will provide unique opportunities to identifying the exact nature of the triggering of VHE flares once the overall geometry of the components (disks, tori and jets) of the two cores has been well constrained. Quasi-periodicities or annual variations can also be induced by a precession of the jet or instabilities affecting it on a large scale or by evolutions or instabilities in the accretion disk. The continuation and improvement of regular multispectral monitoring should make it possible to identify the different cases.

The growing interest in the transient universe and the development of multi-lambda and multi-messenger alert networks, as well as the use of data-driven detection techniques, deep learning and anomaly detection, allowing the possible anticipation of the onset of flares or outbursts, will be particularly fruitful for studies at shorter time scales (Sadeh 2020). Indeed, variations of a few months to a few minutes are directly related to changes in the properties of the gamma emission zone (size, density, speed and acceleration, energy, magnetic field). They will benefit from much better observational coverage, which will make it possible to quantify the evolutions and to model them according to the location of the VHE zone and the particle acceleration processes. The most intense episodes of flares will survey the fastest VHE variabilities from a few hours to a few minutes or even seconds, below the typical transit time $\sim 3h\ (M_{BH}/10^9 M_\odot)$ of light over a Schwarzschild radius and their possible periodicities or quasi-periodicities. This should make it possible to compare the VHE emission scenarios from the magnetosphere or nuclear jets, constrain Doppler factors and explore different sources of fast variability such as intermittent electric fields in gaps, magnetic reconnections, shocks or turbulence.

Different types of transient phenomena associated with AGNs and black holes could also be identified at VHE as tidal disruption events (TDE[20]), which offer the possibility of observing the most minimalist scenario of the AGN activity cycle and the co-evolution of the two accretion-ejection phases live. The detection of possible evaporation signals from primordial black holes would provide information on the properties of the early universe and the existence of low-mass black holes in our

20. A tidal disruption event is when a star or a cloud is disturbed by a tidal effect near the massive black hole of a galactic nucleus and induces an accretion phenomenon, sometimes coupled with the production of a jet, thus achieving a kind of transient mini blazar model.

cosmos. The discovery of electromagnetic counterparts to the events observed by gravitational wave detectors, and LISA, in particular, over the next decade, which will detect waves from massive black holes, should make it possible to deepen the study of gamma phenomena in the vicinity of black holes, whose intrinsic mass and spin properties, as well as the stellar environment and the capture rates of compact stars, can be determined and will set more precise boundary conditions at the base of the jets.

What are the AGNs emitting at VHE? Dedicated observations of already listed AGNs and sky surveys will indicate whether other types of AGN are emitting at VHE beyond blazars and radio galaxies, depending on redshifts and activity levels. The sequence of blazars could be consolidated, or, on the contrary, invalidated by the detection of new types of sources, which will also impact the models which aim to interpret it. A fundamental question will be to verify whether the relativistic jet phenomenon always underlies the VHE emission, and if all VHE gamma-loud AGNs are strong radio-loud sources or if some radio-quiet AGNs are also detected in the VHE gamma-ray domain, as observed in pulsar physics. The case of our galactic center, which can be considered as a low luminosity AGN, could already offer such an example, if the VHE source detected in its direction is well associated with Sgr A *, which will be verified. It will also be necessary to examine to what extent the relativistic Doppler enhancement is necessary for the detection of an AGN at VHE, or if the sensitivities achieved with the new generation of instruments allow the detection of AGN winds, or non-relativistic jets, or jets that are not oriented towards the Earth, and determine the conditions, so that a radio-loud AGN is also a VHE gamma-loud source (whether it is detected or not depending on instrument performance).

The analysis of AGN populations should make it possible to identify whether the "basic building block" of the VHE emission is generally the same for all types of sources, in particular, for blazars and radio galaxies, and to test whether the VHE properties of AGNs appear to be consistent with the classifications adopted so far, to what extent they support the AGN unification theories and how they may account for the diffuse extragalactic background detected at VHE.

What is the origin of the dichotomy between radio-loud and radio-quiet AGNs? This question is critical for the understanding of AGNs, their unification and their lifecycle and asks to identify the respective role of massive black holes and their cosmic environment in the development of jets. Obtaining large samples of gamma-loud AGNs should shed new light on this subject and show whether a new dichotomy between gamma-loud and gamma-quiet AGNs should be considered and to what extent it overlaps (or not) with the radio dichotomy. It is likely that initially the sensitivity performance limits this study to populations of radio-loud AGNs, and to blazars in particular. Beyond the distance and the total luminosity, it will be possible to study whether there are other decisive parameters that determine whether a blazar is a VHE emitter or not, and to verify whether the VHE emission and fast varying flares in particular are a marker of the black hole's spin and of very specific properties of its magnetosphere, if they can appear in the absence of very bright radio jets, or if they are a systematic consequence inherent in the presence of relativistic radio jets. If advances in VLBI with projects such as the EHT succeed in identifying the nature of the jets formed at the heart of AGNs, it will be possible to test what the production mechanisms of the jets that induce VHE emissions are and whether these extreme radiations require the extraction of an ultra-relativistic jet or not by the Blandford–Znajek mechanism. Better knowledge of the acceleration and radiation mechanisms of particles at VHE should also provide new constraints on the magnetic field and help us to test whether its influence is decisive in the phenomenon of AGN jet production or not.

In addition, the VHE emission from the base of the jets can be used to deduce the power in the nuclear jet. This energy balance can be compared, on the one hand, with the accretion rate to test the existence of a direct link between accretion and ejection according to different types of AGN and accretion regimes or not, and, on the other hand, to the energy balance of jets observed on a large scale to constrain various properties of jets such as their radial structure and their content in relativistic and thermal plasmas. This should help us to better understand how the jets interact with their environment and how they are inserted in the cycle of the large-scale inflows and outflows of matter at the galactic level which influence the properties and evolution of AGN cores and host galaxies.

4.6.4. *AGNs, partners of our technological societies*

AGNs, the new recognized borders of our cosmos at the edge of the non-return boundaries of the horizon of massive black holes, are the most distant and brightest perennial sources in our universe, commonly used by astronomers as light beacons to probe their lines of sight and explore intergalactic space–time. In the VHE field, we have seen, for example, how the presence of the extragalactic diffuse background could absorb gamma-ray photons from AGNs by photon–photon interaction and the creation of electron–positron pairs. This absorption effect depends on the energy of the gamma photons and the redshift of the source and is part of the observed spectra. So far, it seems compatible with direct measurements of the infrared background and counts of galaxies at the origin of this background, but is rather in favor of the minimum value indicated by the counts and of a universe as transparent as possible to VHE photons, according to our current knowledge. The observation of AGN populations at different redshifts with a new-generation instrument such as CTA will offer new perspectives on the evolution of galaxies by making it possible to measure the diffuse extragalactic background radiation spectrum from the mid-ultraviolet to the far infrared at redshift $z \sim 0$ with an accuracy of 10% on the total flux and to characterize its evolution up to $z \sim 1$ (Mazin *et al.* 2013; Zech *et al.* 2019). It will also provide a new independent method for the determination of the Hubble constant, the value of which is involved in the overall description of this absorption phenomenon.

The existence of a very weak but non-zero intergalactic magnetic field can also be probed down to values as low as 10^{-24} G by researching substantial time delays in arrival times of primary photons emitted by variable blazars beyond TeV and secondary photons up to a few GeV, created by the effect of pair creation cascades, deflected by the magnetic field and therefore delayed during their propagation towards the Earth. The detection of diffuse halos due to electron–positron pairs around AGNs, achievable with CTA, would, conversely, be the signature of a higher intergalactic magnetic field that is greater than 10^{-16} G (Sol *et al.* 2013; Zech *et al.* 2019). Such results would provide new clues on the still-unknown origin of the cosmic magnetic field, which may have appeared in the primordial universe from electromagnetic quantum fluctuations amplified during inflation, or post-inflation during fundamental force decoupling transitions, or have been generated during the formation of large structures by dynamo amplification of very localized magnetic seeds due to young AGNs.

Following a fairly similar observational approach, the future data expected at VHE on AGN populations could greatly improve the current limits established on the effects of fundamental physics (Doro *et al.* 2013; Zech *et al.* 2019). This is how the search for small time delays as a function of photon energy in the arrival of VHE signals from blazars in an eruptive state at different redshifts makes it possible to search for a possible dispersion of the speed of light propagation in a vacuum, which would be a Lorentz invariance violation signature predicted by certain models of quantum gravity at the Planck scale. Another clue to a possible Lorentz invariance violation could also be obtained if the universe turns out to be more transparent than expected at VHE. Indeed, a Lorentz invariance violation can modify the interaction threshold of the pair creation process and reduce the opacity of the universe at VHE due to the diffuse extragalactic background, which would push the gamma horizon towards the highest redshifts. The existence of axion-like particles (ALP) is also likely to greatly reduce the absorption phenomenon by the diffuse extragalactic background, due to oscillations or photon–ALP conversions that could occur along the lines of sight in the presence of non-zero magnetic fields. Some signs of this phenomenon have been reported without being confirmed at this time. Such a demonstration of the presence of axion-like particles would have multiple consequences, in particular, for dark matter research, where they could be an important component.

Furthermore, due to their extreme properties, AGNs and their jets are of direct interest to our technological societies beyond astronomy, cosmology and physics. Previously stars, which played a fundamental role in maritime navigation, today it is AGNs that are used as cosmic reference marks for the constitution of the celestial reference frames that are essential for space navigation and the precise monitoring of various phenomena, particularly geophysical ones (Charlot *et al.* 2019; Liu *et al.* 2020). The numerous astrometric data acquired for the development of reference frames contribute to increasing statistics on AGN populations, their eruptive phenomena and the evolution of their jet structure, and should participate in the development of a renewed multispectral vision of AGN cores (Petrov *et al.* 2019). On the other hand, any advance in the understanding of the physics of jets and AGNs, and the resulting refined descriptions of their structure and temporal evolution, should ultimately make it possible to further improve the precision of our best current celestial reference frames.

4.7. References

Aab, A., Abreu, P., Aglietta, M. *et al.* (2018). Indication of anisotropy in arrival directions of ultra-high-energy cosmic rays through comparison to the flux pattern of extragalactic gamma-ray sources. *ApJL*, 853, L29.

Aartsen, M.G., Ackermann, M., Adams, J. *et al.* (2018a). Multi-messenger observations of a flaring blazar coincident with high-energy neutrino IceCube-170922A. *Science*, 361, eaat1378.

Aartsen, M.G., Ackermann, M., Adams, J. *et al.* (2018b). Neutrino emission from the direction of the blazar TXS 0506+056 prior to the IceCube-170922A alert. *Science*, 361, 147.

Abdalla, H., Abramowski, A., Aharonian, F. *et al.* (2017). Characterizing the gamma-ray long-term variability of PKS2155-304 with H.E.S.S. and Fermi-LAT. *A&A*, 598, A39.

Abdalla, H., Abramowski, A., Aharonian, F. *et al.* (2018). The gamma-ray spectrum of the core of Centaurus A as observed with H.E.S.S. and Fermi-LAT. *A&A*, 619, id.A71.

Abdalla, H., Adam, R., Aharonian, F. *et al.* (2019). Constraints on the emission region of 3C279 during strong flares in 2014 and 2015 through VHE gamma-ray observations with H.E.S.S. *A&A*, 627, id.A159.

Abdalla, H., Adam, R., Aharonian, F. *et al.* (2020a). Resolving acceleration to very high energies along the jet of Centaurus A. *Nature*, 582, 7812, 356.

Abdalla, H., Adam, R., Aharonian, F. *et al.* H.E.S.S. (2020b). Detection of VHE gamma-ray emission from the quasar PKS 0736+017. *A&A*, 633, A162.

Abdalla, H., Abe, H., Acero, F. *et al.* (2021). Sensitivity of the Cherenkov telescope array for probing cosmology and fundamental physics with gamma-ray propagation. *Journal of Cosmology and Astroparticle Physics*, 2, id.048.

Abdo, A.A., Ackermann, M., Ajello, M. *et al.* (2010). Fermi gamma-ray imaging of a radio galaxy. *Science*, 328, 725.

Abdo, A.A., Ackermann, M., Ajello, M. *et al.* (2011). FERMI large area telescope observations of Markarian 421: The missing piece of its spectral energy distribution. *ApJ*, 736, 131.

Abeysekara, A.U., Archer, A., Benbow, W. *et al.* (2019). Measurement of the extragalactic background light spectral energy distribution with VERITAS. *ApJ*, 885, 150.

Abramowski, A., Acero, F., Aharonian, F. *et al.* (2010). VHE gamma-ray emission of PKS2155-304: Spectral and temporal variability. *A&A*, 520, A83.

Abramowski, A., Acero, F., Aharonian, F. *et al.* (2012a). A multiwavelength view of the flaring state of PKS 2155-304 in 2006. *A&A*, 539, A149.

Abramowski, A., Acero, F., Aharonian, F. *et al.* (2012b). The 2010 VHE gamma-ray flare and 10 years of multiwavelength observations of M87. *ApJ*, 746, id.151.

Acciari, V.A., Aliu, E., Arlen, T. *et al.* (2009). Radio imaging of the VHE gamma-ray emission region in the central engine of a radio galaxy. *Science*, 325, 444.

Acciari, V.A., Ansoldi, S., Antonelli, L.A. *et al.* (2020). Broadband characterisation of the very intense TeV flares of the blazar 1ES 1959+650 in 2016. *A&A*, 638, id.A14.

Acharya, B.S., Actis, M., Aghajani, T. *et al.* (2013). Introducing the CTA concept. *Astroparticle Physics*, 43, 3.

Ackermann, M., Anantua, R., Asano, K. *et al.* (2016). Minute-timescale > 100 MeV gamma-ray variability during the giant outburst of quasar 3C 279 observed by Fermi-LAT in 2015 June. *ApJL*, 824, id.L20.

Aharonian, F. (2000). TeV gamma rays from BL Lac objects due to synchrotron radiation of extremely high energy protons. *New Astronomy*, 5, 377.

Aharonian, F. (2013). Gamma rays at very high energies. In *Astrophysics at Very High Energies: Saas-Fee Advanced Courses 40*, Aharonian, F., Bergström, L., Dermer, C. (eds). Springer.

Aharonian, F., Akhperjanian, A.G., Bazer-Bachi, A.R. *et al.* (2007). An exceptional very high energy gamma-ray flare of PKS 2155-304. *ApJ*, 664, 71.

Aharonian, F., Akhperjanian, A.G., Anton, G. *et al.* (2009). Simultaneous multiwavelength observations of the second exceptional gamma-ray flare of PKS 2155-304 in July 2006. *A&A*, 502, 749.

Aharonian, F., Bergström, L., Dermer, C. (2013). *Astrophysics at Very High Energies: Saas-Fee Advanced Courses 40*. Springer.

Ahnen, M.L., Ansoldi, S., Antonelli, L.A. *et al.* (2018). Extreme HBL behavior of Markarian 501 during 2012. *A&A*, 620, A181.

Akiyama, K., Lu, R., Fish, V.L. *et al.* (2015). 230 GHz VLBI observations of M87: Event-horizon-scale structure during an enhanced very-high-energy gamma-ray state in 2012. *ApJ*, 807, 150.

Akiyama, K., Alberdi, A., Alef, W. *et al.* (2019). The event horizon telescope collaboration. First M87 EHT results I. The shadow of the supermassive black hole. *ApJ L.*, 875, L1.

Aleksic, J., Ansoldi, S., Antonelli, L.A. *et al.* (2014). MAGIC observations and multifrequency properties of the flat spectrum radio quasar 3C 279 in 2011. *A&A*, 567, A41.

Aloisio, R., Blasi, P., De Mitri, I., Petrera, S. (2018). Selected topics in cosmic ray physics. In *Multiple Messengers and Challenges in Astroparticle Physics*, Aloisio, R., Coccia, E., Vissani, F. (eds). Springer.

Anchordoqui, L.A. (2019). Ultra-high-energy cosmic rays. *Physics Reports*, 801, 1.

Asano, K. and Hayashida, M. (2020). Synchrotron gamma-ray emission model of the giant outburst of quasar 3C 279 in 2015 June: Fast reconnection or stochastic acceleration with electromagnetic cascade? *ApJ*, 890, 56.

Balogh, A. and Treumann, R.A. (2013). *Physics of Collisionless Shocks – Space Plasma Shock Waves*. ISSI Scientific Report Series, 12. Spinger.

Begelman, M.C. and Kirk, J.G. (1990). Shock-drift particle acceleration in superluminal shocks – A model for hot spots in extragalactic radio sources. *ApJ*, 353, 66.

Bell, A.R. (1978). The acceleration of cosmic rays in shock fronts – I. *MNRAS*, 182, 147.

Berezinsky, V. and Kalashev, O. (2016). High-energy electromagnetic cascades in extragalactic space: Physics and features. *Physical Review D*, 94, 023007.

Beskin, V.S., Istomin, Ya, N., Pariev, V.I. (1992). Filling the magnetosphere of a supermassive black hole with plasma. *Soviet Astronomy*, 36, 64.

Blandford, R. and Eichler, D. (1987). Particle acceleration at astrophysical shocks: A theory of cosmic ray origin. *Physics Reports*, 154, 1.

Blandford, R., Meier, D., Readhead, A. (2019). Relativistic jets from active galactic nuclei. *ARAA*, 57, 467.

Blumenthal, G.R. and Gould, R.J. (1970). Bremsstrahlung, synchrotron radiation, and Compton scattering of high-energy electrons traversing dilute gases. *Reviews of Modern Physics*, 42, 237.

Böttcher, M. and Baring, M.G. (2019). Multi-wavelength variability signatures of relativistic shocks in blazar jets. *ApJ*, 887, 133.

Böttcher, M. and Reimer, A. (2012). Radiation processes. In *Relativistic Jets from Active Galactic Nuclei*, Böttcher, M., Harris, D.E., Krawczynski, H. (eds). Wiley-VCH.

Böttcher, M., Harris, D.E., Krawczynski, H. (eds) (2012). *Relativistic Jets from Active Galactic Nuclei*. Wiley-VCH.

Broderick, A.E. and Tchekhovskoy, A. (2015). Horizon-scale lepton acceleration in jets: Explaining the compact radio emission in M 87. *ApJ*, 809, id.97.

Cerutti, B., Wernr, G.R., Uzdensky, D.A., Begelman, M.C. (2012). Beaming and rapid variability of high-energy radiation from relativistic pair plasma reconnection. *ApJL*, 754, id.L33.

Cerruti, M., Zech, A., Boisson, C., Emery, G., Inoue, S., Lenain, J.P. (2019). Leptohadronic single-zone models for the electromagnetic and neutrino of TXS 0506+056. *MNRAS*, 483, L12.

Charlot, P., Gabuzda, D., Sol, H. *et al.* (2006). Simultaneous radio-interferometric and high-energy TeV observations of the gamma-ray blazar Mkn 421. *A&A*, 457, 455.

Charlot, P., Jacobs, C.S., Gordon, D. *et al.* (2020). The third realization of the international celestial reference frame by very long baseline interferometry. *A&A*, 644, id.A159.

Cohen, M.H., Cannon, W., Purcell, G.H. *et al.* (1971). The small-scale structure of radio galaxies and quasi-stellar sources at 3.8 centimeters. *ApJ*, 170, 207.

Crinquand, B., Cerutti, B., Philippov, A., Parfrey, K., Dubus, G. (2020). Multidimensional simulations of ergospheric pair discharges around black holes. *Physical Review Letters*, 124, 145101.

Crinquand, B., Cerutti, B., Dubus, G., Parfrey, K., Philippov, A. (2021). Synthetic gamma-ray lightcurves of Kerr black hole magnetospheric activity from particle-in-cell simulations. arXiv:2012.09733. *A&A*, 650, id.A163.

Davis, S.W. and Tchekhovskoy, A. (2020). Magnetohydrodynamics simulations in active galactic nucleus disks and jets. *ARAA*, 58, 407.

Degrange, B. and Fontaine, G. (2015). Introduction to high-energy gamma-ray astronomy. *Comptes rendus physique*, 16, 587–599.

Dermer, C.D. and Menon, G. (2009). *High Energy Radiation from Black Holes*. Princeton University Press.

Diltz, C., Böttcher, M., Fossati, G. (2015). Time dependent hadronic modeling of flat spectrum radio quasars. *ApJ*, 802, 133.

Dominguez, A., Primack, J.R., Rosario, D.J. *et al.* (2011). Extragalactic background light inferred from AEGIS galaxy SED-type fractions. *MNRAS*, 410, 2556.

Doro, M., Conrad, J., Emmanoulopoulos, D. *et al.* (2013). Dark matter and fundamental physics with the CTA. *Astroparticle Physics*, 43, 189.

Eilek, J.A. and Hughes, P.A. (1991). Particle acceleration and magnetic field evolution. In *Beams and Jets in Astrophysics*, Hughes, P. (ed.). Cambridge University Press.

Ellison, D.C., Jones, F.C., Reynolds, S.P. (1990). First-order Fermi particle acceleration by relativistic shocks. *ApJ*, 360, 702.

Feenberg, E. and Primakoff, H. (1948). Interaction of cosmic ray primaries with sunlight and starlight. *Physical Review*, 73, 449–469.

Fegan, D. (2019). *Cherenkov Reflections – Gamma-Ray Imaging and the Evolution of TeV Astronomy*. World Scientific.

Fermi, E. (1949). On the origin of the cosmic radiation. *Physical Review*, 75, 1169.

Fermi, E. (1954). Galactic magnetic fields and the origin of cosmic radiation. *ApJ*, 119, 1.

Fossatti, G., Maraschi, L., Celotti, A., Comastri, A., Ghisellini, G. (1998). A unified view of the spectral energy distribution of blazars. *MNRAS*, 299, 433.

Gangadhara, R.T. and Lesch, H. (1997). On the nonthermal emission in active galactic nuclei. *A&A*, 323, L45.

de Gasperin, F., Vink, J., McKean, J.P. *et al.* (2020). Cassiopeia A, Cygnus A, Taurus A, and Virgo A at ultra-low radio frequencies. *A&A*, 635, A150.

Ghisellini, G., Maraschi, L., Treves, A. (1985). Inhomogeneous synchrotron-self-Compton models and the problem of relativistic beaming of BL Lac objects. *A&A*, 146, 204.

Ghisellini, G., Righi, C., Costamante, L., Tavecchio, F. (2017). The Fermi blazar sequence. *MNRAS*, 469, 255.

Giannios, D., Uzdensky, D.A., Begelman, M.C. (2009). Fast TeV variability in blazars: Jets in a jet. *MNRAS*, 395, L29.

Ginzburg, V.L. and Syrovatskii, S.I. (1964). *The Origin of Cosmic Rays*, H. Massey (trans.), D. ter Haar (ed.). Pergamon Press.

Ginzburg, V.L. and Syrovatskii, S.I. (1965). Cosmic magnetobremsstrahlung (synchrotron radiation). *ARAA*, 3, 287.

Goldoni, P., Pita, S., Boisson, W. *et al.* (2021). Optical spectroscopy of blazars for the Cherenkov telescope array. *A&A*, 650, id.A106.

Gonzalez, W. and Parker, E. (eds) (2016). *Magnetic Reconnection – Concepts and Applications*. Astrophysics and Space Science Library 427, Springer.

Hardcastle, M.J., Lenc, E., Birkinshaw, M. *et al.* (2016). Deep Chandra observations of Pictor A. *MNRAS*, 455, 3526.

Hardcastle, M.J., Williams, W.L., Best, P.N. *et al.* (2019). Radio-loud AGN in the first LoTSS data release – The lifetime and environmental impact of jet-driven sources. *A&A*, 622, id.A12.

Hayashida, M., Nalewajko, K., Madejski, G.M. *et al.* (2015). Rapid variability of blazar 3C 279 during flaring states in 2013–2014 with joint Fermi-LAT, NuSTAR, SWIFT, and ground-based multi-wavelength observations. *ApJ*, 807, 79.

Hervet, O., Boisson, C., Sol, H. (2015). Linking radio and gamma-ray emission in AP Librae. *A&A*, 578, 69.

Hervet, O., Boisson, C., Sol, H. (2016). An innovative blazar classification based on radio jet kinematics. *A&A*, 592, 22.

Hillas, A.M. (1984). The origin of ultra-high-energy cosmic rays. *ARAA*, 22, 425.

Hillas, A.M. (2013). Evolution of ground-based gamma-ray astronomy from the early days to the Cherenkov telescope arrays. *Astroparticle Physics*, 43, 19–43.

Hinton, J.A. and Hofmann, W. (2009). Teraelectronvolt astronomy. *ARAA*, 47, 523–565.

Hirotani, K. and Okamoto, I. (1998). Pair plasma production in a force-free magnetosphere around a supermassive black hole. *ApJ*, 497, 563.

Inoue, S. and Takahara, F. (1996). Electron acceleration and gamma-ray emission from blazars. *ApJ*, 463, 555.

Inoue, S., Ribo, M., Bernardini, E. *et al.* (2019). Key science project: Transients. In *Science with the Cherenkov Telescope Array* , The CTA Consortium (ed.). World Scientific.

Istomin, Y. and Sol, H. (2009). Acceleration of particles in the vicinity of a massive black hole. *Astrophysics and Space Science*, 321, 57.

Jackson, J.D. (1975). *Classical Electrodynamics*, 2nd edition. John Wiley & Sons.

Jaffe, W., Meisenheimer, K., Röttgering, H. *et al.* (2004). The central dusty torus in the active nucleus of NGC 1068. *Nature*, 429, 47.

Jennison, R.C. and Das Gupta, M.K. (1953). Fine structure of the extra-terrestrial radio source Cygnus I. *Nature*, 172, 996.

Jeter, B., Broderick, A.E., Gold, R. (2020). Differentiating disc and black hole driven jets with EHT images of variability in M87. *MNRAS*. 493, 5606.

Jimenez-Rosales, A., Dexter, J., Widmann, F. *et al.* (2020). (GRAVITY collaboration), Dynamically important magnetic fields near the event horizon of Sgr A*. *A&A*, 643, A56.

Jones, F.C. (1968). Calculated spectrum of inverse Compton-scattered photons. *Physical Review*, 167, 1159.

Jorstad, S. and Marscher, A. (2016). The VLBA-BU-BLAZAR multi-wavelength monitoring program. *Galaxies*, 4, 47.

Kardashev, N.S. (1962). Nonstationariness of spectra of young sources of nonthermal radio emission. *Soviet Astronomy*, 6, 317.

Kelner, S.R., Aharonian, F.A., Khangulyan, D. (2013). On the jitter radiation. *ApJ*, 774, 61.

Kim, J.-Y., Krichbaum, T.P., Lu, R.-S. (2018). The limb-brightened jet of M87 down to the 7 Schwarzschild radii scale. *A&A*, 616, A188.

Kim, J.-Y., Krichbaum, T.P., Broderick, A.E. *et al.* (2020a). Event horizon telescope imaging of the archetypal blazar 3C 279 at an extreme 20 microarcsecond resolution. *A&A*, 640, id.A69.

Kim, D.-W., Trippe, S., Kravchenko, E.V. (2020b). Investigating the connection between gamma-ray activity and the relativistic jet in 3C 273 during 2015–2019. *A&A*, 636, A62.

Kirk, J.G. (1994). Particle acceleration. *Plasma Astrophysics. Saas-Fee Advanced Course 24*, Kirk, J.G., Melrose, D.B., Priest, E.R. (eds). Springer.

Königl, A. (1981). Relativistic jets as X-ray and gamma-ray sources. *ApJ*, 243, 700.

Krawczynski, H., Hughes, S.B., Horan, D. *et al.* (2004). Multiwavelength observations of strong flares from the TeV blazar 1ES 1959+650. *ApJ*, 601, 151.

Larionov, V.M., Jorstad, S.G., Marscher, A.P. *et al.* (2020). Multiwavelength behaviour of the blazar 3C279: Decade-long study from gamma-ray to radio. *MNRAS*, 492, 3829.

Lemoine, M., Pelletier, G., Revenu, B. (2006). On the efficiency of Fermi acceleration at relativistic shocks. *ApJ*, 645, L129.

Lemoine, M., Pelletier, G., Vanthieghem, A., Gremillet, L. (2019). Physics of relativistic collisionless shocks III: The suprathermal particles. *Physical Review*, 100, id.033210.

Leverett Jr., D. (1956). Modified fermi mechanism for the acceleration of cosmic rays. *Physical Review*, 101, 351.

Levinson, A. and Blandford, R. (1995). Pair cascades in extragalactic jets II. The beamed X-ray spectrum. *ApJ*, 449, 86.

Lewis, T.R., Finke, J.D., Becker, P.A. (2019). Electron acceleration in blazars: Application to the 3C 279 flare on 2013 December 20. *ApJ*, 884, 116.

Lister, M.L., Aller, M.F., Aller, H.D. *et al.* (2016). MOJAVE.XIII. Parsec-scale AGN jet kinematics analysis based on 19 years of VLBA observations at 15 GHz. *The Astronomical Journal*, 152, 12.

Liu, N., Lambert, S., Zhu, Z. *et al.* (2020). Systematics and accuracy of VLBI astrometry: A comparison with Gaia data release 2. *A&A*, 634, A28.

Longair, M.S. (1994). *High Energy Astrophysics*, 2nd edition. Cambridge University Press.

Mannheim, K. (1993). The proton blazar. *A&A*, 269, 67.

Marcowith, A., Ferrand, G., Grech, M., Meliani, Z., Plotnikov, I., Walder, R. (2020). Multi-scale simulations of particle acceleration in astrophysical systems. *Living Reviews in Computational Astrophysics*, 6, id.1.

Marscher, A.P. (1999). The compact jets of TeV blazars. *Astroparticle Physics*, 11, 19.

Marscher, A.P., Jorstad, S.G., D'Arcangelo, F.D. *et al.* (2008). The inner jet of an active galactic nucleus as revealed by a radio-to-gamma-ray outburst. *Nature*, 452, 966.

Mastichiadis, A., Florou, I., Kefala, E., Boula, S., Petropoulou, M. (2020). A roadmap to hadronic supercriticalities: A comprehensive study of the parameter space for high-energy astrophysical sources. *MNRAS*, 495, 2458.

Mazin, D., Raue, M., Behera, B. *et al.* (2013). Potential of EBL and cosmology studies with the CTA. *Astroparticle Physics*, 43, 241.

Mazin, D., Gerard, L., Ward, J.E. *et al.* (2019). Key science project: Extragalactic survey. In *Science with the Cherenkov Telescope Array*, The CTA Consortium (ed.). World Scientific.

Medvedev, M.V. (2000). Theory of "jitter" radiation from small-scale random magnetic fields and prompt emission from gamma-ray burst shocks. *ApJ*, 540, 704.

Melzani, M., Walder, R., Folini, D., Winisdoerffer, C., Favre, J.M. (2014a). Relativistic magnetic reconnection in collisionless ion-electron plasmas explored with particle-in-cell simulations. *A&A*, 570, A111.

Melzani, M., Walder, R., Folini, D., Winisdoerffer, C., Favre, J.M. (2014b). The energetics of relativistic magnetic reconnection: Ion-electron repartition and particle distribution hardness. *A&A*, 570, A112.

Meyer, M., Scargle, J.D., Blandford, R.D. (2019). Characterizing the gamma-ray variability of the brightest FSRQ observed with the Fermi LAT. *ApJ*, 877, 39.

McKinney, J.C. (2005). Total and jet Blandford-Znajek power in the presence of an accretion disk. *ApJ*, 630, L5.

Mooney, S., Quinn, J., Callingham, J.R. *et al.* (2019). Blazars in the LOFAR Two-Metre Sky Survey first data release. *A&A*, 622, id.A14.

Morris, P.J., Potter, W.J., Cotter, G. (2019). The feasibility of magnetic reconnection powered blazar flares from synchrotron self-Compton emission. *MNRAS*, 486, 1548.

Morrison, P. (1958). On gamma-ray astronomy. *Il Nuovo Cimento*, 7, 858–865.

Mücke, A. and Protheroe, R.J. (2001). A proton synchrotron blazar model for flaring in Markarian 501. *Astroparticle Physics*, 15, 121.

Müller, C., Kadler, M., Ojha, R. *et al.* (2011). Dual-frequency VLBI study of Centaurus A on sup-parsec scales – The highest-resolution view of an extragalactic jet. *A&A*, 530, L11.

Nalewajko, K., Giannos, D., Begelman, M.C., Uzdensky, D.A., Sikora, M. (2011). Radiative properties of reconnection-powered minijets in blazars. *MNRAS*, 413, 333.

Nalewajko, K., Sikora, M., Rozanska, A. (2020). Orientation of the crescent image of M87*. *A&A*, 634, id.A38.

de Naurois, M. (2015). Mazin, ground-based detectors in very high energy gamma-ray astronomy. *Comptes rendus physique*, 16(6–7), 610–627.

Neronov, A. and Aharonian, F.A. (2007). Production of TeV gamma radiation in the vicinity of the supermassive black hole in the giant radio galaxy M 87. *ApJ*, 671, 85.

Nishikawa, K.-I., Hardee, P., Richardson, G. *et al.* (2003). Particle acceleration in relativistic jets due to Weibel instability. *ApJ*, 595, 555.

Nishikawa, K.-N., Dutan, I., Köhn, C., Mizuno, Y. (2021). PIC methods in astrophysics: PIC simulations of relativistic jets, submitted to living reviews in computational astrophysics. arXiv:2008.02105v2.

Ohsawa, Y. (2014). Ultrarelativistic particle acceleration in collisionless shock waves. *Physics Reports*, 536, 147.

Osmanov, Z. (2010). Is very high energy emission from the BL Lac 1ES 0806+524 centrifugally driven? *New Astronomy*, 15, 351.

Osmanov, Z., Rogava, A., Bodo, G. (2007). On the efficiency of particle acceleration by rotating magnetospheres in AGN. *A&A*, 470, 395.

Owen, F., Biretta, J., Eilek, J. (1999). M87 as a radio galaxy, press release. *AAS Meeting*.

Parfrey, K., Philippov, A., Cerruti, B. (2019). First-principles plasma simulations of black hole jet launching. *Physical Review Letters*, 122, 035101.

Parizot, E. (2003). Accélération des particules : les mécanismes de Fermi. *Physique et Astrophysique du Rayonnement Cosmique, Ecole CNRS de Goutelas XXVI*, Parizot *et al.* (eds).

Park, N. (2018). Status of ground based gamma-ray observations. *35th International Cosmic Ray Conference*, PoS ICRC2017, 116.

Paliya, V., Diltz, C., Böttcher, M., Stalin, C.S., Buckley, D. (2016). A hard gamma-ray flare from 3C 279 in 2013 December. *ApJ*, 817, 61.

Petrov, L., Kovalev, Y.Y., Plavin, A.V. (2019). A quantitative analysis of systematic differences in positions and proper motions of Gaia DR2 with respect to VLBI. *MNRAS*, 482, 3023.

Pita, S., Goldoni, P., Boisson, C. *et al.* (2014). Spectroscopy of high energy BL lacertae objects with X-shooter on the VLT. *A&A*, 565, id.A12.

Pohl, M., Hoshino, M., Niemiec, J. (2020). PIC simulation methods for cosmic radiation and plasma instabilities. *Progress in Particle and Nuclear Physics*, 111, id.103751.

Prandini, E. (2017). Highlights from TeV extragalactic sources, review talk at the conference "VHE Phenomena in the Universe". *Moriond*, arXiv:1706.01670.

Protheroe, R.J., Bhat, C.L., Fleury, P., Lorentz, E., Teshima, M., Weekes, T.C. (1997). Very high energy gamma rays from Markarian 501. *25th International Cosmic Ray Conference (ICRC)*, Durban, arXiv:astro-ph/9710118.

Punch, M., Akerlof, C.W., Cawley, M.F. *et al.* (1992). Detection of TeV photons from the active galaxy Markarian 421. *Nature*, 358, 477.

Quinn, J., Akerlof, C.W., Biller, S. *et al.* (1996). Detection of gamma rays with E > 300 GeV from Markarian 501. *ApJ*, 456, L83.

Rani, B., Jorstad, S.G., Marscher, A.P. *et al.* (2018). Exploring the connection between parsec-scale jet activity and broadband outbursts in 3C 279. *ApJ*, 858, 80.

Rieger, F.M. (2019). An introduction to particle acceleration in shearing flows. *Galaxies*, 7, 3.

Rieger, F.M. and Aharonian, F.A. (2008). Variable VHE gamma-ray emission from non-blazar AGNs. *A&A*, 479, L5.

Rieger, F.M. and Duffy, P. (2006). A microscopic analysis of shear acceleration. *ApJ*, 652, 1044.

Rieger, F.M. and Duffy, P. (2021). Turbulence and particle acceleration in shearing flows. *ApJL*, 907, id.L2.

Rieger, F.M. and Levinson, A. (2018). Radio galaxies at VHE energies. *Galaxies*, 6, 116.

Romano, P., Böttcher, M., Foschini, L. *et al.* (2020). Prospects for gamma-ray observations of narrow-line Seyfert 1 galaxies with the CTA – II. Gamma-gamma absorption in the broad-line region radiation fields. *MNRAS*, 494, 411.

Romero, G.E., Boettcher, M., Markoff, S., Tavecchio, F. (2017). Relativistic jets in active galactic nuclei and microquasars. *Space Science Reviews*, 207, 5.

Rybicki, G.B. and Lightman, A.P. (1979). *Radiative Processes in Astrophysics*. John Wiley & Sons.

Rybicki, G.B. and Lightman, A.P. (2004). *Radiative Processes in Astrophysics*, 2nd edition. John Wiley & Sons.

Sadeh, I. (2020). Data-driven detection of multimessenger transients. *ApJL*, 894, id.L25.

Sikora, M., Begelman, M., Rees, M. (1994). Comptonisation of diffuse ambient radiation by a relativistic jet: The source of gamma rays from blazars? *ApJ*, 421, 153.

Sironi, L. and Spitkovsky, A. (2014). Relativistic reconnection: An efficient source of non-thermal particles. *ApJL*, 783, L21.

Sironi, L., Keshet, U., Lemoine, M. (2015). Relativistic shocks: Particle acceleration and magnetization. *Space Science Review*, 191, 519.

Sol, H. (2018). Active galactic nuclei at very high energies: Cosmic accelerators and probes of space-time. *Journal of Astrophysics and Astronomy*, 39, id.52.

Sol, H., Zech, A., Boisson, C. *et al.* (2013). Active galactic nuclei under the scrutiny of CTA. *Astroparticle Physics*, 43, 215.

Sturm, E., Dexter, J., Pfuhl, O. *et al.* (2018). (Gravity collaboration), Spatially resolved rotation of the broad-line region of a quasar at sub-parsec scale. *Nature*, 563, 7733, 657.

Summerlin, E.J. and Baring, M.G. (2012). Diffusive acceleration of particles at oblique, relativistic, magnetohydrodynamic shocks. *ApJ*, 745, 63.

Thorne, K.S., Price, R.H., MacDonald, D.A. (1986). *Black Holes: The Membrane Paradigm*. Yale University Press.

Tramacere, A., Massaro, E., Taylor, A.M. (2011). Stochastic acceleration and the evolution of spectral distributions in synchro-self-Compton sources: A self-consistent modeling of blazars' flares. *ApJ*, 739, id.66.

Treumann, R.A. and Baumjohann, W. (2015). Spontaneous magnetic reconnection – Collisionless reconnection and its potential astrophysical relevance. *The Astronomy and Astrophysics Review*, 23, 4.

Valverde, J., Horan, D., Bernard, D. *et al.* (2020). A decade of MWL observations of the TeV blazar 1ES 1215+303: Extreme shift of the synchrotron peak frequency and long-term optical-gamma-ray flux increase. *ApJ*, 891, id.170.

Villard, P. (1900). Sur la réflexion et la réfraction des rayons cathodiques et des rayons déviables du radium, et sur le rayonnement du radium. *Comptes Rendus de l'Académie des Sciences. CXXX*, 1010–1012 and 1178–1179.

Walker, R.C., Hardee, P.E., Davies, F.B. *et al.* (2018). The structure and dynamics of the subparsec jet in M87 based on 50 VLBA observations over 17 years at 43 GHz. *The Astronomical Journal*, 855, 128.

Weekes, T.C., Cawley, M.F., Fegan, D.J. *et al.* (1989). Observation of TeV gamma-rays from the Crab Nebula using the atmospheric Cherenkov imaging technique. *ApJ*, 342, 379.

Werner, G.R., Philippov, A.A., Uzdensky, D.A. (2019). Particle acceleration in relativistic magnetic reconnection with strong inverse Compton cooling in pair plasmas. *MNRAS*, 482, L60.

Zech, A., Cerruti, M., Mazin, D. (2017). Expected signatures from hadronic emission processes in the TeV spectra of BL Lacertae objects. *A&A*, 602, id.A25.

Zech, A., Mazin, D., Biteau, J. *et al.* (2019). Key science project: Active galactic nuclei. *Science with the Cherenkov Telescope Array*, The CTA Consortium (ed.). World Scientific, 231.

Zhang, B.T., Petropoulou, M., Murase, K., Oikonomou, F. (2020). A neutral beam model for high-energy neutrino emission from the blazar TXS 0506+056. *ApJ*, 889, id.118.

5

Galactic Bulge–Black Hole Co-evolution, Feeding and Feedback of AGNs

Françoise COMBES
LERMA, Paris Observatory, France

Since the 1990s, we have known that there is a super-massive black hole in every galaxy, and that its mass is proportional to the mass of the bulge. To better understand how these black holes were formed, in symbiosis with their galaxies, we will look at their demography, the scaling relations between properties of black holes and host galaxies, and their evolution in Hubble time. Observations at high angular resolution now allow us to enter the black hole sphere of influence, see the molecular tori evoked in the AGN unification paradigm, and understand the feeding processes of black holes. These are often accompanied by feedback processes, which moderate the formation of galaxies.

5.1. Co-evolution: mass of black holes and bulges

5.1.1. *How can we measure the mass of black holes?*

If it is relatively easy to measure the mass of the black hole in our own Galaxy, the Milky Way (see Chapter 2); this is because of its proximity, 8 kpc. For galaxies belonging to the Local Group, it is already more difficult; the closest spiral, Andromeda, is ~ 800 kpc away, so 100 times farther. And for the galaxies closest to

For a color version of the figures in this chapter, see www.iste.co.uk/combes/nuclei.zip.

Active Galactic Nuclei,
coordinated by Françoise COMBES.

our environment, in our local super-cluster Laniakea, the distance is around 10–80 Mpc, up to 10,000 times farther. It was only with the angular resolution of the Hubble Space Telescope (HST, 0.05-0.1''), launched in 1990, that it was possible to measure the velocity of stars and gas very close to the black hole, in its sphere of influence. The sphere of influence is defined as the area where the gravity of the black hole dominates that of the bulge, that is, its radius is $r_s = G\ M_{BH}/\sigma_v^2 = 10–50$ pc according to the mass of the galaxy. At a distance of 20 Mpc, 0.1'' is equivalent to 10 pc and we can solve the sphere of influence of the black hole. Later adaptive optics on ground-based telescopes made it possible to achieve comparable resolutions. To determine the mass of the black hole, we must measure the velocity V of stars or gas at a certain radius R. The mass will be proportional to the product V^2 R.

One of the first relationships, linking the mass of the black hole and that of the bulge, was established by Magorrian *et al.* (1998), who observed 36 galaxies with the HST. They had already demonstrated a slightly different relationship for elliptical galaxies with or without cores. The core refers to the distribution of light in galaxies: if the density function $\rho(R)$ is a power law (with slope 1-2), there is no core, but if the law $\rho(R)$ is a plateau, we say that there is a core. The existence of cores could be due to the action of binary black holes, formed in the merger of galaxies that form elliptical galaxies. During a major merger of two galaxies, with each possessing a super-massive black hole, the dynamical friction causes the two galaxies to spiral towards each other, as well as the black holes, which begin to orbit fairly quickly in the center of the newly formed elliptical galaxy. Before merging, the pair of black holes must lose their relative angular momentum by dynamical friction on the stars of the spheroid: these stars gain energy, and are evacuated from the center, which creates a deficit of light in the center, and a core (Kormendy and Bender 2009, 2013). According to this scenario, galaxies with a core are the most massive, with the least rotation, and appear to have been formed by merging galaxies without gas. The coreless elliptical galaxies were undoubtedly formed by merging gas-rich galaxies; through dissipation, the gas fell towards the center and formed stars, creating an excess of light. These galaxies are less massive, and have more rotation.

Several tracers were used to determine the mass of the black hole: for the kinematic method, the velocity of stars, gas, and even masers in radio-astronomy (see Figure 5.1). The mass of the black hole M. is approximately 0.5% of the mass of the bulge. What matters is only the mass of the bulge or spheroid, not the total mass, nor the mass of the disk for spiral galaxies nor the mass of the dark matter halo (see Kormendy *et al.* 2011a, 2011b).

The method based on the kinematics of stars is quite uncertain, because the stars in the center of elliptical galaxies are not only animated by rotation, but mainly by anisotropic velocity dispersion. The result then depends on the modeling, to determine the degree of anisotropy (σr, σθ, σϕ dispersions), since the observations of the line-of-sight velocity are degenerated. The gas velocity does not suffer from this problem; however, the gas is in general less massive, and easier to disturb, for example by the feedback of the active nucleus, hence the existence of outflows. The maser method is particularly precise and exact. A maser emission (the equivalent of a microwave laser) is a coherent emission of molecules (OH, H_2O, etc.), where the molecules are all excited and emit simultaneously, in a stimulated manner. This can occur in regions of star formation, or the accretion disks around black holes, when the line of sight corresponds to a tangent to the disk seen edge-on, the place where molecules are crowded in the line of sight (see Figure 5.2). The emission is then amplified, and can even be detected at a very high angular resolution, of around a milliarcsecond, by very long baseline interferometry (VLBI). A hundred galaxies have been observed in this way.

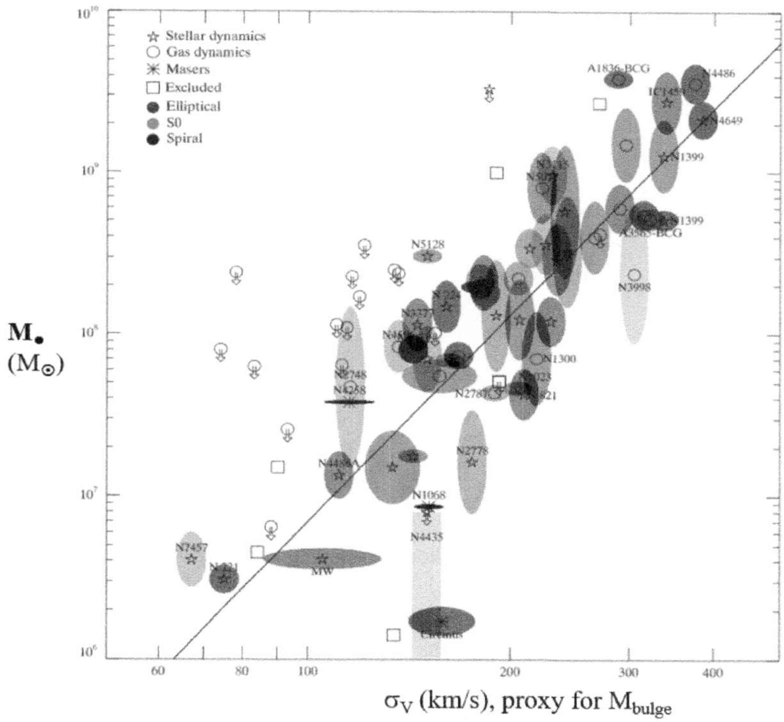

Figure 5.1. *Relationship between the mass of the central black hole M• and the mass of the bulge, or the velocity dispersion (σ_V) in the center of the galaxy. The different symbols indicate the method of determination, and the colors indicate the categories of host galaxies (source: Gültekin et al. (2009))*

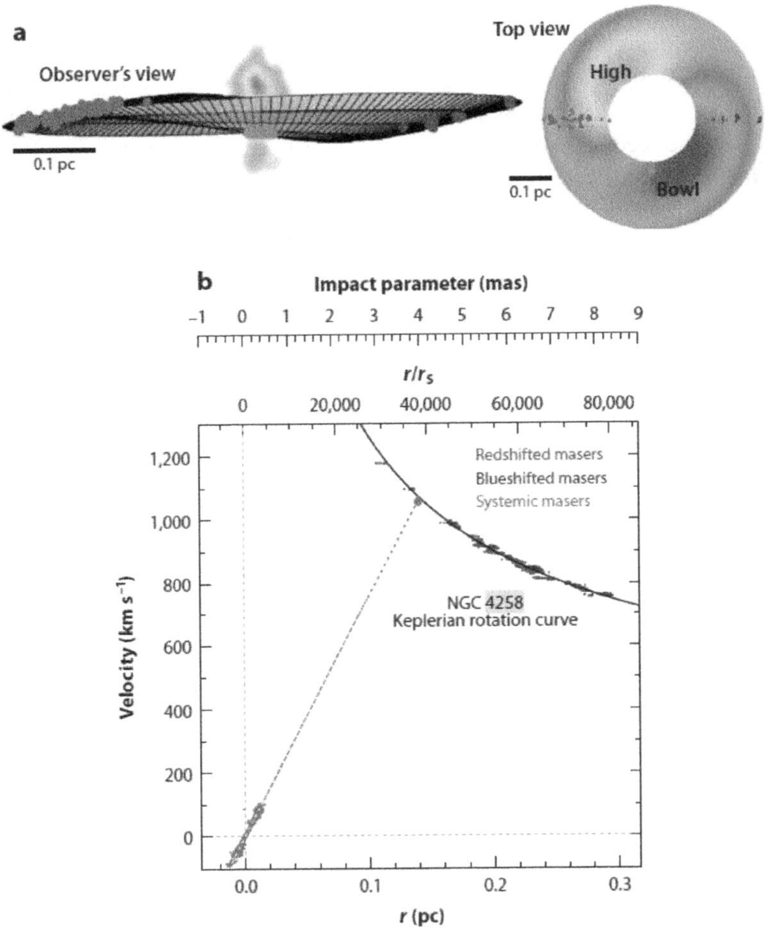

Figure 5.2. *The galaxy NGC 4258 and its H_2O masers, discovered by Greenhill et al. (1995): (a) Schema of the disk seen edge-on, showing a warp of the plane, with the radio jet perpendicular to it. In the face-on view on the right, we see the relief in colors (top in yellow, bottom in blue), and the positions of the maser emissions, according to their relative velocities (positive, zero and negative, red, green and blue points, respectively). (b) The velocities are reported on the bottom of the rotation curve, which is very close to a Keplerian curve (expected around a point mass of $4\ 10^7\ M_\odot$), with some small uncertainties due to the de-projection. The scale is in parsec at the bottom, and also in milliarcsecond (mas) and Schwarzschild radius at the top (or radius of the horizon of the black hole $r_S = 2\ GM_\bullet/c^2$). In green, the almost zero velocity is due to the masers drifting in the plane of the sky, we can then detect their proper motion, they move from right to left in 12 years, over an area of ± 4° viewed from the center (source: Moran (2008))*

The previous methods require angular resolution and can only be effective for nearby galaxies, whether the nucleus is active or not. For more distant galaxies, it is still possible to measure the mass of the black hole if the nucleus is active, and temporal and spectral resolution can replace spatial resolution. As long as the continuum emitted by the quasar or AGN varies, the variation propagates through the accretion disk into the Broad Line Region (BLR), very close to the black hole. The technique is called "reverberation mapping". The BLR responds up to radius R to brightness variations of the central continuum, which variably photoionizes it. Responses are seen in the lines after a time ~R/c shorter than the dynamic time tdyn = R/V (see Figure 5.3).

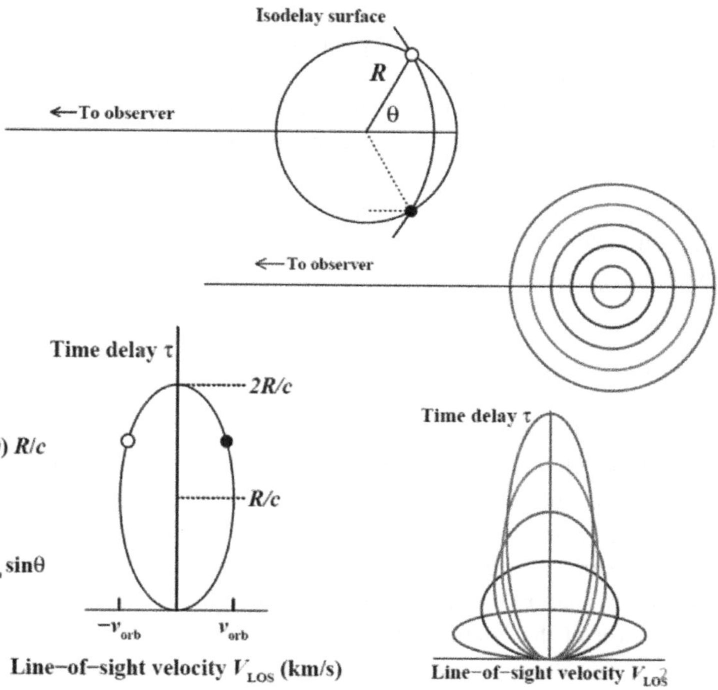

Figure 5.3. *Principle of reverberation mapping. For a signal coming from radius R, the delay after a burst of the continuum coming from the origin is $\tau(R,\theta) = (1 + \cos\theta)$ R/c. Each line from the BLR (Hα, Hβ, etc.) has an optimum emission radius, symbolized by the various colors. Each angle θ corresponds to a Doppler velocity towards the observer V_{LOS}. We therefore observe the bottom-right diagram*

Blandford and McKee (1982) first used the term RM "reverberation mapping", for this method. A significant number of AGNs (about 60) have been studied over time, thanks to numerous observations and long-term monitoring of variations, and the masses of their black holes have been precisely measured (e.g. Peterson *et al.* 2004). Today, reverberation mapping is the most accurate "luxury" method for a distant AGN. It can be done using robotic telescopes. It is then used to calibrate the L-R relationship and Luminosity-Size of the BLR region, and thus deduce the masses of black holes from a much larger number of AGNs.

The relationship between quasar luminosity and BLR size is now well established by a power law of slope 2, that is, $L \propto R^2$. This relationship can be explained if we assume that all bright AGNs have the same degree of ionization. The luminosity at a given frequency is proportional to the number of photons, which itself is proportional to the ionizing flux multiplied by the area R^2, and by the particle density or electron density n_e. This amounts to assuming that U and n_e vary little from one AGN to another. Once the luminosity of the AGN (and therefore the size R of the BLR) and the width of the lines ΔV of the BLR region have been measured, we can apply the gravitational equilibrium, $M_{BH} = \Delta V^2 R/G$, within a projection factor of the velocity, because the orientation of the accretion disk is not known.

5.1.2. *Bulge dispersion velocities*

Soon enough, it was noted that there was a much better relationship between the mass of the black hole and the central velocity dispersion of the bulge (Ferrarese and Merritt 2000). The velocity dispersion at the center is generally a function of the mass of the spheroid, a scaling relation called the Faber–Jackson relation for elliptical or early-type galaxies (Faber and Jackson 1976). It is the equivalent of the Tully–Fisher relation for spiral galaxies, where the visible mass is proportional to the 4th power of the rotational velocity, M (stars + gas) $\propto V^4$ (see McGaugh *et al.* 2000).

In Figure 5.4, it is easy to see that the relation involving the velocity dispersion is less scattered. Note, however, that the effective velocity dispersion σe is defined as an average of the total kinetic energy, weighted at each radius by the corresponding luminosity (and therefore, mass). More precisely, the square of the dispersion σe^2 is equal to the mean of $V^2(r) + \sigma^2(r)$, weighted by the luminosity at radius r. In the case where the mass of the black hole is determined by the kinematics of stars or gas, the two variables, black hole mass and dispersion are not entirely independent, which may partly explain the weaker scatter of the relationship.

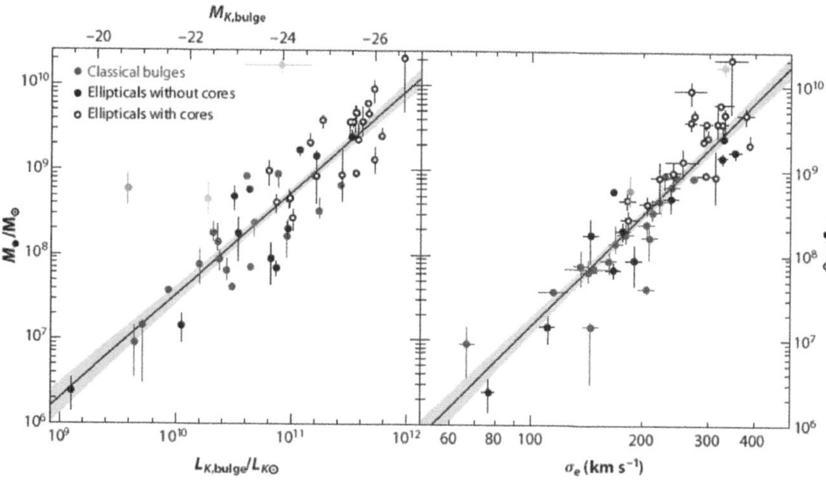

Figure 5.4. *Comparison of the relation between M_{BH} and total luminosity, with the relation M_{BH} and velocity dispersion σe. A distinction is made between classical bulges of spirals (in red) and ellipticals (in black). Empty circles are ellipticals with cores. For larger masses of black holes, M_{BH} becomes almost independent of σe, at least for ellipticals with a core (source: Kormendy and Ho (2013))*

In addition, we can see in Figure 5.4 that there is a small distinction between ellipticals with or without a core. This is already found in the Faber–Jackson relation: if the coreless ellipticals do verify the law $L \propto \sigma_v^4$, the velocity dispersion is much lower for galaxies with cores, and the relation becomes rather $L \propto \sigma_v^8$. The interpretation is that the galaxy mergers that form the most massive ellipticals result in the merger of black holes, which in their coalescence heat and evacuate matter from the center, reducing the velocity dispersion (Kormendy and Bender 2013). The mass of the bulge then becomes independent of sigma, and the M_{BH}–sigma relation saturates at large masses. Elliptical galaxies with cores have box shapes, and do not spin, while galaxies without a core have disk shapes and more rotation. These observations are fully consistent with the essential role of black hole mergers in core formation.

Another distinction can be made between classical bulges, which are believed to be the result of minor mergers of galaxies and pseudo-bulges, resulting from the secular evolution of spiral disks. Figure 5.5 shows that the relationship between the mass of the black hole and the mass of the bulge mainly concerns classical bulges. However, the relationship becomes very scattered at small mass. This could come from a determination problem, because for small masses, the black hole sphere of influence is very small, and requires very high spatial resolution. Also, the activity of nuclei is weak and can be confused with other activities, such as star formation. The mass of nuclear disks, whether stellar or gaseous, becomes comparable to the mass of the black hole.

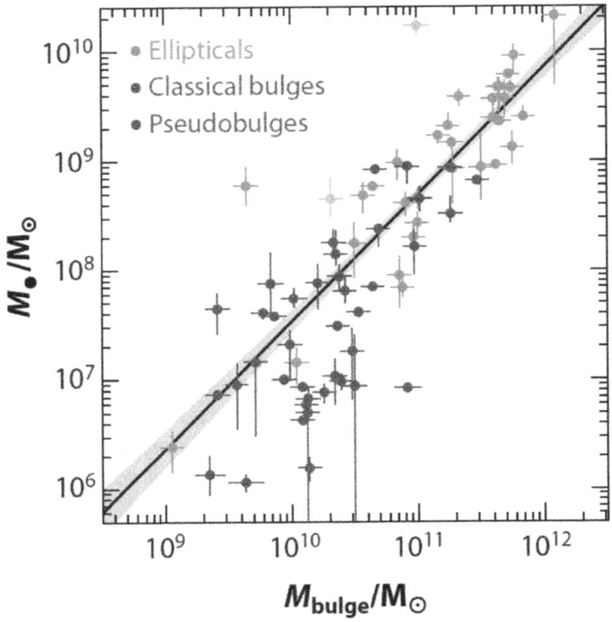

***Figure 5.5.** The proportionality between the mass of the black hole and the mass of the bulge is less scattered for massive spheroids, early-type galaxies. For galaxies of lower mass, the relation is with "classical" bulges and not pseudo-bulges of barred spiral galaxies, and since it is difficult to separate them, the relation is much more scattered (source: Kormendy and Ho (2013))*

For spiral galaxies, other black hole mass indicators have been proposed, such as the pitch angle of the spiral arms (Davis *et al.* 2014), or the Sérsic index (Mutlu-Pakdil *et al.* 2016). The relationships are empirical, but could be justified by the indirect correlation between the tightening of the spiral arms and the stability of the disk, which is increased by a massive bulge.

It is even more difficult to measure the mass of a possible black hole, of intermediate mass, in dwarf galaxies. However, a certain number of dwarfs have an active nucleus, detected thanks to its X-ray, optical/infrared or radio emission (see the review by Reines and Comastri (2016)). It is difficult to distinguish AGNs in dwarfs, however, as the diagnostics can also apply to a region of starburst

(Hainline *et al.* 2016). The relationship between the mass of the black hole and the mass of the bulge seems to continue, even at low masses (Baldassare *et al.* 2016).

5.1.3. *Special cases and exceptions: obese black holes*

There are exceptions to the general law, for example a number of systems have black holes that are too light for their velocity dispersion. The vast majority of these are interacting galaxies that will merge into one. The velocity dispersion has already increased, but the black holes have not yet merged, and the final system will reverse the relationship in less than a billion years.

There is also a whole category of objects where the black hole is obese, compared to the host galaxy. These galaxies are in clusters, and they are often the brightest galaxy in the cluster. These galaxies are like cannibals in the center of the cluster, they merge with many small companions, and most notably, swallow hot gas before it has even had time to form stars. This process could cause the black hole to grow faster than the bulge. One examples of an obese black-hole can be found in the NGC 1277 galaxy, within the Perseus cluster. It is a lenticular galaxy, possessing a rotating disk, and a massive bulge. van den Bosch *et al.* (2012) measured a giant black hole mass of 1.7 10^{10} $M_{\odot}$, or more than half of the mass of the bulge, instead of the standard ratio of 0.5%. This large value was questioned by Emsellem (2013), reducing the mass of the black hole by a factor of 4, which was evaluated using stellar kinematics, and proposing a higher bulge mass, and perhaps a velocity disturbance due to a bar in the line of sight. The mass of the black hole was, however, confirmed using molecular gas kinematics (Scharwächter *et al.* 2016).

Other examples can be found in the Perseus cluster, and also in the cores of other clusters. If the ratio of the mass of the black hole to the mass of the bulge is too large, it would appear to be due to a deficit of star formation in the bulge, rather than an excess of black hole mass. In clusters, star formation can be easily stopped, as the very hot gas from the cluster acts as a wind that sweeps away the gas from galaxies. Likewise, galaxies can be suffocated if they can no longer replenish themselves with cold gas, as the filaments of matter and gas are heated and destroyed at the entrance of the cluster. Observations support the scenario, because the stellar population in these galaxies is very old. These could be "relic" galaxies, where all of the stars were formed at the beginning of the universe, and which could then have acquired gas to fuel their black hole, but without forming new stars (see Trujillo *et al.* (2014)).

The difficulty in determining the mass of black holes using stellar dynamics is well illustrated by the example of our nearest spiral galaxy: Andromeda or M31. Its proximity makes it a very favorable object, because with the HST, we have excellent angular resolution. At a distance of 780 kpc, 0.1 arcsec represents 0.37 parsec, and the kinematics can be observed well within the sphere of influence (r~20 pc) of the black hole, which is approximately 10^8 $M_\odot$. However, there are difficulties which arise precisely because of this proximity. We can see that the dynamics of the stars in the center are not in equilibrium. There is an eccentric nuclear disk of stars, 2 pc in diameter. The black hole is not in the center of the disk. This asymmetry is probably generated by a density wave, which rotates with a high angular velocity.

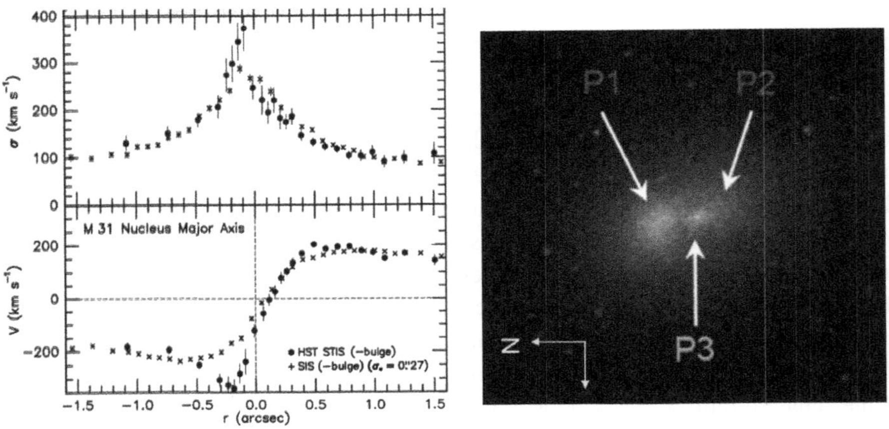

Figure 5.6. *The black hole in the center of our neighboring galaxy M31. On the left: the stellar kinematics, obtained with the HST by Bender et al. (2005), are presented, with the dispersion at the top, reaching a maximum of 400 km/s, and the rotation along the major axis at the bottom. On the right: the photometry of the HST is presented, showing the three light components of the eccentric nuclear stellar disk, according to Lauer et al. (2012). The black hole is in P2, which is not the brightest point*

Figure 5.6 shows the arrangement of the eccentric nuclear stellar disk. The brightest point on the disk is P1. The black hole, detected by the maximum of the velocity dispersion, is in P2. The separation between P1 and P2 is 0.5'' = 1.8 pc. The rotational velocity is V = 200 km/s, with an orbital period of 50,000 years. P3 is a very young blue star cluster, no more than 200 million years old. The kinematic major axis of the velocity field does not correspond to the photometric major axis, symbolized by the segment P1-P2. The tilt of the stellar disk is uncertain, but would rather be 55°, quite different from that of the M31 galaxy, which is very tilted (i = 77°). The system should not be stable, and our observation of this state would be highly unlikely. For example, star clusters should merge, unless there is a dynamical mechanism to lengthen the lifespan of such a distribution.

This mechanism is probably a wave of symmetry m = 1 (Tremaine 1995); these are the most frequent waves to develop in a Keplerian disk, that is, dominated by a point mass. N-body simulations have shown that such a wave is compatible with the observations (Bacon *et al.* 2001). In the simulations, the nuclear disk of M31 is subject to a natural mode, with a very slow rotational pattern speed (3 km/s/pc, period of two million years) which can be maintained for over a thousand dynamical times, or periods of rotation. The resulting morphology and kinematics can reproduce the photometry of the nuclear disk observed in M31 and the average stellar velocity, with its asymmetries. They require a central mass concentration (the black hole) and also a cold stellar disk of 20–40% of the mass of the black hole. Such a slow mode could be excited when interstellar clouds from the outermost gaseous disk infiltrate towards the center. Nuclear disks formed from accumulated gas are possible candidates for the precursors of these types of structures, and may be common in central regions of galaxies. Peiris and Tremaine (2003) simulated a massless, non-self-gravitating nuclear disk. They tried several inclinations and alignments with the main disk of the Andromeda galaxy. Good agreement with observations is only obtained when the stellar disk is not aligned.

5.1.4. *High redshift black holes*

Does the relationship between the masses of the black hole and bulge persist during evolution, and when was it established? A large number of quasars have been observed at large z values, even greater than 6, and the dynamics of the host galaxy could be determined by the CO and [CII] lines, which are redshifted into the millimeter range (observable with ALMA or NOEMA interferometers). The mass of their black hole is determined by optical observation of the spectrum of their BLR region, and the relationship between brightness and size of the BLR region, described earlier in this chapter. For the most part, the mass of high z black holes is greater than expected from their dynamical mass, and from the relationship obtained locally (see Figure 5.7, Venemans *et al.* (2016)). This surprising result could come from the great uncertainties when estimating the dynamical mass. It is not possible to distinguish the bulge from the disk, for example, and the mass of the black hole is compared to the total dynamical mass of the host. Therefore, the mass of the bulge could only be less, which would increase the distance from the local relation. Another uncertainty is that of the host galaxy morphology. The inclination of the rotating molecular disk is not well known and the result is only valid statistically. Estimating the mass of the black hole, using the brightness-size relationship of the BLR region, also has a factor of uncertainty due to unknown inclination. In most systems, it is assumed that the [CII] or CO lines are centered on the systemic velocity of the host galaxy, and can be reliably used to calculate dynamical mass.

However, the broad MgII optical emission lines are systematically shifted towards the blue. The average shift is 500 km/s, but can go up to 1,700 km/s. It is interpreted as an outflow due to feedback from the central AGN; the symmetrical red-shifted region is too obscured by dust to be detected. However, even though the dynamical mass is not well estimated, it is possible to have a lower limit with the gas mass, from the two lines CO and [CII] and also from the emission of the dust. In these large-z systems, the gas fraction is often greater than 50%, and the dynamical mass can only be underestimated by a factor of 2 at most. The masses of the black holes are then three to four times higher than expected according to the local relation.

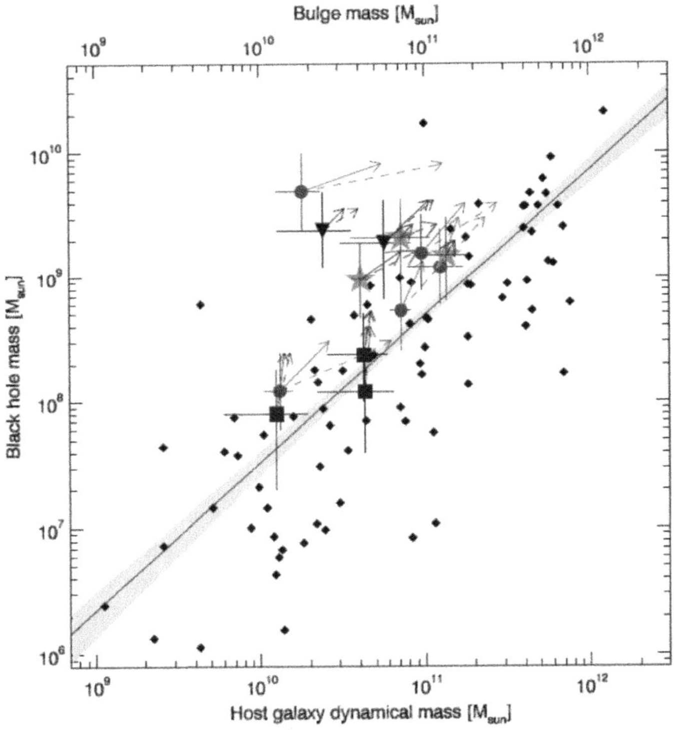

Figure 5.7. *The mass of the black hole as a function of the dynamical mass of distant galaxies (z = 6) hosting quasars, and for local galaxies, as a function of the mass of the bulge (top scale). The black diamonds represent local galaxies (Kormendy and Ho, 2013). Their relationship is underlined by the shaded area. Large-colored symbols are high z quasars, which overwhelmingly have excess black hole masses. According to the accretion rate of matter deduced from the luminosity of the quasar, and taking into account the observed star formation rate, it is possible to extrapolate the trajectory of the points (arrows) over the next 50 million years (source: Venemans et al. (2016))*

5.2. Interpretations: cause or effect?

As presented at the beginning of this chapter, there is a strong relationship between the masses of black holes and the masses of bulges, locally. There is also a simultaneity of evolution between star formation, or the mass growth of bulges, and the activity of nuclei, or the mass growth of black holes. These two combined relationships suggest a symbiotic formation of galaxies with their black holes.

5.2.1. *Evolution of star formation and the accretion of black holes*

The distribution of active star-forming galaxies showed a very strong evolution with the redshift. The majority of the stars we see in local galaxies were formed in the first half of the universe's life. A parallel evolution is observed for the activity of the nuclei. While bright quasars are quite rare locally, they were much more common at the beginning of the universe. The activity of nuclei is a tracer of mass accretion by super-massive black holes at the center of each galaxy, and we can use the continuity argument (Soltan 1982) to associate the growth rate of black holes to the number and total luminosity of quasars over time. For this, we use the average radiation efficiency of an AGN: any mass m swallowed by a black hole will radiate $\varepsilon = 10\%$ of its mass energy mc^2, and thus contribute to the brightness of the quasar. There is also an efficiency term, because some nuclei may have reduced activity, for the same rate of accretion (e.g. advection-dominated accretion flow (ADAF), see Chapter 3). For this black hole mass growth rate, the overall rate and the merging of galaxies leading to the merging of black holes, without an increase in their combined mass, does not count.

How can we monitor the evolution of black holes over time? The main source of information comes from studying the active nuclei as a function of redshift, in order to establish the all-z luminosity function, and obtain the mass of black holes from the spectrum of the BLR and the brightness-size relation of the BLR. If the mass accretion rate is dmacc/dt, the radiation efficiency is ε and the luminosity of the AGN is $L = \varepsilon$ dmacc/dt c^2. In this accretion, the mass of the black hole increases by $dM_\bullet/dt = (1-\varepsilon)$ dmacc/dt, and the link between the increase in mass of black holes and the brightness of the AGN is $dM_\bullet/dt = (1-\varepsilon)/\varepsilon \, L/c^2$.

Normally, this inventory would account for the mass function of black holes today at $z = 0$, based on the relation $M_\bullet$-M_{bulge} and the bulge mass function, regardless of nucleus activity. Yet the comparison does not match, there is twice the mass of black holes today compared to the integrated luminosity of all AGNs, and it must be assumed that the mass accretion by black holes is not always radiatively efficient (Fabian and Rees 1995). It is also possible that much of the radiation from AGNs is obscured dust, and has escaped observation.

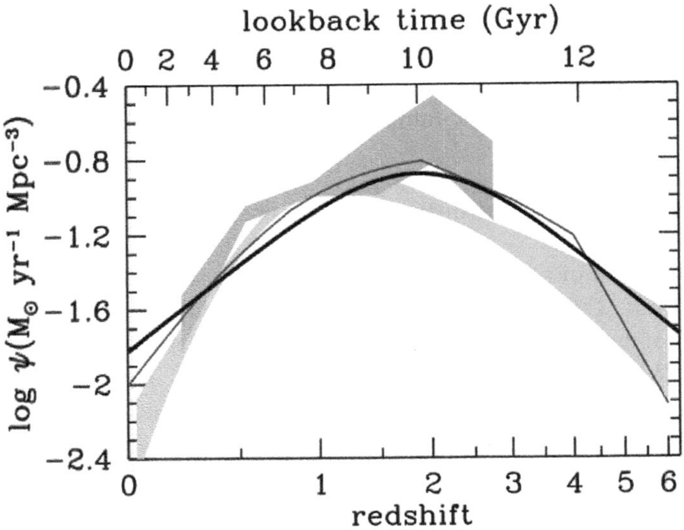

Figure 5.8. *The cosmic evolution of star formation density is the thick black curve. The density increases to a maximum of z = 2, then decreases by a factor of ~ 10 until today. The activity of quasars and AGNs follows a parallel evolution: the red curve and the green area are two estimates using the X-ray indicator; the blue area is derived from the infrared domain. The rate of mass accretion on black holes has been multiplied by 3300 to enter the same scale as the density of star formation (source: Madau and Dickinson (2014))*

Another more global method is to observe the accumulated background radiation, especially in X-rays, by AGNs during evolution.

Figure 5.9 shows the entire spectrum of cosmic background radiation. The energy is dominated by microwave background radiation coming from the Big-Bang, then the optical and infrared background radiation coming from the radiation of stars in the galaxies, and from dust heated by young stars and also by AGNs. X-ray radiation is dominated by AGNs, but the spectrum of harder radiation indicates that much of this radiation is obscured by dust, so it is re-radiated in the far infrared or sub-millimeter range, at lower frequency. X-rays can detect obscured AGNs, but up to a column density of 10^{24} cm^{-2}. Beyond that, for higher column densities, even the X-rays do not get out, these are so-called "Compton-thick" sources. To make an inventory of all AGNs, all of these corrections must be made.

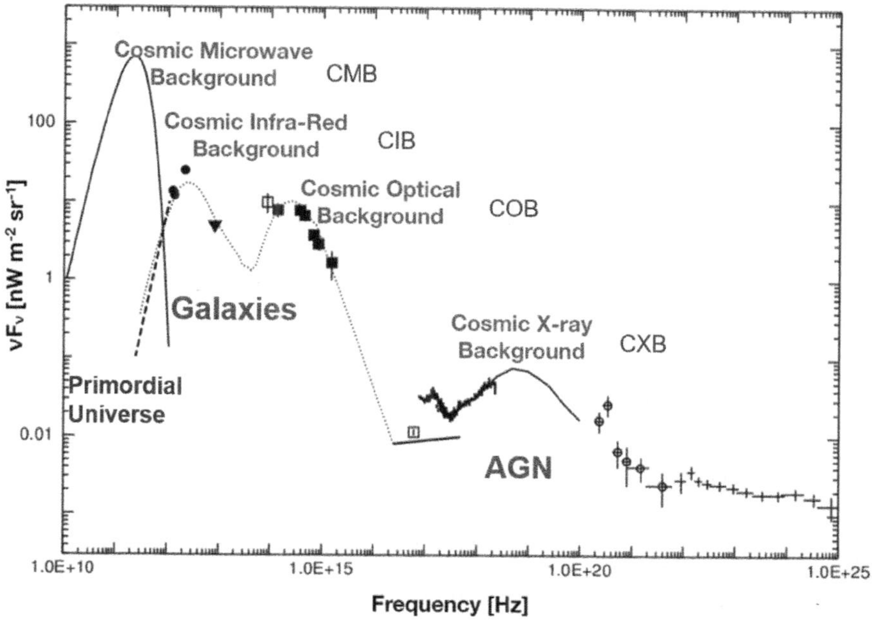

Figure 5.9. *The spectral density of cosmic energy, from radio waves to the highest gamma frequencies. Representation in $\nu F\nu$ weighs the radiation with its energy, and with the log-log scale, a horizontal line corresponds to radiation of equal power per decade of energy (source: Hasinger (2000))*

5.2.2. Downsizing, formation of AGNs coupled with the formation of stars

The evolution of the luminosity functions of AGNs reveals an anti-hierarchical effect called "downsizing" (Cowie *et al.* 1996). This effect has also appeared for star formation: the most active galaxies today are the less massive galaxies, the late-type spirals and the irregulars. The brightness of the most active galaxies continues to decrease over time. However, in the standard cosmological model, which is hierarchical, the objects that form first are the smallest, they then merge to form the largest, but these are dark matter halos. As for black holes, the most massive ones appear earlier in the universe, and are now silent, while activity continues in Seyfert galaxies, in spiral types.

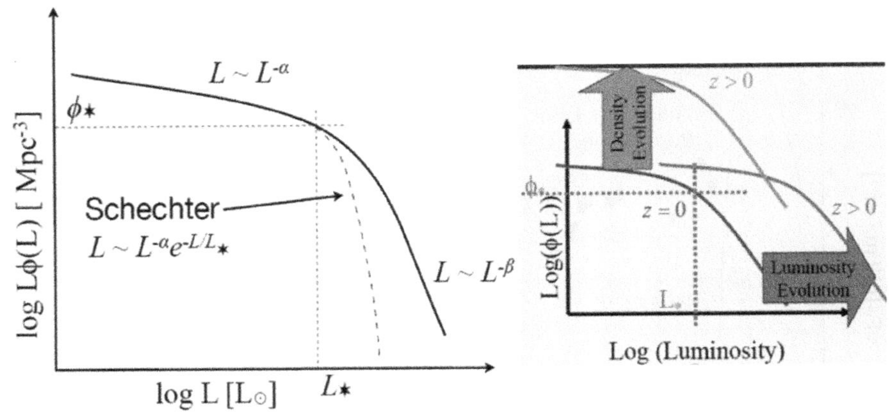

Figure 5.10. *Left: The AGN luminosity function is a double power law, α for low luminosities (L<L*) and β for higher values. It can be written as $\Phi(L) = \Phi(L^*)/[(L/L^*)^\alpha + (L/L^*)^\beta]$, with $\alpha < 1$ and $\beta > 1$. It is very different from the luminosity function of galaxies for the brightest objects. Right: the evolution of the luminosity function with the redshift breaks down into an evolution of the luminosity (L* increases), and also the density of AGNs (Φ* increases)*

The luminosity function of AGNs can be described by a double power law, with two slopes α and β (see Figure 5.10). It is different from the luminosity function of galaxies, which is a Schechter law, first a power law of slope α at low luminosity (L<L*), then an exponential law, in exp (-L/L*). The brightness function of AGNs varies significantly with redshift, both in brightness and density (see Figure 5.10). The two effects cannot be separated. On the contrary, the two slopes α and β also change with the redshift, and that is what causes the downsizing. Observations show that AGNs of higher masses and luminosities form earlier (anti-hierarchical formation), this implies coupled evolutions of luminosity and density, and an evolution of slopes α and β with z. A model reproducing the X-ray, optical and infrared observations, therefore in bolometric luminosity of AGNs, clearly shows the anti-hierarchy and the variation of slopes α and β (see Figure 5.11). The AGN activity maximum is well reproduced at a redshift equal to 2, as shown in Figure 5.8.

Regarding the mass of black holes, the evolution is well reproduced, assuming an average radiation efficiency ε = 0.04-0.16, with an accretion rate such that the active phase of AGNs occurs with a luminosity equal to 0.1–1.7 of the Eddington luminosity. The lifespan of these active phases varies between 1.5 10^8 years for large black hole masses (greater than 10^8 $M_\odot$), and 4.5 10^8 years for the smallest (Marconi *et al.* 2004). For the anti-hierarchy, massive black holes are formed in the deepest potential wells, which form first. The lightest black holes form in shallow wells. They are formed later in the universe, and moreover, they are more sensitive to the feedback phenomenon. They therefore take much longer to grow.

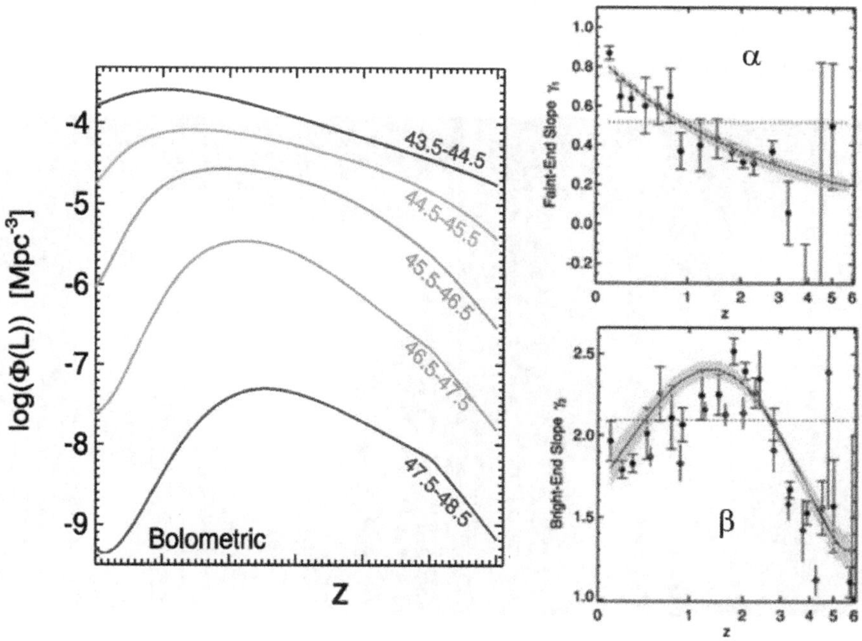

Figure 5.11. *On the left, the various curves correspond to a given bolometric luminosity, in erg/s (only the log is indicated). For the brightest AGNs, the maximum occurs at a greater redshift than for weaker AGNs. This is the anti-hierarchical evolution (downsizing). In this model, it is necessary that the slopes of the brightness function of the quasars, α and β, evolve as a function of z, as shown on the right (source: Hopkins et al. (2007))*

In the 1990s, during the first far infrared sky survey using the IRAS satellite, a new phase in the evolution of galaxies was discovered, a huge burst of star formation: these objects are called "starburst galaxies". There was no knowledge of them before, as their immense star-forming activity is almost entirely obscured by dust, and emerges in far infrared (Sanders and Mirabel 1996). They are ultra-luminous infra-red galaxies, or ULIRGs ($L_{IR} > 10^{12}$ $L_{\odot}$). At these very strong luminosities, almost all of the objects are the result of interacting or coalescing galaxies. In the interaction, the tidal torques were able to drive enormous amounts of interstellar gas towards the center of the galaxies, and produce bursts of star formation. In these very compact areas, the dust is concentrated, and re-emits the light from buried young stars in the far infrared. The gas is not entirely consumed by star formation, and can also fuel central black holes. The ULIRG phase could thus be a precursor of the quasar activity phase. The correspondence between the luminosities of extreme starbursts and quasars is compatible with this hypothesis. The activity phase of the quasar is shorter as the feedback is more violent, which corresponds to the observations.

Figure 5.12. *The advanced coalescing two-galaxy system, NGC6240 (HST image, left), is an ULIRG, a starburst galaxy. It is the prototype of starbursts buried in dust and molecular gas. The near infrared emission of the H_2 molecule could be detected, and at high angular resolution, two nuclei appear, emitting strongly in X-rays, indicating that the merger is not complete. The powerful and hard X-rays confirm that both nuclei are AGNs. Two iron atom recombination lines (Kα), in fluorescence near the two black holes, were also detected (source: Komossa et al. 2003))*

However, there are still some degeneracies when interpreting the luminosity function of quasars. Two models reproduce the observed luminosity function: either there are many short-lived quasars, or there are rare long-lived quasars. A constraint on the parameters is already given by the mass function of black holes established locally. Mass accretion must be relatively shared among all galaxies during evolution, since black holes with extreme masses (greater than 10^{10} $M_{\odot}$) are not often observed. However, there are several free parameters, not only the radiation efficiency ε, but also the Eddington rate: hence, does the growth of black holes mainly take place near the maximum set by the radiation pressure (see Chapter 1)? The luminosity function of AGNs in the two-slope power laws α and β has been interpreted on the one hand, (α) by more regular growth away from the Eddington rate, and on the other, (β) by more intermittent growth at the Eddington rate, during galaxy mergers. These would be the same galaxies seen in these two emission regimes, but in different phases of their evolution. Since both low and high luminosity AGNs come from the same population, a wide range of brightness comes from a small domain of the mass of galaxies and black halos. The consequence is that the degree of agglomeration of quasars, in groups or clusters, should not depend on their luminosity. An attempt at clarification could be made by comparing the agglomeration rate of AGNs with that of dark matter distribution. If quasars are activated by merging galaxies, the maximum attainable luminosity is correlated with the mass of the halo, but not the instantaneous luminosity. The detailed study of the simulated light curves, the degree of correlation with the mass of the halo, and the comparison with the agglomeration of black halos and galaxies showed that the agglomeration rate of AGNs had very little dependence on luminosity, which is compatible with the predictions of the model (Lidz *et al.* 2006). Quasars cluster less than elliptical (red) galaxies, but more than spiral (blue) galaxies. They form an intermediate population.

5.2.3. *Feedback: how can the black hole influence the bulge?*

Many models propose to explain the excellent correlation between the masses of black holes and the masses of bulges through the AGN feedback phenomenon: as soon as interstellar gas arrives towards the center, it can both form stars and fuel a black hole, but as soon as the latter is active, its energy will heat the gas or eject it mechanically, preventing any subsequent star formation and therefore limiting the mass of the bulge. Yet, the radius of the black hole sphere of influence $R_{infl} = GM_\bullet/<V^2>$, defined as a function of the average velocity or dispersion velocity $<V^2>$ of the bulge at radius R, is extremely small; approximately 20 pc, much smaller than the radius of the bulge (of the order of kpc). More precisely, the observed mass correlation is $M_\bullet \sim 0.002\ M_{bulge}$, with the mass of the bulge equal to $M_{bulge} \sim 5 <V^2> R_{bulge}/G$, that is, $R_{infl} \sim 10^{-2}\ R_{bulge}$. The volume of influence of the black hole is therefore 10^{-6} times the volume of the bulge. Under these conditions, it is difficult to consider a gravitational exchange of information.

However, when we calculate the energy returned to the rest of the galaxy as the black hole grows, accreting at a dmacc/dt rate, with an average efficiency of $\varepsilon \sim 0.1$ and an AGN luminosity of $L = \varepsilon$ dmacc/dt c^2, the growth of the corresponding black hole mass is $dM_\bullet/dt = (1-\varepsilon)$ dmacc/dt. The energy expended during the growth of the black hole is therefore $E_c = \varepsilon/(1-\varepsilon)$ M. c^2. This can be compared with the gravitational energy of the bulge $E_{bulge} \sim M_{bulge} <V^2>$. It appears that the growth energies of the black hole and the bulge are in the ratio $E_c/E_{bulge} = \varepsilon/(1-\varepsilon)$ (M./M_{bulge}) $c^2/<V^2> \sim 400$!

The active nucleus deploys enough energy to destroy the bulge. Therefore, the whole problem lies in determining what fraction of this energy will be radiated outwards without affecting the galaxy, and what fraction will be coupled with the disk or the bulge, thus justifying an interaction between the two, and a possible moderation. We will now identify the various possible models of coupling between AGNs and galaxies, depending on the brightness of the AGN.

5.2.4. *Feedback: radiative mode and quasar wind, or mechanical mode of the radio jet*

One of the fairly obvious modes of interaction between an AGN and the galaxy is through the wind evaporating from the accretion disk, when the luminosity of the quasar is very high, near the Eddington limit. Note that Eddington luminosity is obtained when the radiation pressure compensates for gravity, and the gas can no longer flow freely and feed the black hole. This Eddington limit is $L_{Edd} = 3.3 \times 10^{13}$ $L_\odot$ ($M/10^9$ $M_\odot$), for a black hole of 10^9 $M_\odot$, for example. A black hole, with matter distributed spherically around it, will not be able to radiate beyond it. This is possible, however, if the material is distributed on a disk. The accretion rate corresponding to this limit is 20 $M_\odot$/year ($M/10^9$ $M_\odot$).

In the case of very bright quasars, it is therefore common for a wind of matter to rise above the disk (see Figure 5.13). For all of the others, the possible mechanism would rather be mechanical, through driving the jet. Note, however, that radio jets are only observed in 10% of all AGNs.

The presence of significant radio jets (radio loud quasars) is anti-correlated with the brightness of the quasar. More precisely, the ratio between radio and optical brightness R = Lrad/LB is a decreasing function of the Eddington ratio λ = Lacc/LEdd (see Ho (2002), Figure 5.14). This is verified by both categories of AGNs, "radio-loud" and "radio-quiet", even though the ratio R is always higher for the first sequence. A saturation of the parameter R occurs at low accretion rates, for $\lambda < 10^{-3}$, for low-luminosity AGNs, such as FRI among the radio sources, or even Seyfert galaxies and LINERs.

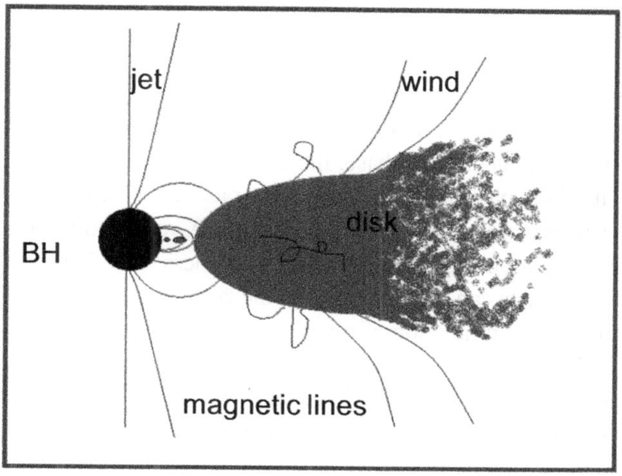

Figure 5.13. *Wind of matter emitted by the accretion disk, for very bright quasars near the Eddington limit. The wind is generated by the radiation pressure, aided by the magnetic pressure. The emission of the jet is closer to the black hole, and takes place, for example, per the Blandford–Znajek mechanism (see previous chapters)*

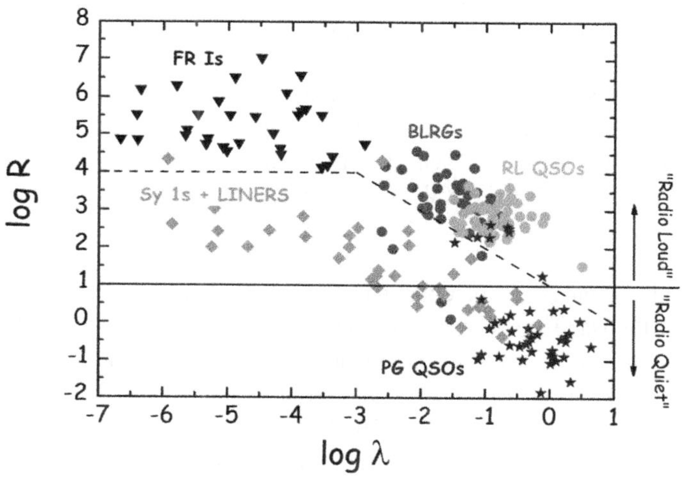

Figure 5.14. *Anti-correlation between the ratio R of the radio and optical luminosities R = Lrad/LB, and the Eddington ratio of the accretion rate λ = Lacc/LEdd. The various colors correspond to the various AGN categories: the brilliant "PG, Palomar-Green" radio-silent quasars are in blue, the radio-loud or FRII quasars are in blue-cyan, the weaker radio sources (FRI) are in black, the Seyfert and LINERs are in green and the BLR galaxies with broad lines are in red (source: Sikora et al. (2007))*

This anti-correlation can be explained by the various phases of black hole accretion, depending on its accretion rate compared to the Eddington rate (e.g. Fender *et al.* 1999). In the very active, high-luminosity phase, matter forms a thin accretion disk, which radiates efficiently, and electrons are not emitted. The disk mostly emits soft X-rays. In the weak accretion phase, the disk is truncated until it no longer exists at all. Instead, the hot material, which does not radiate effectively, forms a plasma ring; the relativistic electrons then transform the soft X-rays into hard X-rays. Hence, there is usually an anti-correlation between brightness and X-ray energy. In the low-light phase, radio jets can form, starting at the corona, which is the base of the jet. The corona favors the accumulation of magnetic flux near the black hole, thanks to its 3D geometry (Sikora and Begelman 2013). It would be favored by an episode of gas accretion or fusion between galaxies. It is thus explained that the hard X-rays are correlated with radio emission. Likewise, the most powerful radio sources are elliptical galaxies, formed by the fusion of spiral galaxies.

5.2.4.1. *The radiative or quasar mode*

For the radiative feedback mode, simple models have attempted to predict the ejection effectiveness of matter, based on the characteristic shock waves, well known in supernova models. Figure 5.15 schematically shows the different spheres, the adiabatic (reverse) shock wave, the contact discontinuity between wind and the interstellar medium, and the shock wave advancing in the galaxy.

Winds of almost relativistic speed near the black hole were observed by their iron lines in X-rays (Tombesi *et al.* 2011). These streams of matter are called Ultra-Fast Outflows (UFOs). Their speeds are detected thanks to the lines of the highly ionized iron Fe XXV/XXVI in absorption. The speeds are typically c/30 to c/10.

Further in the galaxy, molecular gas flows are observed, at speeds of several hundred to a thousand km/s (e.g. Feruglio *et al.* 2010; Cicone *et al.* 2014). These observations were at first a surprise: how do we explain that molecules remain in the waves? The shock waves should have destroyed the molecules. The gas is heated in the shocks to temperatures of 10^6 to 10^7 K, enough to photo-dissociate hydrogen molecules and their tracers. However, the already shocked molecular gas, between the contact discontinuity and the shock wave, cools very efficiently, within a time scale well under a million years (Zubovas and King 2014). It becomes multi-phasic, with a phase fragmented into dense clouds, by Rayleigh–Taylor instabilities, which induces the formation of stars. This formation sometimes induces a luminosity comparable to that of an AGN, if it is greater than a rate of 100 $M_{\odot}$/year! In this case, it becomes very difficult to distinguish molecular flows due to starburst or AGN. The criterion of velocity can sometimes be used to decide. AGNs can eject gas at very high velocities, greater than the escape velocity, which is rarely the case with starburst galaxies (except for dwarf galaxies).

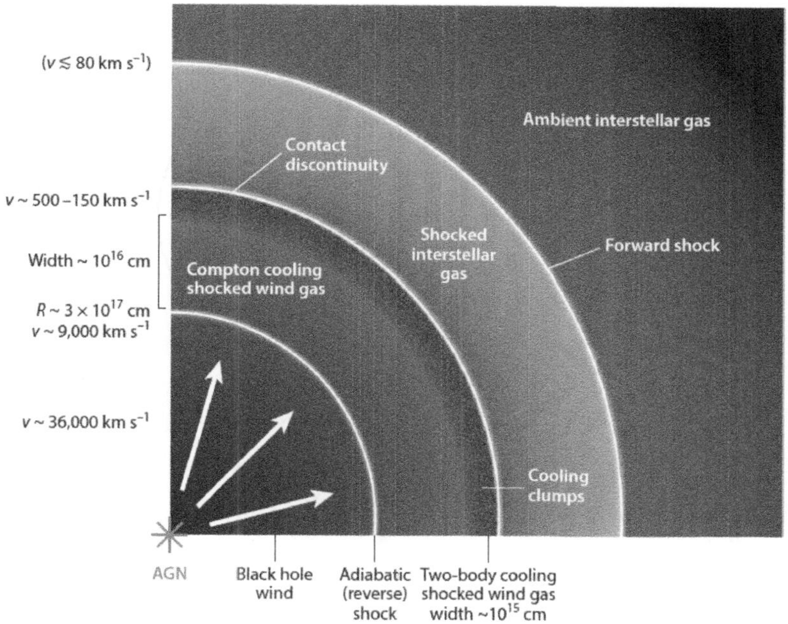

Figure 5.15. *Diagram of the wind emitted by a high luminosity AGN, by radiation pressure, in the quasar radiative mode. A flow of plasma at very high speed (~c/10) produces a shock wave in the ambient interstellar gas. A reverse shock returns to the quasar, with the contact discontinuity surface between the two, which delimits the interstellar gas which has already undergone the shock wave, consisting of wind cooled at a more moderate speed*

There are several cases, as shown in Figure 5.16, depending on whether the cooling in zone "b" of the shocked wind is effective or not, the shock energy is conserved or not, and how far the interstellar gas shell is ejected. If the energy is conserved, then we can observe a momentum in the molecular gas much greater than the initial momentum in the AGN wind. This case occurs for very fast winds, that is, V_{in}>10,000 km/s, with few radiative losses (Faucher-Giguère and Quataert 2012). The molecular gas velocity Vs, and flow dMs/dt, then receives a momentum boost. If there is almost conservation of energy, to a factor of 2, we can write dMs/dt Vs^2 ~1/2 dMin/dt Vin^2. And the quasar wind momentum dMin/dt Vin is approximately LAGN/c. The molecular outflow momentum is dPs/dt = dMs/dt Vs, and that of wind is dPin/dt = dMin/dt Vin. By conserving energy we then have the ratio of moments: Ps/Pin = ½ Vin/Vs; for Vin= 30 000 km/s, and Vs~300 km/s, the ratio Vin/(2 Vs) can reach up to 50!

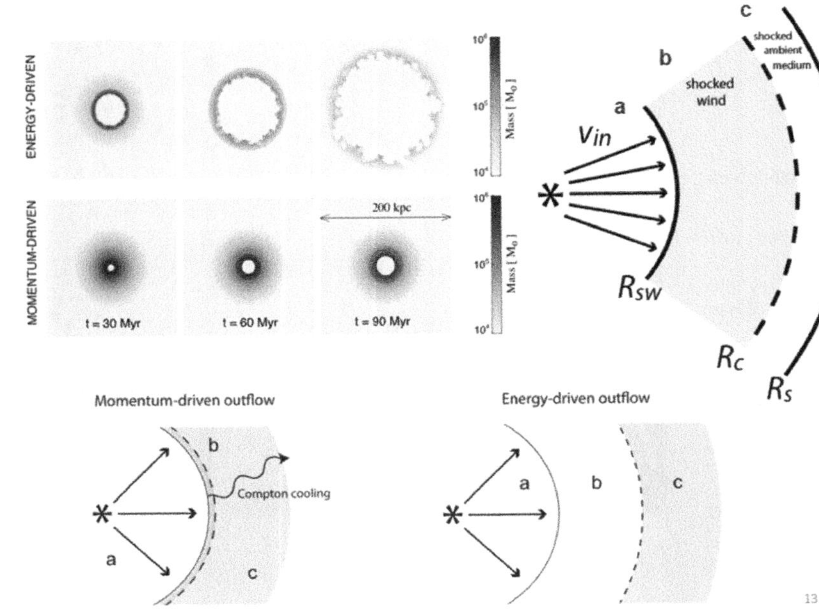

Figure 5.16. *Diagram (top right) of the various outflow phases around an AGN in the radiative phase (quasar mode). R_c is the discontinuity surface, R_s the shock in the interstellar medium, R_{SW} the reverse shock towards the relativistic wind (source: Faucher-Giguère and Quataert 2012). Zone "a" is characterized by the inflow velocity V_{in}, then zone "b" of the shocked wind, which can be considerably reduced in volume, if the cooling is effective, as in the hypothesis of a momentum conserved flow, which does not conserve energy. The "b" zone remains important, in the case where energy is conserved, as shown by the simulations by Costa et al. (2014). At the top, in orange, the gas shell extends very far radially, in the case where energy is conserved. At the bottom, in blue, the extension of the gas is much smaller, because energy is not conserved. Each panel corresponds to a 200 kpc square, for a black hole with mass 10^8 $M_\odot$, which emits the Eddington luminosity. The stages of the simulation are 30, 60 and 90 million years (source: Costa et al. (2014))*

This explains why the molecular flow momentum is much greater than LAGN/c. Among the various molecular flows observed in recent years, statistics have shown that more energy is conserved. In fact, the flow momentum is on average 20 times greater than that of the AGN wind (see Figure 5.17, Cicone *et al.* (2014)).

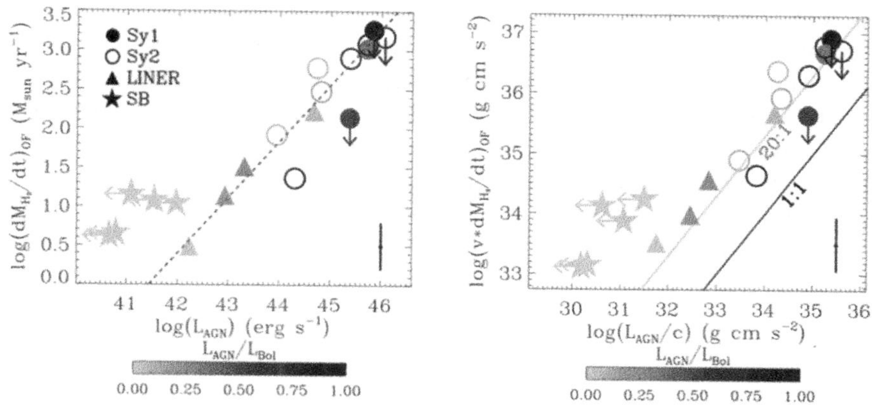

Figure 5.17. *Left: molecular outflow rate, as a function of the AGN bolometric luminosity (L_{AGN}), for several objects, identified by symbols: circles = Seyfert galaxies, triangles = LINERs, stars = starburst galaxies. The color indicates the relationship between the luminosity of the AGN and the bolometric luminosity of the host galaxy (L_{AGN}/Lbol), according to the color palette shown below. Only galaxies with an AGN verify the correlation (not starbursts). On the right: the molecular flow momentum is plotted as a function of the quasar wind momentum L_{AGN}/c, using the same symbols and colors (source: Cicone et al. (2014))*

5.2.4.2. *The mechanical mode, or radio mode*

When AGN is in a low-luminosity phase, the only energetic manifestation is that of the radio jet, if it exists. The jet can drag out interstellar gas, if the coupling with the disk is sufficient. This is often the case, because the jet is not perpendicular to the plane of the galaxy, as one might think. Instead, the jet is perpendicular to the accretion disk, but the latter is quite independent and decoupled from the orientation of the large-scale disk. The sizes are very different; the molecular tori are of the order of parsec, the galactic disks of the kilo parsec, and the timescales are the same (dynamic times different by a factor of 1,000). Even gravitationally, the potential felt by the accretion disk and the molecular torus is quasi-spherical, Keplerian under the influence of the central black hole, and does not know the orientation of the large-scale disk. Very often the jet is observed sweeping the gas of the galaxy (e.g. Garcia-Burillo *et al.* 2014; Dasyra *et al.* 2015).

The radio mode is characterized compared to the quasar mode, not only by the luminosity of the AGN, but also by the collimation of the molecular outflow. Sometimes the latter looks like an extremely fine jet (Aalto *et al.* 2016), and only a radio jet can cause it, even though it has not yet been observed. On the contrary, the impact of a highly collimated radio jet does not seem likely to be very significant, if its action is only to dig a channel in the interstellar medium and thus escape the galaxy. However, it is not, because step by step, the jet heats all of the surrounding gas, and because of Kelvin–Helmholtz instabilities, has a cross section much larger than its own. It then forms a cocoon, the impact of which is considerable on the galaxy, both in energy and amount of momentum transferred. The simulation of Figure 5.18 shows the formation of this cocoon, in a fractal medium representative of the interstellar medium, and its porosity (Wagner and Bicknell 2011).

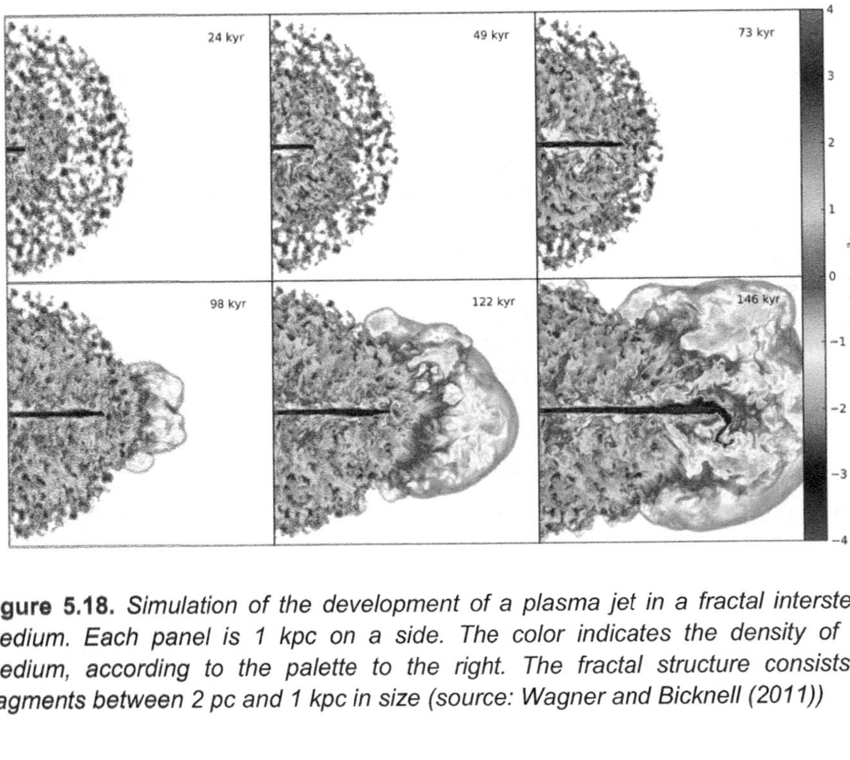

Figure 5.18. *Simulation of the development of a plasma jet in a fractal interstellar medium. Each panel is 1 kpc on a side. The color indicates the density of the medium, according to the palette to the right. The fractal structure consists of fragments between 2 pc and 1 kpc in size (source: Wagner and Bicknell (2011))*

5.3. Black hole fueling processes

5.3.1. *Problem of angular momentum, stellar processes*

To feed the central black hole, the matter must lose a lot of angular momentum. As the latter is proportional, per unit of mass, to the velocity and to the radius, and the velocity varies very little, it is the radius that determines the orders of magnitude. For matter near the center (~ 1 kpc), it will be necessary to arrive on the accretion disk, whose radius is typically 10^{-2} pc, therefore a loss of a factor 10^5. What is the inflow of matter necessary for a luminous quasar? A quasar can emit Vs; up to $L = 10^{13}\ L_\odot$, that is, 10^{39} W (10^{46} erg/s). With a radiation efficiency of $\varepsilon = 10\%$, the luminosity is written as $L = dM/dt\ c^2\ \varepsilon$, we therefore need an accretion rate of $dM/dt = 1.7\ (0.1/\varepsilon)\ (L/10^{39}\ W)\ M_\odot$/year. Assuming typical duty cycles of 100 million years, the black hole would have to swallow $2\ 10^8\ M_\odot$, meanwhile, which is a notable fraction of the gas in a galaxy. This matter is not in the immediate vicinity, and it will take a dynamic mechanism to exchange the angular momentum at the kpc scale in the galaxy.

To bring in the gas in such a short time, and cause it to lose its angular momentum, torques are required, and therefore tangential forces, produced by disk disturbances. If the disk remains axisymmetric, the gravitational forces are only radial, and there are no torques. Asymmetries can be bars, spirals, or interactions between galaxies. The formation of these disturbances has been described in the "Galaxies" volume of this series; in this chapter, we will describe and quantify the angular momentum exchanges.

First, let us consider the matter in the immediate environment of the black hole that could initially fuel it: the stars.

Stars lose mass over their lifetime, on average for a star cluster, at a rate of 10^{-11} $M_\odot$/year, per $M_\odot$ of star. If there is a nuclear star cluster in the few central parsecs, then it is possible to have enough gas that has already lost its angular momentum to form the star cluster. The formation of this nuclear cluster could be due to a merger of galaxies, capable of forming a cluster of $4\ 10^9\ M_\odot$ and therefore provide $\sim 10^7\ M_\odot$ in one activity cycle, almost simultaneously with the starburst (Norman and Scoville 1988). However, in this starburst formation, the direct gas supply may be dominant. The stars could also help fuel the black hole when there is no gas. They can be destroyed by tidal forces, reduced to gas, and then swallowed by the black hole accretion disk. For this, the mass of the black hole M_{BH} must be less than the Hills mass, of $3\ 10^8\ M_\odot$. This limit arises from the fact that the average density $\langle\rho\rangle$ of a black hole inside its horizon (of radius $Rh = 2\ G\ M_{BH}/c^2$) is a decreasing function of its mass: $\langle\rho\rangle \propto M_{BH}/R^3 \sim 1/M_{BH}^2$. This density becomes equal to the average density inside a star in the main sequence, for $M_{BH} = 3\ 10^8\ M_\odot$. For more massive black holes, stars are no longer destroyed by the tidal effect outside the horizon, but inside, and no longer contribute to AGN activity (Hills 1975). It should also be taken into account that stars in the vicinity of a black hole can be "swollen" or "bloated", therefore more fragile and sensitive to mass loss. The stars could also be giants. There is also another process that could fuel the black hole: collisions between stars, in the dense stellar cluster existing in the nucleus. The velocities of stars as they approach the black hole are relativistic, and collisions can tear off a lot of mass, or even destroy stars (Rauch 1999). Collisions also have the advantage of redistributing angular momentum, and thus refilling the loss cone, through the distribution of low angular momentum stars, which have already been swallowed by the black hole. Collisions between stars are now a preferred way to form intermediate-mass black holes in dense star clusters (Reinoso *et al.* 2018; Sakurai *et al.* 2019).

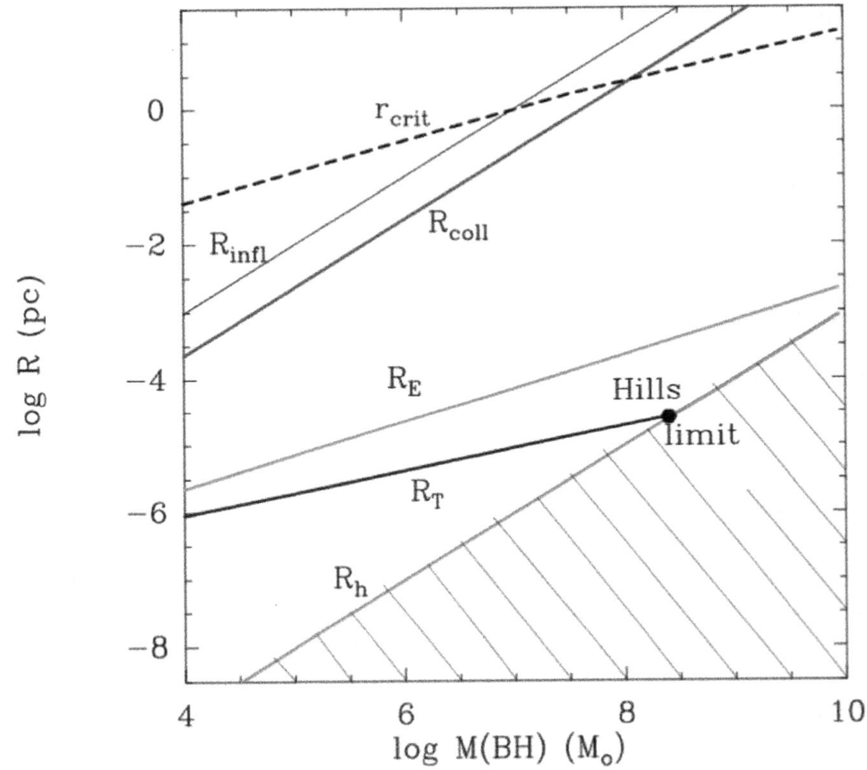

Figure 5.19. *Characteristic radii corresponding to different physical phenomena, depending on the mass of the black hole. From top to bottom, R_{infl}, the radius of the black hole sphere of influence, R_{coll}, the collision radius, R_E, the Eddington radius, R_T, the tidal radius, and R_h, the black hole horizon or Schwarzschild radius (see definitions in text). The loss cone effects for stars are significant within the critical radius r_{crit} (dotted lines)*

The processes to fuel the super-massive black hole with stars alone are necessary for AGNs active at z = 0 in galaxies with no gas available. It is useful to quantify them to determine some characteristic scales, as illustrated in Figure 5.19. The radius R_{infl} corresponds to the black hole sphere of influence, $R_{infl} = GM_{BH}/\langle V^2 \rangle$, already defined at the beginning of this chapter, depending on the velocity dispersion of the bulge <V>. The collision radius R_{coll}, is the one below which collisions release gas, that is, the velocity of stars around the black hole $(GM_{BH}/r)^{1/2}$ is comparable to the escape speed v∗ of a typical star with mass M* and radius r*, $v^* = (GM*/r*)^{1/2}$. The Eddington radius R_E is the radius below which a star receives more energy than its own Eddington luminosity, $R_E \sim r*(M_{BH}/M*)^{1/2}$. From this moment, their envelopes can evaporate or form "bloated" ("swollen") stars, more fragile to mass loss. The tidal radius R_T is that below which a star can be destroyed by tidal forces from the black hole, it can be calculated as the Roche radius, where the density of a star equals the average density of the black hole: $R_T \sim r*(M_{BH}/M*)^{1/3}$. Figure 5.19 quite successfully shows the Hills mass when R_T becomes equal to the horizon of the black hole. There is then no more activity in the nucleus when the black hole swallows a star, but on the other hand, its mass still grows!

5.3.2. *Fueling from gas, gravitational torques*

To evacuate the angular momentum of the gas outwards, torques are required. Could these be viscous torques? Interstellar gas in a galactic disk actually has very low viscosity. It is either very dispersed (atomic gas of average density 1 particle per cm^3), or fragmented into dense molecular clouds, which collide barely once or twice per rotation. Clouds cannot be viewed as fluids, but as ballistic particles, with some dissipation due to collisions. In the thin accretion disk around the black hole, there are very significant viscous torques. Viscosity is modeled as a force proportional to pressure. The magnitude of the viscous torques is such that they can reduce gas to a radius of 4 pc in a billion years. Beyond that, the timescales will be much too large for viscosity to be efficient (e.g. Pringle 1981). Viscous torques are negligible for galactic disks at the kilo parsec scale; however, the disks are subject to gravitational instabilities. The instabilities will produce disturbances, such as spiral arms and bars. The instabilities heat the disk, and create the dispersion of streaming motion velocities, which stabilize the stars (Toomre parameter $Q >> 1$). The gas is dissipative and radiates this energy. The Toomre parameter reverts to 1 for gas, which may be subject to other instabilities, and so continuously. These instabilities, constantly renewed, transfer mass and angular momentum, and are at the origin of an effective kinematic viscosity (Lin and Pringle 1987).

The nature of spiral and barred instabilities has been described in the "Galaxies" volume (see Chapter 4). The stellar bar of a galaxy is a density wave that can generate spiral waves in the gas component. The stars follow in their families of periodic orbits that are parallel or perpendicular to the bar, changing 90° with each Lindblad resonance. The gas flow lines also gradually rotate 90° at each resonance, following a spiral wave. The gravity torques exerted by the bar on the gas spiral are summarized in Figure 5.20. They change sign at each resonance.

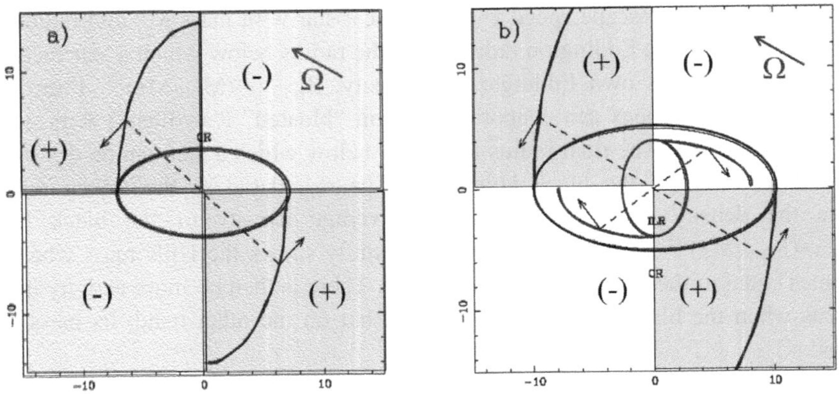

Figure 5.20. *The stellar bar (here horizontal) divides the plane of the disk into four quadrants: in pink, the gravitational torques of the bar on the gas are positive, and in blue, the torques are negative. The direction of the torques is oriented by the direction of rotation in the galaxy (here the direct direction). The spiral arms generated by the barred potential on the gas are in the pink quadrants outside the corotation (ends of the bar), and in the blue quadrants between corotation and Internal Lindblad Resonance (ILR)*

It is evident from the nature of the gravitational torques that the gas will be driven from the corotation outwards, where it will accumulate at the Outer Lindblad Resonance (OLR). At resonances, the gas settles in rings, circular or elongated, which are symmetrical about the bar: thus, no torque is applied to it, and the gas accumulates to form stars. Inside the corotation, the torque is negative, and the gas will accumulate in an internal resonance ring or ILR.

To quantify the time scales of these angular momentum transfers, it is possible to calculate the gravitational torques directly on the observed galaxies. The gravitational potential of a spiral galaxy, mainly due to stars, is obtained by projecting a near infrared image. The forces are calculated at each pixel of the disk by derivation, assuming the disk is thin, isothermal, of a constant height scale equal to ~1/12th of the radial scale of the image (Garcia-Burillo *et al.* 2005). On the torque map, which is broken down into four quadrants around the bar, the contours of the molecular gas overlap, as shown in Figure 5.21. It is easy to see whether the gas is concentrated in the positive or negative regions. By taking the azimuthal average, we obtain the angular momentum transfer rate at each radius.

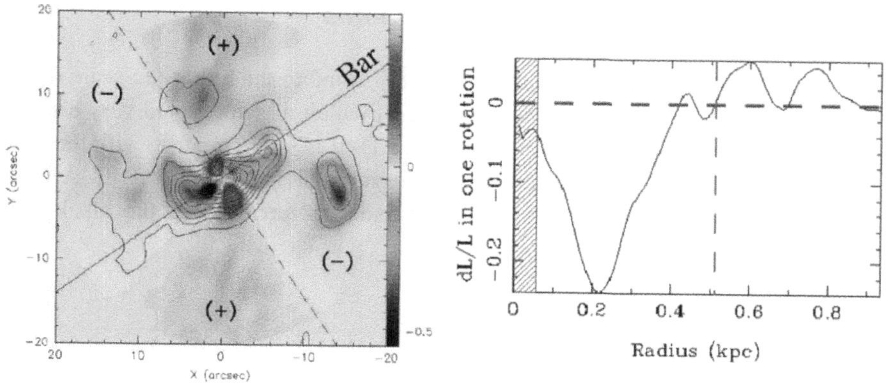

Figure 5.21. *Calculation of the gravitational torques in an observed spiral galaxy. Left: color map of torques in the de-projected galaxy. The major axis of the galaxy is horizontal. The bar separates the space into four quadrants of positive (red) or negative (blue) torques. The contours represent the observed molecular gas density. On the right: from the two-dimensional map of the torques (figure on the left), the azimuthal average of the torques exerted on the gas, weighted by the density of the gas in each pixel, is calculated for each radius. The torque is equal to the change in angular momentum dL/dt, and the unit of the vertical axis is chosen as the relative change in L in one rotation, at each radius. The torque is negative at the center, and will transfer 30% of the angular momentum of the gas in one rotation (source: Casasola et al. (2011))*

The calculation carried out on around 30 nearby galaxies with an active nucleus, but of low luminosity (Seyfert or LINER) showed that at the scale of ~100 pc, only a third of the galaxies were in the active nucleus feeding phase (Garcia-Burillo and Combes 2012). The rest of the time, the gas is stalled at the ILR and forms stars in a ring. From the fraction of negative torques, we can deduce a fairly short feeding phase for black holes, of around 10^7 years. Once the mass has accumulated sufficiently towards the center and into the ring, the next step is the decoupling of a secondary bar, or a nested nuclear structure. Mass accumulation increases the rotational velocity at the center and the rate of precession Ω–κ/2, which is close to bar pattern speed. Two ILR resonances occur, weakening the primary bar, as the periodic orbits between the two ILRs are perpendicular to the bar. A faster secondary bar decouples in the center, inside the smaller ILR, as shown by numerical simulations. A certain number of nested structures can even follow one another, of symmetry m = 2 and m = 1 closer to the black hole (Keplerian potential, see Hopkins and Quataert (2010)).

5.3.3. *Sphere of influence and molecular torus*

The series of nested structures must certainly come to an end at the scale of the molecular torus, of size 1–10 pc, the existence of which was suggested very early in the unification paradigm of AGNs (see Chapter 1). It is only recently, with the very high resolution and sensitivity of millimeter interferometers (ALMA, NOEMA), that it has been possible to detect molecular tori and structures within the black hole sphere of influence.

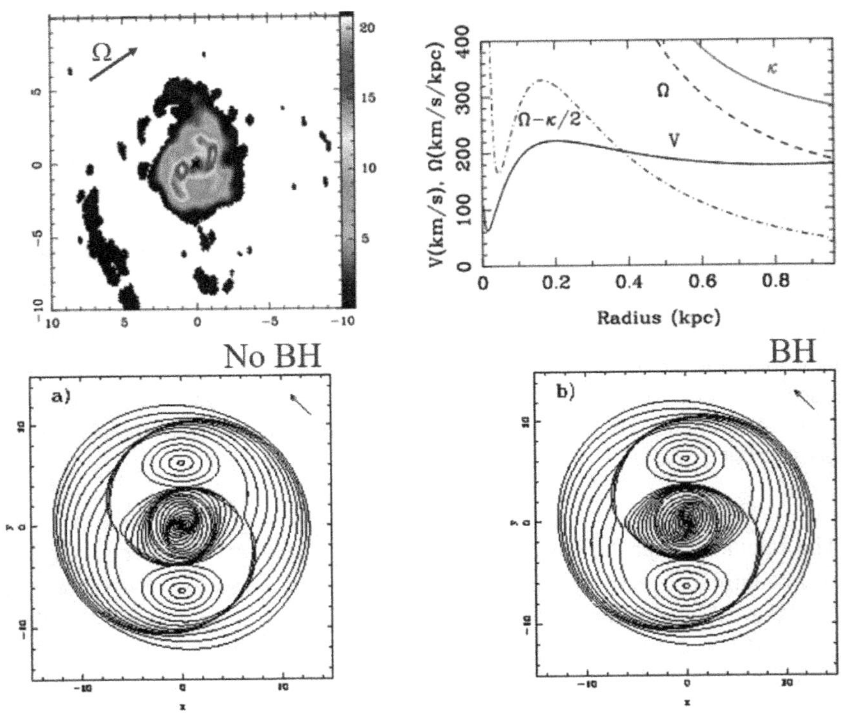

Figure 5.22. *Detection of a "trailing" nuclear spiral at the core of NGC 1566 (Combes et al. 2014). Top left: map of the CO line (3-2) observed with ALMA, in the projected galaxy plane (axes are in arcseconds, 1" = 35 pc). Top right: rotational velocity V, $\Omega = V/r$ and epicyclic frequency κ in NGC1566. The blue curve $\Omega - \kappa/2$ is the precession rate of elliptical orbits. This rate goes up in the black hole sphere of influence to within 50 pc. Bottom: representation of the gas flow lines in the bar rotating frame (the bar is horizontal). At the corotation, the gas only travels through an epicycle at the Lagrange points L4 and L5, on the vertical of the center. In the absence of a black hole, the orbital precession velocity $\Omega - \kappa/2$ does not increase towards the center, and the nuclear spiral structure is "leading". It is "trailing" with a black hole (source: Buta and Combes (1996))*

As one enters the black hole sphere of influence, the precession velocity of the elliptical orbits, which drives the gas streamlines, increases, as the frequencies Ω and κ vary in a Keplerian manner as $r^{-3/2}$. When the gas is precipitated towards the center by the gravitational torques, its rate of precession first decreases, which causes the spiral to proceed in the leading direction, then increases, which causes the spiral to reverse in the trailing direction. This has been observed for the galaxy NGC 1566, as shown in Figure 5.22 (Combes *et al.* 2014). The winding sense of the spiral controls the sign of the gravitational torques, as shown in Figure 5.20. Without the black hole, the spiral is leading and the torque is positive: the gas accumulates towards the ILR (inner resonance). On the contrary, in the presence of a black hole, the torque is negative, and the gas will feed the nucleus.

A similar trailing nuclear structure was also observed in NGC 613 and NGC 1808 (Audibert *et al.* 2019, 2021). The precise calculation of the gravitational torques has confirmed the feeding of the black hole, with a timescale of around four million years (see Figure 5.23).

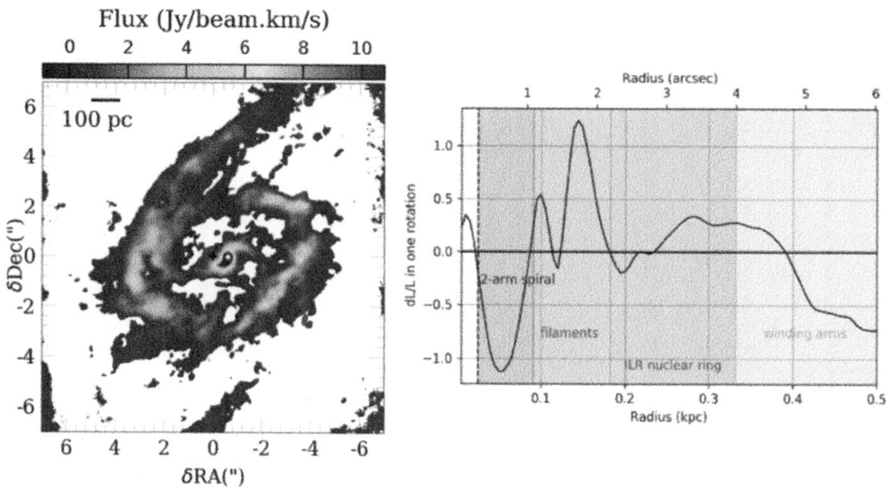

Figure 5.23. *ALMA map of the CO line (3-2) at the center of NGC 613, showing the detection of a trailing nuclear spiral inside the ring at ILR resonance (1" = 83 pc). On the right, the calculation of the gravitational torques shows that inside 100 pc, the gas loses its angular momentum in one rotation (source: Audibert et al. (2019))*

The nuclear spiral structures observed are the last before the molecular torus. Just inside these spirals, compact circumnuclear disks have been observed, which appear to be decoupled by their morphology and kinematics. Most often, their orientation and inclination on the sky plane are different from those of the disk of the host galaxy (Combes *et al.* 2019). These components correspond to what is expected, to account for the obscuration of certain active nuclei, depending on their orientation in the line of sight. It is not clear from spatial resolution whether they are really torus-shaped with a molecular gas deficiency in the center, except in a few rare cases. The radius of molecular tori varies from 6 to 27 pc, and their mass from 0.7 to 3.9 10^7 $M_\odot$. The torus is generally centered on the continuum source detected in millimeter waves, which is essentially synchrotron emission from AGNs.

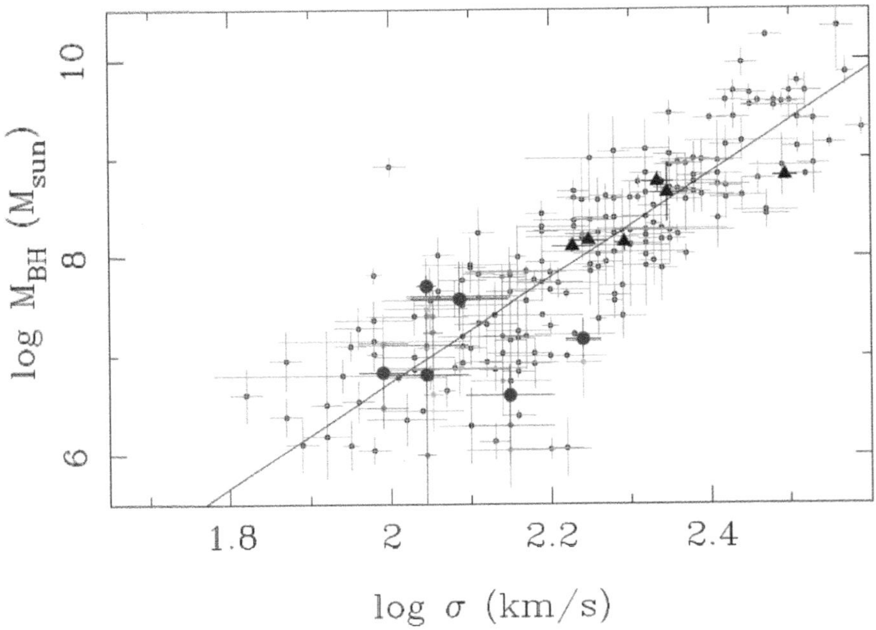

Figure 5.24. *Measurement of the mass of black holes by the molecular gas kinematics method in the black hole sphere of influence. The red circles represent spiral galaxies with low-luminosity AGNs (Combes et al. 2019), and the blue triangles represent the similar measurement in elliptical galaxies with higher spheroid masses (WISDOM project, Davis et al. 2018). The gray symbols come from van den Bosch (2016), adjusted by a straight line of slope 5.35. Estimates of black hole mass in spiral galaxies using different methods (red circles) are also shown in green, turquoise and pink*

It is no surprise that the circumnuclear disk (or torus) is decoupled from the rest of the galaxy. Indeed, inside the black hole sphere of influence, when the potential becomes Keplerian and spherical, the gas no longer has any information about the orientation and flattening of the rest of the galaxy. However, there is still a memory of the original angular momentum of the gas. Although, the phenomenon of star formation and associated feedback (ejection of supernovae, fountain effect above the plane, and subsequent gas accretion) randomizes the new orientations of the gas (e.g. Emsellem *et al.* 2015).

With the molecular gas being observed in the black hole sphere of influence, in relative dynamic equilibrium, it is possible to deduce the mass of the black hole, by using the kinematic method. This method is all the more welcome since the other methods do not give very precise results for low spheroid masses and low black hole masses. Scattered measurements may reflect intrinsic dispersion, but more galaxies observed in molecular gas at high angular resolution are needed to conclude (see Figure 5.24).

5.4. AGN feedback – efficiency

5.4.1. *The need for feedback*

The feedback phenomenon is invoked in the standard cosmological scenario ΛCDM to avoid forming galaxies that are too massive. As shown in Figure 5.25, the mass function of the observed galaxies is very far from what might be expected, from the mass function of the dark matter halos, assuming that the baryon fraction is equal to the universal fraction ($f_b = 17\%$) in each halo. The observed baryon deficit is particularly significant for small masses, where the feedback from supernovae is supposed to be very efficient, given the shallow depth of the potentials of these dwarf galaxies. For the most massive galaxies, feedback from supernovae is not efficient, and the role of AGNs is critical. The simulations succeed in suppressing the formation of very massive galaxies and it remains to verify the efficiency of this feedback in the observations.

Feedback from AGNs is particularly necessary for the most massive galaxies. However, these are the ones that have the most massive black holes too, and for which it is not certain that outward manifestations of energy are possible. Indeed, a quasar radiates efficiently from a thin accretion disk, which can no longer exist beyond a certain black hole mass, which we will define now. Indeed, the outer limit of the accretion disk at the origin of the AGN emission, is a function almost independent of the mass R_{disk}~0.01 pc ~ 2000 AU. Beyond that, the disk cannot subsist, the medium becomes self-gravitating and forms stars. On the contrary, the radius of the horizon is proportional to the mass of the black hole M_{BH}, Rh~ 20 ($M_{BH}/10^9$ $M_\odot$) AU, and the last stable orbit is 3 Rh for a non-rotating black hole and 0.5 Rh for maximum rotation (a = 1). In any case, the disk can only exist between this last stable orbit, and R_{disk}, and no longer exists for a black hole mass greater than 5 10^{10} $M_\odot$, assuming typical parameters (King 2016). Black holes can still increase their mass, but they will be invisible, with no feedback phenomenon possible.

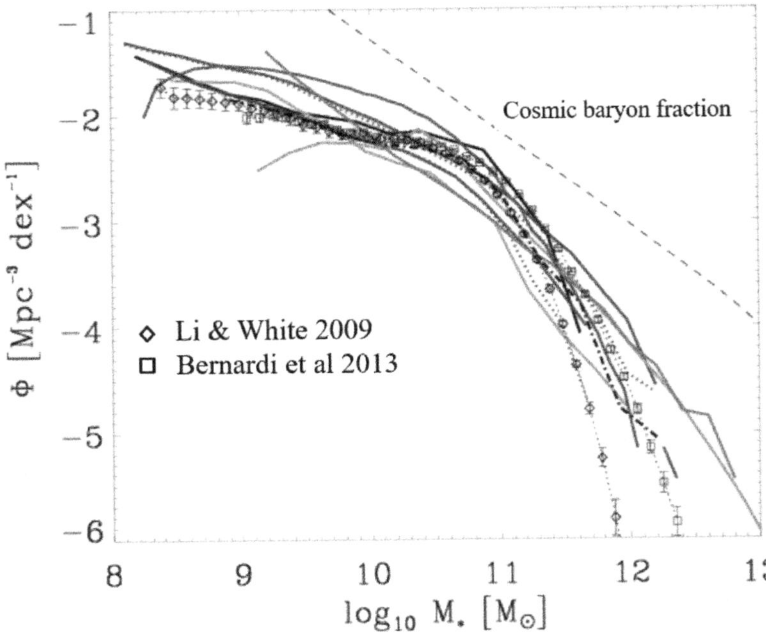

Figure 5.25. *Observed mass function of galaxies, rhombuses and squares with error bars (Li and White 2009; Bernardi et al. 2013) compared with cosmological simulations using the feedback effects of star formation and AGNs, calibrated on the observations (various color curves, corresponding to several simulations). The top dashed curve is the theoretical mass function, based on the mass function of dark matter halos, assuming the universal baryon fraction, without feedback (source: Naab and Ostriker (2017))*

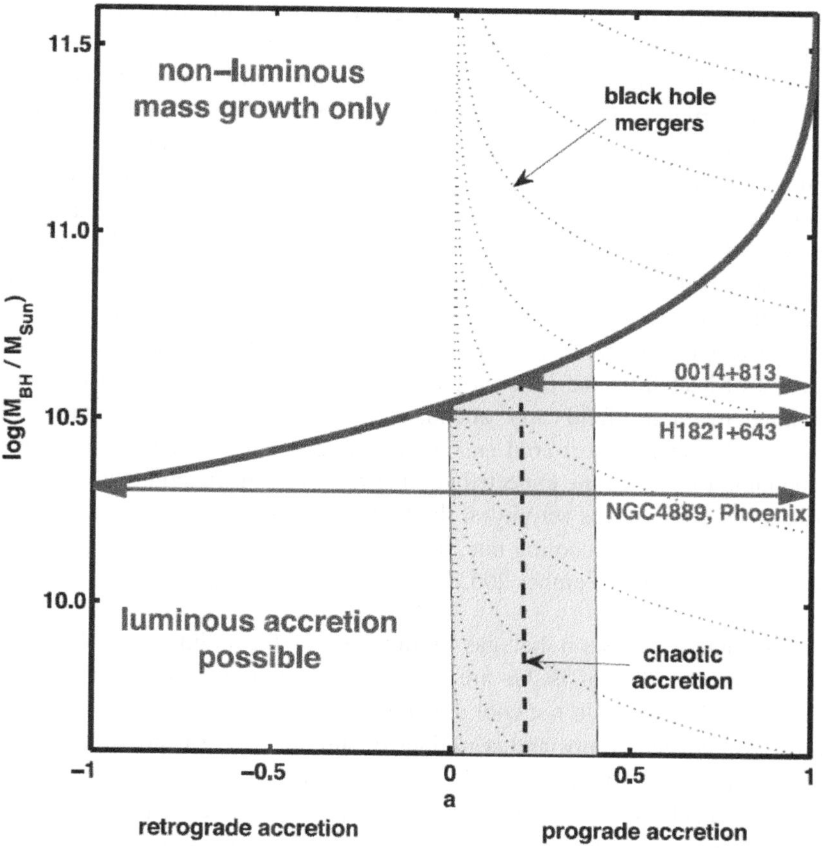

Figure 5.26. *Calculation of the maximum mass that the black hole of an AGN can have (curve in blue). A black hole can always increase its mass beyond, but this will be done without luminosity (and therefore without feedback). The abscissa is the dimensionless parameter "a", characterizing the spin of the black hole. The parameter a is between 0 and 1 (maximum rotation). Here, negative "a" values indicate retrograde accretion of matter with respect to the black hole spin. Three of the highest measured black hole mass values are shown in red. We see that the two highest values can already constrain the spin of the black hole. The dotted lines show the statistical effects of black hole mergers on their mass growth (source: King (2016))*

5.4.2. *Prototype: cooling flows*

Clusters and groups of galaxies are bathed in a very hot gas, visible in X-rays, the temperature of which can reach 100 million degrees for the richest clusters. This gas is in virial equilibrium in the cluster potential well, and the average timescale of its radiative cooling is greater than the Hubble time. Yet, the cooling rate is proportional to the square of the density, because the radiation of the ionized gas occurs when particles collide. At the center, where the density is higher, the cooling time t_{cool} becomes less than the Hubble time, typically $t_{cool} \sim 300$ million years. The gas must therefore cool, lose its pressure support and fall towards the central galaxy. The latter, called a Brightest Cluster Galaxy (BCG), is a very massive galaxy, often a cannibal, which swallows the satellites spiraling towards the center by dynamical friction. Most of the time, it is also an AGN. Gas accretion rates up to 1000 $M_{\odot}$/year are predicted, but are not observed. The stars that this cold gas should have formed are not detected either. Many hypotheses have been proposed, such as the conduction of heat from the external parts towards the center to avoid its cooling, a blocking magnetic field and abundance anomalies. However, since 2002, astronomers began to detect cold and molecular gas in the center of the clusters, corresponding to 10% of the expected cooling rate, and understood that this flow of gas was at the origin of the activity of the BCG nucleus. The resulting AGN is able to limit and moderate the cooling rate, by using the associated feedback phenomena (Edge 2001; Salomé and Combes 2003; Peterson and Fabian 2006).

The impact of AGN on hot gas cooling is via the radio mode. In relaxed, cool-core clusters, 70% of BCGs are radio sources, while there are only 20% in other clusters. The latter do not cool down, because they are dynamically disturbed by collisions between groups and accretions of sub-clusters. It is possible that all of the clusters spend some time in the relaxed and disturbed phases, cool-core or not. In the case of feedback, the radio jet of the central elliptical galaxy digs quite remarkable cavities in the hot cluster gas, up to 100 kpc from the center (see Figure 5.27). The cavities fill with plasma from the jet, forming a cocoon. This relativistic and non-thermal gas (synchrotron emission) is less dense than the hot gas in the cluster (thermal emission). They are in pressure equilibrium, but the less dense rises by Archimede forces. Cavities are bubbles that move, rise through buoyancy, and move away from the center. They occur in each phase of AGN activity, the jet of which is intermittent. Therefore, it is possible to see several cavities or bubbles in the hot gas (up to three sets, because then the bubbles disperse). The periods of activity of the jet are around 40 million years, similar to the rise time of the bubbles up to 20 kpc.

The cold molecular gas ($T \sim 20$ K) is observed far from the center, concentrated just outside the wall of the cavities (see Figure 5.27). It is possible that hot low entropy gas has been entrained by the bubbles and cools just above the bubble wall, where the gas is compressed, and the cooling time t_{cool} becomes about 10–20 times the free-fall time t_{ff} (McCourt *et al.* 2012). Alternatively, it can be considered that the gas infall time t_{infall} is much greater than the free-fall time, due to the turbulence and the hot gas entrained by the bubbles, and the fact that $t_{cool}/t_{infall} \sim 1$ is the factor triggering the cooling (McNamara *et al.* 2016). Simulations have shown that when t_{cool}/t_{ff} drops below 10–20, the hot gas is subject to thermal instabilities, which make the gas turbulent and inhomogeneous. The density can be higher than average in places and cause cooling (Gaspari *et al.* 2012). As soon as the gas cools, it fragments, becomes dense and molecular, loses pressure support and should free-fall in the form of filaments. In fact, the gas velocities observed by the CO lines (H_2 tracer) show a much lower speed, showing that the gas is slowed by the dynamical ram pressure of the hot gas and/or the magnetic field (Olivares *et al.* 2019). The filaments are spectacular in ionized gas, mapped by Hα (see Figure 5.27). Ionized gas is perfectly correlated with molecular gas, and can be considered as the interface between molecular filaments and the very hot ambient environment.

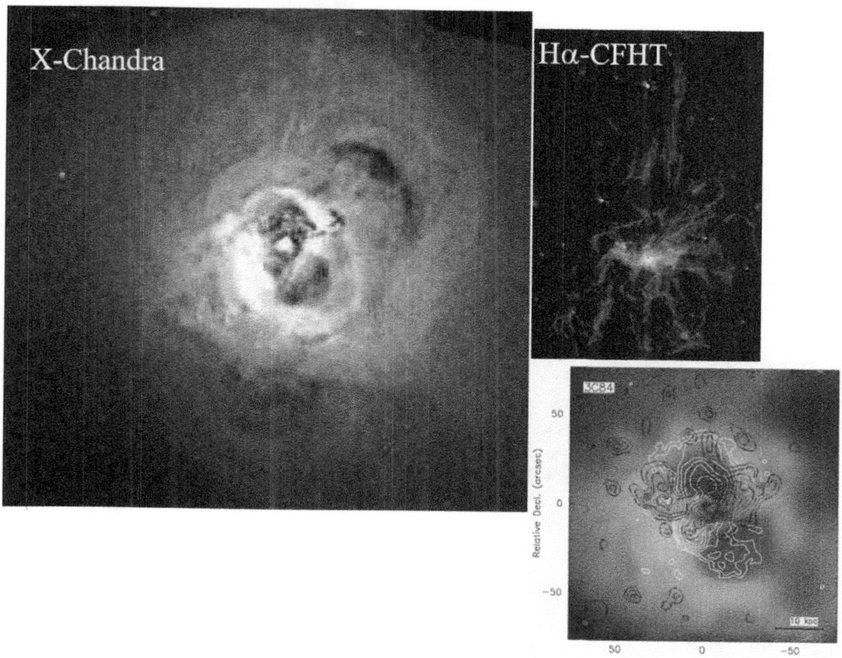

Figure 5.27. *The cooling flow in the Perseus cluster. Left: the X-ray image obtained by the Chandra satellite (Fabian et al. 2003). The cavities dug into the hot gas are spectacular. Top right: at the same scale, the Hα image obtained with the CFHT telescope by Gendron-Marsolais et al. (2018). Bottom right: on the blue and yellow X-ray image, the contours of the radio continuum of the VLA (radio jet) are superimposed in white and those of the molecular gas are in black (line of the CO molecule (2-1), IRAM-30m, Salomé et al. (2006)). The radio jet fills the cavity, as the cold molecular gas concentrates on the edges of the cavities, falling towards the center*

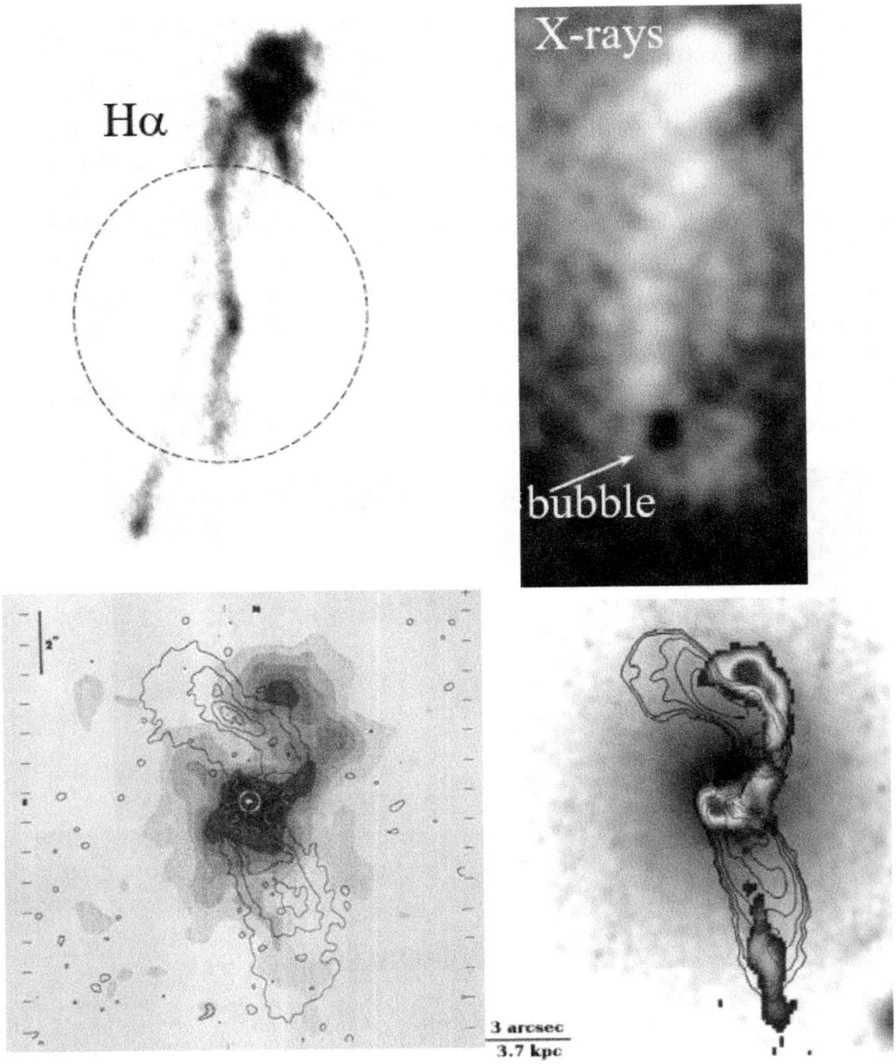

Figure 5.28. *The cooling trail cluster, Abell 1795. Top left: the BCG is not in the center of the cluster, symbolized by the dashed circle, with a radius of 22 kpc, but in the north, 33 kpc from the center (Cowie et al. 1983). Top right: the trail is also very visible in X-rays with the Chandra satellite (same scale). Bottom left: zoomed in image of the central part of the galaxy, the contours of the VLA continuum are superimposed on the Hα image in gray (van Breugel et al. 1984). Bottom right: color image at the same scale of the CO(2-1) map obtained with ALMA, superimposed on a grayed HST image (Russell et al. 2017)*

From the energy balance perspective, the AGN feedback model to moderate the cooling flow is quite compatible with the orders of magnitude provided by observations. It can be seen that the radiative losses of gas in the cooling zones are compensated by the thermal energy plus the mechanical energy of cavity expansion (PV). The cavities are emptied of hot gas, and the X-ray emission only comes from the shell surrounding the cavities. After de-projection, the surface brightness on the cavities is compatible with the emission from the wall alone. The mass of the shells is comparable to the mass ejected from the cavities. The X-ray gas was pushed out of the cavities by the plasma ejected by the radio source and compressed into the shells (see McNamara *et al.* (2016)).

Another phenomenon will be taken into account in the cooling of the gas moderated by the feedback of the central AGN. Clusters are frequently disturbed by the accretion of a group, or the merger with another cluster. In the relaxation that follows, the off-center BCG can oscillate around the center, with a period of approximately 300 million years, comparable to the gas cooling time. Thus, it can form a cooling trail behind the BCG, very visible in Hα, as shown in the case of Abell 1795 in Figure 5.28. Once again, the molecular gas is very correlated with the ionized gas, and is formed around the cavities dug by the radio jet (Russell *et al.* 2017). These motions generated by the interaction dynamics contribute to turbulence and inhomogeneities, which promote cooling far from the center of the BCG.

5.4.3. *Molecular outflows*

In the case of AGNs at the center of clusters, it is clear that the feedback phenomenon is essential to moderate the cooling of the hot gas and avoid the formation of too massive central galaxies. Is the feedback from low-luminosity AGNs at the center of nearby spiral galaxies (Seyfert or LINER) as efficient as, and well-coupled to the host galaxy? In section 5.2, we showed how statistically molecular outflows are correlated with the AGN luminosity, and that the radio mode was often most effective at low luminosities ($L < L_{Edd}$), provided that the radio jet is partly oriented in the plane of the host galaxy, and can sweep and drag the molecular gas.

This is indeed the case with a large number of nearby galaxies, such as NGC 1377, illustrated in Figure 5.29. This galaxy is very special, because its centimetric radio emission is not detected. However, there is a very collimated molecular outflow, which cannot come only from star formation and the feedback of supernovae. The molecular jet even reverses: it approaches, then moves away from the observer on either side of the disk (with alternating velocities shifted towards blue and then towards red). This reversal indicates a precession, with a jet moving in a plane close to the plane of the sky. This precession, which could be at an angle of approximately 10°, is not compatible with star formation feedback, but demonstrates the existence of a radio jet, below the detection threshold. The galaxy NGC 1068, Seyfert 2 prototype, due to obscuration, also has a jet in the plane of the galaxy (e.g. Garcia-Burillo *et al.* 2014). Either way, the AGN feedback is effective in delaying star formation by heating and displacing the available gas. However, the displaced quantities are quite small (a few percent at most) compared to the total gas content of the galaxy, and furthermore, the velocities are far from the escape velocity, so the gas will become available again to form other stars.

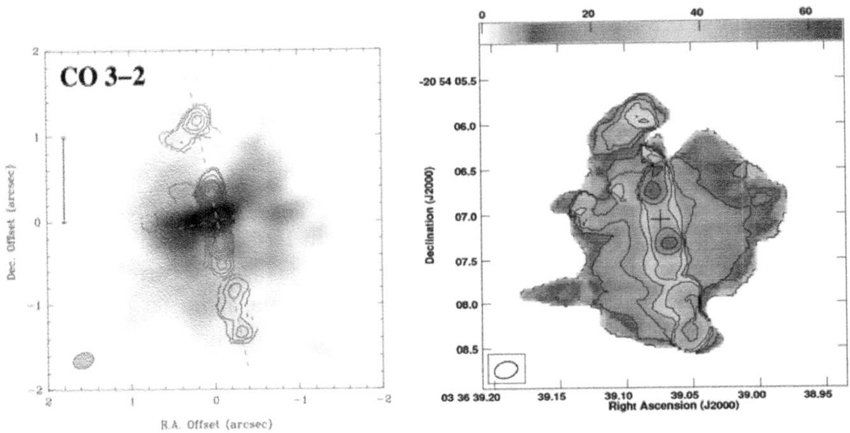

Figure 5.29. *NGC 1377 seen by ALMA. Left: the integrated CO(3-2) emission for velocities close to the systemic velocity (±30 km/s) is presented in gray. The emission in the approaching velocities is indicated by the contours in blue (-80 to -150 km/s), and receding velocities in red (80 to 150 km/s). The vertical bar indicates a scale of 100 pc. The ellipse at the bottom gives the angular resolution. Right: color map of the velocity dispersion (see palette) contours range from 4.4 to 57 km/s. The velocity dispersion underlines the molecular jet (source: Aalto et al. (2016))*

For nearby galaxies, it seems difficult to demonstrate that AGN feedback is efficient in stopping star formation. On the contrary, there are many statistics to show an excellent correlation between starburst formation and nucleus activity, and the molecular gas content may only be more fragile for strong luminosities (e.g. Fiore *et al.* 2017). Perhaps it could be a matter of timescales, the activity of the nuclei could be shifted compared to a starburst coming from a galaxy merger, but the common period is still long, up to 1.5 billion years (Volonteri *et al.* 2015). Nearby AGN host galaxies have quite normal star-forming properties, even in their central regions (Casasola *et al.* 2015). In numerical simulations, depending on the algorithms and resolutions used, the feedback from AGNs can be very efficient (Nayakshin 2014), or not efficient at all (Roos *et al.* 2015), sometimes the feedback is even positive (Gaibler *et al.* 2012; Bieri *et al.* 2015).

5.5. Conclusion

The relationship between masses of black holes and host spheroids is now well established, and interpreted as the co-evolution and symbiosis between black holes and their galaxies. The exceptions to this relationship are very indicative of the mechanisms behind this symbiosis. The simultaneous growth of the bulge and the black holes comes naturally from the gas accretion that feeds both, and also from the mergers of galaxies. Limiting and moderating star formation by AGN feedback has often been proposed to explain the relationship between the masses.

The feedback mechanisms can be broken down into two categories: (1) the quasar, or radiative mode, exerted by relativistic winds, coming from the brightest AGNs; (2) the radio, or mechanical mode, exerted by radio jets, coming from AGNs of low luminosities relative to their Eddington luminosity.

Molecular outflows are very frequently observed around nearby AGNs, with velocities ranging from 200 to 1200 km/s in the most extreme cases, typically ejecting 10^8 $M_\odot$ of gas. The mass ratio between the rate of ejected gas and the rate of star formation can range from 1 to 5. Statistically, the outflows conserve energy, and their angular momentum is increased compared to that of the wind, by a factor ~ 20. However, the feedback from AGNs, if it allows a little delay in star formation, does not seem efficient in stopping it completely.

The radio mode is very efficient in clusters, to moderate the cooling of hot gas in the vicinity of the central massive galaxy. This moderation is produced by the action of radio jets, whose non-thermal plasma sculpts cavities into the hot gas in the cluster. The energy of the AGN heats up the gas, but the bubbles that rise by buoyancy also generate the cooling of the gas, which forms molecular filaments that fall back on the black hole to feed it, thus closing the feedback cycle.

5.6. References

Aalto, S., Costagliola, F., Muller, S. *et al.* (2016). *A&A*, 590, A73.

Audibert, A., Combes, F., Garcia-Burillo, S. *et al.* (2019). *A&A*, 632, A33.

Audibert, A., Combes, F., Garcia-Burillo, S. *et al.* (2021). *A&A*, 656, A60.

Bacon, R., Emsellem, E., Combes, F. *et al.* (2001). *A&A*, 371, 409.

Baldassare, V.F., Reines, A.E., Gallo, E. *et al.* (2016). *ApJ*, 829, 57.

Bender, R., Kormendy, J., Bower, G. *et al.* (2005). *ApJ*, 631, 280.

Bernardi, M., Meert, A., Sheth, R.K. *et al.* (2013). *MNRAS*, 436, 697.

Bieri, R., Dubois, Y., Silk, J., Mamon, G.A. (2015). *ApJ*, 812, L36.

Blandford, R.D. and McKee, C.F. (1982). *ApJ*, 255, 419.

van den Bosch, R.C.E. (2016). *ApJ*, 831, 134.

van den Bosch, R.C.E., Gebhardt, K., Gültekin, K. *et al.* (2012). *Nature*, 491, 729.

van Breugel, W., Heckman, T., Miley, G. (1984). *ApJ*, 276, 79.

Buta, R. and Combes, F. (1996). *Fund. Cosmic Phys.*, 17, 95.

Casasola, V., Hunt, L.K., Combes, F. *et al.* (2011). *A&A*, 527, A92.

Casasola, V., Hunt, L.K., Combes, F., García-Burillo, S. (2015). *A&A*, 577, A135.

Cicone, C., Maiolino, R., Gallerani, S. *et al.* (2014). *A&A*, 562, A21.

Combes, F., García-Burillo, S., Casasola, V. *et al.* (2015). *A&A*, 565, A97.

Combes, F., García-Burillo, S., Audibert, A. *et al.* (2019). *A&A*, 623, A79.

Costa, T., Sijacki, D., Haehnelt, M.G. (2014). *MNRAS*, 444, 2355.

Cowie, L.L., Hu, E.M., Jenkins, E.B., York, D.G. (1983). *ApJ*, 272, 29.

Cowie, L.L., Songaila, A., Hu, E.M., Cohen, J.G. (1996). *AJ*, 112, 839.

Dasyra, K.M., Bostrom, A.C., Combes, F., Vlahakis, N. (2015). *ApJ*, 815, 34.

Davis, B.L., Berrier, J.C., Johns, L. *et al.* (2014). *ApJ*, 789, 124.

Davis, T.A., Bureau, M., Onishi, K. *et al.* (2018). *MNRAS*, 473, 3818.

Edge, A. (2001). *MNRAS*, 328, 762.

Emsellem, E. (2013). *MNRAS*, 433, 1862.

Emsellem, E., Renaud, F., Bournaud, F. *et al.* (2015). *MNRAS*, 446, 2468.

Faber, S.M. and Jackson, R.E. (1976). *ApJ*, 204, 668.

Fabian, A.C. and Rees, M.J. (1995). *MNRAS*, 277, L55.

Fabian, A.C., Sanders, J.S., Allen, S.W. *et al.* (2003). *MNRAS*, 344, L43.

Faucher-Giguère, C.-A. and Quataert, E. (2012). *MNRAS*, 425, 605.

Fender, R., Corbel, S., Tzioumis, T. (1999). *ApJ*, 519, L165.

Ferrarese, L. and Merritt, D. (2000). *ApJ*, 539, L9.

Feruglio, C., Maiolino, R., Piconcelli, E. *et al.* (2010). *A&A*, 518, L155.

Fiore, F., Feruglio, C., Shankar, F. *et al.* (2017). *A&A*, 601, A143.

Gaibler, V., Khochfar, S., Krause, M., Silk, J. (2012). *MNRAS*, 425, 438.

Garcia-Burillo, S. and Combes, F. (2012). *JPhCS*, 372, a2050.

Garcia-Burillo, S., Combes, F., Schinnerer, E. *et al.* (2005). *A&A*, 441, 1011.

Garcia-Burillo, S., Combes, F., Usero, A. *et al.* (2014). *A&A*, 567, A125.

Gaspari, M., Ruszkowski, M., Sharma, P. (2012). *ApJ*, 746, 94.

Gendron-Marsolais, M., Hlavacek-Larrondo, J., Martin, T.B. *et al.* (2018). *MNRAS*, 479, L28.

Greenhill, L.J., Jiang, D.R., Moran, J.M. *et al.* (1995). *ApJ*, 440, 619.

Gültekin, K., Richstone, D.O., Gebhardt, K. *et al.* (2009). *ApJ*, 698, 198.

Hainline, K.N., Reines, A.E., Greene, J.E., Stern, D. (2016). *ApJ*, 832, 119.

Hasinger, G. (2000). *LNP*, 548, 423.

Hills, J.G. (1975). *Nature*, 254, 295.

Ho, L.C. (2002). *ApJ*, 564, 120.

Hopkins, P.F. and Quataert, E. (2010). *MNRAS*, 407, 1529.

Hopkins, P.F., Richards, G.T., Hernquist, L. (2007). *ApJ*, 654, 731.

King, A. (2016). *MNRAS*, 456, L109.

Komossa, S., Burwitz, V., Hasinger, G. *et al.* (2003). *ApJ*, 582, L15.

Kormendy, J. and Bender, R. (2009). *ApJ*, 691, L142.

Kormendy, J. and Bender, R. (2013). *ApJ*, 769, L5.

Kormendy, J. and Ho, L.C. (2013). *ARA&A*, 51, 511.

Kormendy, J., Bender, R., Cornell, M.E. (2011a). *Nature*, 469, 374.

Kormendy, J., Bender, R., Cornell, M.E. (2011b). *Nature*, 469, 377.

Lauer, T.R., Bender, R., Kormendy, J. *et al.* (2012). *ApJ*, 745, 121.

Li, C. and White, S.D.M. (2009). *MNRAS*, 398, 2177.

Lidz, A., Hopkins, P.F., Cox, T.J. *et al.* (2006). *ApJ*, 641, 41.

Lin, D.N.C. and Pringle, J.E. (1987). *MNRAS*, 225, 607.

Madau, P. and Dickinson, M. (2014). *ARA&A*, 52, 415.

Magorrian, J., Tremaine, S., Richstone, D. *et al.* (1998). *AJ*, 115, 2285.

Marconi, A., Risaliti, G., Gilli, R. *et al.* (2004). *MNRAS*, 351, 169.

McCourt, M., Sharma, P., Quataert, E., Parrish, I.J. (2012). *MNRAS*, 419, 3319.

McGaugh, S.S., Schombert, J.M., Bothun, G.D., de Blok, W.J.G. (2000). *ApJ*, 533, L99.

McNamara, B.R., Russell, H.R., Nulsen, P.E.J. *et al.* (2016). *ApJ*, 830, 79.

Moran, J.M. (2008). *ASPC*, 395, 87.

Mutlu-Pakdil, B., Seigar, M.S., Davis, B.L. (2016). *ApJ*, 830, 117.

Naab, T. and Ostriker, J.P. (2017). *ARA&A*, 55, 59.

Nayakshin, S. (2014). *MNRAS*, 437, 2404.

Norman, C. and Scoville, N.Z. (1988). *ApJ*, 332, 124.

Olivares, V., Salomé, P., Combes, F. *et al.* (2019). *A&A*, in press arXiv1902.09164.

Peiris, H.V. and Tremaine, S. (2003). *ApJ*, 599, 237.

Peterson J.R. and Fabian A.C. (2006). *Phys. Rep.*, 427, 1.

Peterson, B.M., Ferrarese, L., Gilbert, K. M. *et al.* (2004). *ApJ*, 613, 682.

Pringle, J.E. (1981). *ARA&A*, 19, 137.

Rauch, K.P. (1999). *ApJ*, 514, 725.

Reines, A.E. and Comastri, A. (2016). *PASA*, 33, 54.

Reinoso, B., Schleicher, D.R.G., Fellhauer, M. *et al.* (2018). *A&A*, 614, A14.

Roos, O., Juneau, S., Bournaud, F., Gabor, J.M. (2015). *ApJ*, 800, 19.

Russell, H.R., McNamara, B.R., Fabian, A.C. *et al.* (2017). *MNRAS*, 472, 4024.

Sakurai, Y., Yoshida, N., Fujii, M.S. (2019). *MNRAS*, 484, 4665.

Salomé, P. and Combes, F. (2003). *A&A*, 412, 657.

Salomé, P., Combes, F., Edge, A.C. *et al.* (2006). *A&A*, 454, 437.

Sanders, D.B. and Mirabel, I.F. (1996). *ARA&A*, 34, 749.

Scharwächter, J., Combes, F., Salomé, P. *et al.* (2016). *MNRAS*, 457, 4272.

Sikora, M. and Begelman, M.C. (2013). *ApJ*, 764, L24.

Sikora, M., Stawarz, L., Lasota, J-P. (2007). *ApJ*, 658, 815.

Soltan A. (1982). *MNRAS*, 200, 115.

Tombesi, F., Cappi, M., Reeves, J.N. *et al.* (2011). *ApJ*, 742, 44.

Tremaine, S. (1995). *AJ*, 110, 62.

Trujillo, I., Ferré-Mateu, A., Balcells, M. *et al.* (2014). *ApJ*, 780, L20.

Venemans, B.P., Walter, F., Zschaechner, L. *et al.* (2016). *ApJ*, 816, 37.

Volonteri, M., Capelo, P.R., Netzer, H. *et al.* (2015). *MNRAS*, 452, L6.

Wagner, A.Y. and Bicknell, G.V. (2011). *ApJ*, 728, 29.

Zubovas, K. and King, A. (2014). *MNRAS*, 439, 400.

6

Binary Black Holes in the Gravitational Wave Universe

Monica COLPI

University of Milano-Bicocca, Milan, Italy

The discovery by the LIGO Scientific Collaboration and Virgo Collaboration of the first source of high-frequency gravitational waves, GW150914, depicting the coalescence of two stellar black holes (Abbott *et al.* 2016), heralded the birth of a new era of exploration of the universe and led to a paradigm shift in our understanding of the collapse of very massive stars. Also, a new revolution is at our doorstep with the launch in the early 2030th of LISA, the Laser Interferometer Space Antenna that will survey for the first time the low-frequency Gravitational Wave Universe with sufficient sensitivity to grant us access to an immense cosmological volume. LISA will detect the inspiral and merger of the first massive black holes and help shed light onto how these objects formed and grew giant, to explain (i) the yet unknown origin of the early quasars outshining when the universe was less than one billion years old, and (ii) the evolution of the dormant supermassive black holes that inhabit the bright galaxies of today's universe. LISA will detect the teeming population of black holes of 10^{4-6} $M_\odot$ across all cosmic epochs, exploring the low-mass tail of the supermassive black hole. Prospected multi-messenger observations of merging massive black holes in gas-rich environments will let us also probe for the first time the physics of accretion in violently changing spacetimes, and possibly test the emergence of powerful jets in the imminent vicinity of a merger. The link between stellar and supermassive black holes is discussed to offer a view on how future facilities such as LISA and the

For a color version of the figures in this chapter, see www.iste.co.uk/combes/nuclei.zip.

Active Galactic Nuclei,
coordinated by Françoise COMBES.

Einstein Telescope will let us explore black hole birth processes across their entire mass spectrum. This chapter aims to describe the "astrophysical black holes" from a new perspective, that is, as powerful sources of gravitational waves (GWs), and to highlight the profound link between black hole evolution and galaxy assembly.

6.1. Introduction

According to general relativity, GWs are ripples in the fabric of spacetime propagating with speed equal to the speed of light. They are produced by non-symmetric, accelerated mass motions and travel almost unimpeded throughout the universe, carrying unique information on the physical properties at the source. The extreme accelerations experienced by colliding black holes in the imminent vicinity of the merger make them the loudest sources of GWs in the universe (Sathyaprakash and Schutz 2009).

Astrophysical black holes are described uniquely by their mass $M_{\rm BH}$ and spin vector, defined so that the modulus of the black hole angular momentum is $J_{\rm BH} = \chi_{\rm BH} G M_{\rm BH}^2/c$. While the mass is unconstrained and can vary from the Planck mass to arbitrarily large values, the dimensionless spin $\chi_{\rm BH}$ is limited to vary between [0,1], to avoid the occurrence of a naked singularity when $\chi_{\rm BH} > 1$.

The mass and spin vector determine all the observational features of black holes as electromagnetic (EM) and GW sources in different astrophysical settings. The mass and spin fix the dimension of the event horizon, the evanescent spherical surface below which not even light can escape, and the position of the last innermost stable circular orbit (ISCO). For accreting black holes, ISCO indicates the position of the inner rim of an equatorial accretion disk fueling the black hole, and for two coalescing black holes, the approximate distance below which the inspiraling black holes start their radial infall towards coalescence ending with the formation of a new, more massive black hole. It is close to ISCO that the bulk of the electromagnetic and gravitational waves is emitted, as gravity there provides a huge reservoir of energy. Binary black holes surrounded by a mini-disk fueled by a larger disk can jointly emit electromagnetic and gravitational waves.

GWs carry exquisite information on the black hole masses and spin, including spin orientation. Thus, detecting the GW signal from a population of both stellar and massive binary black holes as a function of redshift will provide unique insight on their origin and evolution, conveying rich new information about the history of star and structure formation in our expanding universe. In addition, studying their GW signals, that is, their phase and amplitude, will let us explore the nature of gravity in the strong field, the dynamical sector and its propagation properties (Berti *et al.* 2015; Barack *et al.* 2019).

Detecting the GW signal from merging massive black holes is the key science objective of the ESA-led space mission LISA (Amaro-Seoane *et al.* 2017; Colpi *et al.* 2019), the first interferometer in space that will detect low-frequency GWs, around 40 μHz and 0.1 Hz, emitted by binaries weighing between a few $10^4\,M_\odot$ and a few $10^7\,M_\odot$. Third-generation ground-based interferometers, such as the Einstein Telescope and Cosmic Explorer (Punturo *et al.* 2010; Sathyaprakash *et al.* 2012, 2019; Reitze *et al.* 2019), will detect high-frequency GWs, from 3 Hz to 5 kHz, emitted by the cosmic population of stellar black holes. Opening the GW window will shed light on the potential link between the stellar and supermassive black holes (Valiante *et al.* 2016) and how gravitational collapse develops on all scales, among stars and galaxies.

6.2. Black holes in the cosmic landscape

Astrophysical black holes[1] come in two different flavors: *stellar* and *supermassive* black holes.

Stellar black holes (sBHs) are widespread in all the galaxies of the universe, resulting from the core-collapse of very massive stars. The high masses of the sBHs revealed by LIGO-Virgo, up to $\sim 50\,M_\odot$ (Abbott *et al.* 2019), hint formation sites of low metallicity Z, below 0.5% of the solar value $Z_\odot = 0.0196$ (Heger *et al.* 2000; Spera and Mapelli 2017). Furthermore, for isolated black holes, theory predicts the existence of a high-mass gap, between about $60\,M_\odot$ and $150\,M_\odot$ (Whalen *et al.* 2013). In this window, pair-instabilities during oxygen burning lead to the complete disruption of the progenitor star. sBHs heavier than $150\,M_\odot$ can form at $Z < 1\%\,Z_\odot$, if the initial mass function of stars extends further out, up to hundreds of solar masses. Figure 6.1 is a diagram showing the mass spectrum of sBHs, the pair-instability gap strip and potential zones where beyond-gap sBHs may exist (we note that this gap is filled by black holes resulting from the coalescence of sBHs). Whilst editing this chapter, new gravitational wave signals from merging stellar black holes have been discovered by the LIGO, Virgo and KAGRA scientific collaborations (Abbott *et al.* 2021), and among these GW190521, the heaviest stellar black hole known as of today, of $\sim$ 150 $M_\odot$ (Abbott *et al.* 2020). This coalescence involved two black holes of 85 and 66 $M_\odot$, with the heavier in the nominal pair-instability gap, challenging stellar evolution models, and suggesting new avenues for the formation of such large stellar black holes.

Supermassive black holes (SMBHs) are observed as luminous QSOs and AGN, powered by accretion of gas (Merloni 2016), or as massive dark objects (MDOs) at the center of quiescent galaxies (Kormendy and Ho 2013). The SMBH mass spectrum currently observed extend from about $5 \times 10^4\,M_\odot$ up to about $1.2 \times 10^{10}\,M_\odot$, as illustrated in Figure 6.2. There is observational evidence that SMBHs underwent substantial cosmic evolution driven by two main mechanisms: accretion and, to a minor extent, mergers. Both processes are intimately linked with the evolution of their

1. In this chapter, we hypothesize that all EM sources powered by extreme gravity, such as AGN, quasars and X-ray sources, host Kerr black holes as described by general relativity.

host galaxy, which is in turn ruled by the complex process of galaxy formation and assembly. Recovering the properties of massive black holes at birth and tracking their irreversible evolution towards higher masses and possibly larger spins is one of the fascinating problems of contemporary astrophysics.

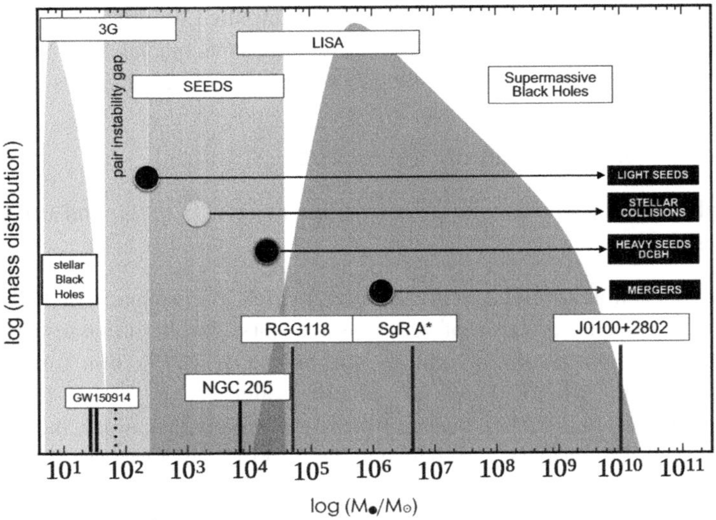

Figure 6.1. *Diagram illustrating the black hole mass spectrum encompassing the whole astrophysical relevant range, from sBHs to SMBHs, through the unexplored (yellow-beyond-gap and light-blue) zone where seed black holes of intermediate mass are expected to form and grow. Vertical black lines (dotted) on the left denote the two sBH (final sBH mass) in GW150914 (Abbott et al. 2016); the IMBH in the low-mass early-type galaxy NGC205 (Nguyen et al. 2019), and in the dwarf galaxy RGG118 (Baldassare et al. 2015); the SMBH in the Milky Way (SgR A*) (Genzel et al. 2010), and in the quasar J0100+2802 (the heaviest SMBH ever observed (Wu et al. 2015)). The mass distribution of sBHs, drawn from the observations of the Galactic sBH candidates (Özel et al. 2010), has been extended to account for the high-mass tail following the discovery of GW150914. The minimum (maximum) sBHs are arbitrarily set equal to* $3\,M_\odot$ *(*$60\,M_\odot$*), and the theoretically predicted pair-instability gap is depicted as a narrow gray strip. The SMBH distribution has been drawn scaling their mass according to the local galaxy mass function and SMBH correlation (Kormendy and Ho 2013). The decline below* $\sim 10^5\,M_\odot$ *is arbitrary. Dots illustrate seeds whose origin is indicated in the black frames*

SMBHs are considered to be the relic of those active black holes outshining as quasars as early as redshift $z \sim 7.5$ (Valiante *et al.* 2017; Bañados *et al.* 2018) and as AGN during the peak of the star formation rate (Merloni 2016). Their contribution to the local mass density, dominated by SMBH of mass $> 10^8\,M_\odot$ is found of the order of $\rho_\bullet^{BH} \sim 2 \times 10^5\,M_\odot\,Mpc^{-3}$ (Marconi *et al.* 2004). Further observations indicate that AGN contributed to the rise of a cosmic X-ray background resulting mainly from unresolved and obscured AGN of mass ~ 10^{8-9} M$_{\odot}$ in the redshift interval $0.5 < z < 3$. As gas accretion leads to mass deposition on the black hole, the light present in the X-ray astrophysical background mirrors the mass increment experienced by SMBHs over cosmic history due to accretion. This density increment is found to be close to $\Delta\rho^{\rm BH}_{\rm light} \sim 3.5 \times 10^5$ M$_{\odot}$ pc^{-3} (Merloni 2016).

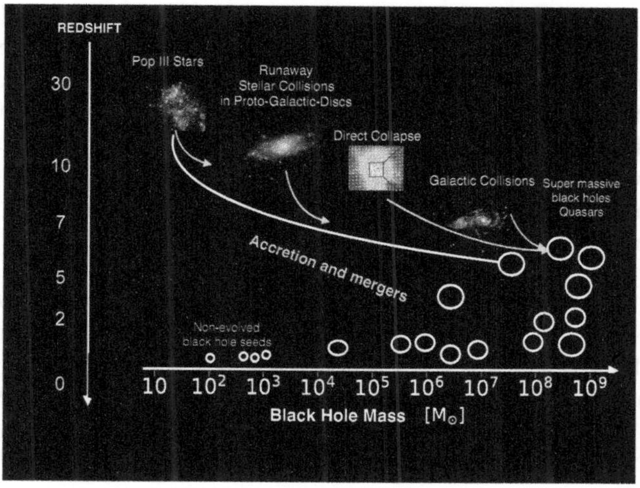

Figure 6.2. *Scenarios of black hole seed formation and evolution. In the early universe, seeds could form from Population III stars, from runaway stellar collisions in dense nuclear star clusters in proto-galactic disks, from the direct collapse of dense, unstable gas clouds illuminated by a UV background, or in lower-redshift gas-rich galactic collisions. In environments favorable for their growth, seeds acquire mass via episodes of intense accretion and mergers up to* 10^{8-9} M$_{\odot}$, *outshining as quasars as early as* $z \sim 7$. *Those seeds that did not grow into SMBHs can instead be found in the local universe as leftover IMBHs and/or as slowly growing massive black holes of* 10^{5-6} M$_{\odot}$ *that are still elusive to electromagnetic observations and that will be detected, for example, by the ESA-led mission Athena*

The close match between these two independent measurements is an indication that radiatively efficient accretion played a large part in the building of SMBHs in galaxies, from redshift $z \sim 3$ to the present. It further indicates that information residing in the initial mass spectrum of the, albeit unknown, black hole ancestor population has been erased during events of copious accretion and mergers along the course of cosmic evolution. Information on the natal spin is also lost: a non-rotating black hole is spun-up to $\chi_{\rm BH} = 1$ after increasing its mass by a factor $\sqrt{6}$ for prograde accretion. Conversely, a maximally rotating black hole is spun-down to null spin by retrograde accretion after increasing its mass by a factor $\sqrt{2/3}$. Thus, after many e-fold increases in the mass, the spin modulus can either reach values close to unity or change erratically, depending on whether accretion episodes occurred in a coherent or chaotic mode. Thus, the spin is a key parameter whose value conveys information on the accretion history (Sesana *et al.* 2014). Furthermore, coalescences can change the spin both in modulus and direction.

Current observations hint in favor of the existence at high redshifts of an underlying population of black holes of a smaller variety that grew along cosmic history through episodes of accretion and merging. They are called *seeds* (Volonteri and Begelman 2010; Valiante *et al.* 2016; Volonteri *et al.* 2021). The concept of seeds has also emerged to explain the appearance of a large number of SMBHs of 10^8 $M_\odot$ or higher, recently discovered at $z \sim 6-7$ (Bañados *et al.* 2018). Seeds are middleweight black holes that form quite early in cosmic history, at redshift z as large as 20, when the universe was only 180 Myr old. They form in metal-poor dark matter halos under rare conditions, and those that evolved to grow giant enable the presence of luminous quasars during the epoch of cosmic reionization.

Different formation channels have been proposed for the seeds. *Light* seeds refer to black holes with masses in the hundreds $M_\odot$ forming in metal-free/poor dark matter halos at redshifts as large as 20–30 (Hirano *et al.* 2015). Seeds of 10^3 $M_\odot$ may also rise from runaway collisions of massive stars in dense star clusters of low metallicity (Devecchi *et al.* 2012), or from repeated mergers of sBHs in star clusters subjected to gas-driven inflows (Lupi *et al.* 2014).

Heavy seeds refer instead to middleweight black holes of about 10^{4-5} $M_\odot$ resulting from the monolithic collapse of massive gas clouds (Umeda *et al.* 2016). Intense ultraviolet radiation from adjacent star-forming regions and large infall rates of metal-poor gas are required to suppress gas cooling and fragmentation in the birth clouds undergoing direct collapse. Alternatively, the formation of heavy seeds may be aided by massive nuclear inflows in major gas-rich galaxy mergers at lower redshifts (Mayer *et al.* 2015).

Often, black holes in the mass range between 100 $M_\odot$ up to $\sim$ 10^5 $M_\odot$ are called *intermediate mass* black holes (IMBH hereon). See again Figure 6.2, which illustrates scenarios for the formation and evolution of black hole seeds to account for the population of high-redshift quasars and SMBHs in today's bright galaxies.

The *Gravitational Wave Universe* is the universe we explore using GWs as messengers (Amaro-Seoane *et al.* 2017; Colpi *et al.* 2019; Sathyaprakash *et al.* 2019). Observing this universe over a wide interval of GW frequencies, from mHz to kHz, will allow us to answer a number of key questions:

– How and when did black hole seeds form?

– Are sBHs the only elementary building blocks of SMBHs that grew through accretion and mergers, or is there a genetic divide between sBHs and SMBHs?

– How fast do seeds grow and pair hand-in-hand with the growth of cosmic structures?

– How do massive black hole binaries form?

– Do massive black holes merge when galaxies merge?

– What is the mass spectrum of astrophysical black holes, and how does it evolve with cosmic time?

– What is the spin distribution and what does it tell us about the evolution of massive black holes?

6.3. Gravitational waves from coalescing black hole binaries

According to GR, a GW is a perturbation of the geometry of spacetime, that is, of the metric tensor $h_{\mu\nu}$ which propagates at the speed of light over a stationary background[2]. The perturbation in the four-dimensional space, orthogonal to the propagation direction, is characterized by only two linearly independent propagating degrees of freedom, called the plus and cross polarizations, denoted as h_+ and $h_\times$, respectively. A single detector can measure a linear combination of the two signals, called *strain amplitude*, $h(t) = F_+h_+(t) + F_\times h_\times(t)$, where F_+ and $F_\times$ are the detector antenna patterns that contain in their expression the two angles identifying the source in the sky (Sathyaprakash and Schutz 2009).

GWs are sourced by non-symmetrically accelerated masses, that is, the charges of gravity, and trace the large-scale coherent motions of cosmic bodies. Thus, there is a similarity with the process of emission of EM radiation from accelerated elementary charged particles. However, no gravitational dipole radiation is emitted as the total linear momentum of a close system is conserved. Hence, to leading order in the multipole expansion, gravitational radiation is *quadrupolar*, where the quadrupole moment is of order $Q \sim \varepsilon Ma^2$ for a source of mass M, dimension a, and non-symmetric deformation ε. Einstein's linearized field equations establish a relation between the second time derivative of Q and the strain amplitude $h \propto \ddot{Q}$: a back of the envelope calculation gives $h \sim (a/d_L)(R_G/a)(u/c)^2$, where $R_G = 2GM/c^2$ is the Schwarzschild radius associated with the mass M, d_L is the source luminosity distance, and u is the velocity of the asymmetric mass motion (G and c denote the gravitational constant and the speed of light).

Binaries have a time-varying quadrupolar motion and thus emit GWs. The frequency of the GW is related to the inverse of the dynamical time associated with the non-symmetric motion, and for a circular binary is, to leading order, twice the

2. For further knowledge on the propagating properties of spacetime and the GW sources, we refer the reader to the excellent book, *Gravitational Waves* by Michele Maggiore, vol. 1, edited by Oxford University Press.

Keplerian frequency: $f_{gw} = (1/\pi)(GM_B/a^3)^{1/2}$, where now a is the semi-major axis of the relative orbit and M_B is the total mass of the binary. In the imminent vicinity of a binary black hole (BBH) merger, a is comparable to about a few GM_B/c^2 and the GW frequency f_{gw} coasts to values scaling with the fundamental frequency $f_o \equiv c^3/(GM_B)$, which depends on the inverse of the total mass of the binary M_B. Thus, merging sBH binaries emit at frequencies much higher than SMBH binaries, according to the ratio of their masses. The associated amplitude of the GW is $h \sim 2GM_B/(c^2 d_L)$, which is a tiny dimensionless number. In typical astrophysical conditions, $h \sim 10^{-22} - 10^{-19}$. Gravitational radiative losses lead the binary to progressively harden as GWs carry away orbital energy and angular momentum. The initial inspiral is a slow adiabatic contraction, which occurs over a timescale much longer than the orbital period. Only near merger, the loss is dramatically fast.

The coalescence of a black hole binary in vacuum has the beauty of being one of the elementary processes in nature. It is a well-posed problem characterized by a relatively small set of intrinsic parameters: the mass ratio $q \equiv m_2/m_1$, where m_1 and m_2 are the masses of the two black holes in the binary, and the individual spins $\mathbf{S}_i$ ($= \chi^i_{bh} G m_i^2/c$ for $i = 1, 2$, respectively). The semi-major axis at large separations (or total energy E_B) and the modulus of the orbital angular momentum, whose direction can be taken as the reference binary axis due to the isotropy of space, define the initial condition. The merger product is a new Kerr black hole of mass M_f and spin $\mathbf{S}_f$. There is no intrinsic scale in vacuum Einstein's gravity. Thus, any solution can be re-scaled knowing the binary mass $M_B = m_1 + m_2$ and mass ratio under examination. Coalescing black holes produce a universal GW signal that contains unique information on the black hole individual masses, spins and luminosity distance. It is as a progression of 3+1 phases:

– The *inspiral* refers to the phase in which the two black holes behave as point masses with orbital velocities that are small compared to the speed of light c. Their dynamics and emission can be described by the post-Newtonian theory (Blanchet 2014). In the advanced stage of the inspiral, the binary is circular, as GWs carry orbital energy faster than angular momentum.

The inspiral of a circular binary has the characteristic shape of a *chirp*, having both the frequency and amplitude slowly sweeping to higher values. If we express the orbital energy as a function of f_{gw}, we have $E_B = -[\pi^{2/3}/2]\nu M_B (GM_B)^{2/3} f_{gw}^{2/3}$, where ν is the *symmetric mass ratio*, $\nu = m_1 m_2/M_B^2 = q/(1+q)^2$ (equal to 1/4 for an equal mass binary and invariant under the change of the mass ratio $q = m_2/m_1$ in q^{-1}). The rate of change of E_B is given by

$$\dot{E}_B = -\frac{32}{5}\frac{c^5}{G}\left(\frac{GM_{\rm chirp}}{c^3}\pi f_{gw}\right)^{10/3} \qquad [6.1]$$

where the right-hand side gives the radiated power in GWs for a circular binary. Equation [6.1] contains the so-called binary *chirp* mass

$$M_{\rm chirp} \equiv \frac{(m_1 m_2)^{3/5}}{(m_1+m_2)^{1/5}} = \nu^{3/5} M_{\rm B} = \frac{q^{3/5}}{(1+q)^{6/5}} M_{\rm B}^{2/5}, \qquad [6.2]$$

that is, a combination of the individual black hole masses, expressed in terms of the symmetric mass ratio ν. Equation [6.1] can be recast to determine, to leading order, the rate of change of the frequency of the GW during the inspiral phase

$$\dot{f}_{\rm gw} = \frac{96}{5}\pi^{3/8}\left(\frac{G M_{\rm chirp}}{c^3}\right)^{5/3} f_{\rm gw}^{11/3}[1+\mathcal{D}^{\rm PN}], \qquad [6.3]$$

where $\mathcal{D}^{\rm PN}$ is the post-Newtonian correction term, which contains information on the mass ratio and spins of the interacting black holes (Blanchet 2014).

As $E_{\rm B}$ and $f_{\rm gw}$ can be expressed in terms of the semi-major axis a, equation [6.1] can be recast to give the rate of change of $\dot{a}$ and thus the *coalescence time*, defined as the time it takes the binary to reach null separation (in the point mass approximation) from an initial value of semi-major axis $a_\circ$, set by "astrophysical" conditions. For a circular binary

$$t_{\rm coal} = \frac{5}{256}\frac{c^5}{G^3}\frac{{a_\circ}^4}{\nu M_{\rm B}^3}. \qquad [6.4]$$

Equation [6.4] shows a steep dependence on $a_\circ$ and can be used either to determine the time to coalescence from birth (that can be millions to billions years long, or even longer) or to infer the time of duration of the GW signal when the source enters the sensitivity band of the detector, typically at $a_\circ \sim nGM_{\rm B}/c^2$ with n being an integer number: $t_{\rm coal} = (5/256)(1/\nu)n^4 GM_{\rm B}/c^3$, which is shorter for lighter black holes. If the binary has an initial eccentricity $e_\circ$, the coalescence time is shorter by a factor $G(e_\circ)(1-e_\circ^2)^{7/2}$, with $G(e_\circ)$ being a function varying from 1 (for $e_\circ = 0$) to 1.80 (for $e_\circ = 1$).

The chirp signal is accompanied by an increase of the amplitude h of the GW. According to general relativity, the GW signal, that is, the plus and cross polarization components of the metric perturbation take the following form:

$$h_+(t) = \frac{4}{d_L}\left(\frac{G M_{\rm chirp}}{c^2}\right)^{5/3}\left(\frac{\pi f_{\rm gw}(t_{\rm ret})}{c}\right)^{2/3}\frac{1+\cos^2(\iota)}{2}\cos[\Phi^+_{\rm N}(t_{\rm ret})] \qquad [6.5]$$

$$h_\times(t) = \frac{4}{d_L}\left(\frac{G M_{\rm chirp}}{c^2}\right)^{5/3}\left(\frac{\pi f_{\rm gw}(t_{\rm ret})}{c}\right)^{2/3}\cos(\iota)\sin[\Phi^\times_{\rm PN}(t_{\rm ret})] \qquad [6.6]$$

where t_{ret} is the retarded time (as waves propagate at the speed of light), Φ_{PN} is the phase computed at the desired post-Newtonian order, d_L is the luminosity distance and ι is the angle between the orbital angular momentum vector and the line of sight. Note how the GW amplitude depends on the chirp mass and frequency: for a given value of $\mathcal{M}_{chirp}$, as the frequency of the GW increases, so does the amplitude of the oscillation. Furthermore, $h_+/h_\times$ directly gives the inclination angle ι. When the binary is seen edge-on ($\iota = \pi/2$), $h_\times = 0$, radiation has + polarization only, since from the observer's point of view, the motion of the two black holes projected on the sky is linear. When $\iota = 0$, the binary is seen face-on/off and the black holes execute a circular motion in the sky. In this case, both polarization components have equal amplitude and are out of phase by $\pi/2$. If a network of detectors enables the measurement of the two independent polarization states of a GW, from the chirp signal (equations [6.3], [6.5] and [6.6]), we can infer the luminosity distance d_L of the source.

Black hole coalescence occurs at any redshift, and thus also at cosmological distances. In these circumstances, the GW signal is affected by the expansion of the universe. If z_s is the cosmological redshift of the source of GWs, it is possible to show that (i) the frequency of the GW f_{gw} as measured in the source frame is related to the frequency f in the observer frame as $f = f_{gw}/(1 + z_s)$; (ii) the chirp mass in the source frame $\mathcal{M}_{chirp}$ is replaced by the redshifted chirp mass $\mathbb{M}_{chirp} = (1 + z_s)\mathcal{M}_{chirp}$; (iii) the source distance is identified with the luminosity distance d_L, which depends on how the universe is expanding; and (iv) the time to coalescence as measured in the observer frame is $\tau_{coal} = (1 + z_s)t_{coal}$. Note that the combination $\mathbb{M}_{chirp}f = \mathcal{M}_{chirp}f_{gw}$ is redshift independent and that the above equations for the chirp and amplitude of the GW in the observer frame can be rewritten by substituting f_{gw} with f, and $\mathcal{M}_{chirp}$ with $\mathbb{M}_{chirp}$, respectively. In general, measuring the luminosity distance does not provide the source redshift which is inferred using the d_L versus z relation from our best model for the cosmic expansion of the universe.

Coalescing BBHs are gravitational standard candles, or, as often said, *standard sirens*. Indeed, if an EM counterpart is associated with and identified with the GW source, it is then possible to have a direct measure of z_s and this allows for an independent measure of the d_L versus z relation, and thus of the Hubble expansion parameter with no use of a cosmic distance ladder (Tamanini *et al.* 2016).

– The *merger* corresponds to the nonlinear plunge-in phase and coalescence. Moving at around one-third of the speed of light, the two highly deformed black hole spacetimes are described in the realm of numerical relativity. The plunge refers to the rapid increase in the magnitude of the inward radial velocity leading to merger. It initiates when the two black holes attain a distance comparable to ISCO, defined here as $R_{ISCO} = 6GM_B/c^2$. The frequency derivative is no longer described by equation [6.3], as the frequency coasts to a finite value when the new BH forms. This maximum frequency in the source frame is close to twice the Keplerian frequency at the ISCO for a binary of total mass $M_{\rm B}$,

$$f_{\rm gw,max} \sim \frac{c^3}{\pi 6^{3/2} G M_{\rm B}} = 4.4 \times 10^3 \left(\frac{\rm M_\odot}{M_{\rm B}} \right) \text{Hz}. \qquad [6.7]$$

The simple scaling $f_{\rm gw,max} \propto M_{\rm B}^{-1}$ shows that detecting coalescing sBHs requires interferometers operating at frequencies 10^5 times higher than those requested to detect massive black holes with $M_{\rm B} = 10^6\,{\rm M_\odot}$.

The merger of BBHs carries tremendous luminosity of the order of $L_{\rm gw} \approx \nu(c^5/G) \sim 3.6 \times 10^{59} \nu$ erg s^{-1} which is independent on $M_{\rm B}$. This huge luminosity is emitted for a short time lapse, in the immediate vicinity of the merger proper. What depends on the mass scale is the total energy radiated away in GWs which can account for 3.8% – 42% of $\nu M_{\rm B} c^2$ depending on the black hole spins and whether the orbit is prograde or retrograde, relative to the spin direction. Fitting formulae instructed by numerical relativity simulations (Barausse *et al.* 2012) help infer the mass and spin of the newly formed black hole. For equal-mass non-spinning BBHs, the final spin is $\chi_{\rm spin,f} = 0.6865$, showing that a merger creates a highly spinning black hole, the spin resulting mainly from the orbital angular momentum at ISCO.

– The *ring-down* is the phase describing the settling of the new black hole into a quiescent, stationary state described by the Kerr family. In this phase, GWs carry away the residual spacetime deformation. The radiation consists of a superposition of quasi-normal modes whose frequencies and decaying timescales depend only on the mass and spin of the black hole, according to the "no-hair" theorem.

– A fourth phase of *gravitational recoil* exists, as the new black hole that forms experiences a kick (Lousto *et al.* 2010). The gravitational recoil emerges when the two black holes are not symmetric, as in this case GWs carry away linear moment. The asymmetry can be due to unequal masses, unequal spins, or a combination of the two. For non-spinning BBHs, the maximum recoil is of $\sim$ 175 km s^{-1} when the mass ratio is $q \sim 0.195$ km s^{-1}. However, generic binaries with in-plane spin components may lead to much higher recoil velocities, up to $\sim 4,000$ km s^{-1}

The GW signal from a coalescing binary carries precious information on both external and internal parameters of the source. The signal is degenerate in combinations of some of these parameters. Still, accurate waveforms inferred from post-Newtonian expansion, numerical relativity, effective one-body and perturbation theories enable us to characterize almost fully the source parameters (Buonanno and Sathyaprakash 2014). Detecting a population of merging black holes across cosmic distances is a unique way to infer how they formed and evolved. This is made possible by measuring their masses, spins and luminosity distance. It is useful to remember that mergers elevate the mass and spin of black holes and thus contribute to their growth.

6.4. The Gravitational Wave Universe

The GW Universe comprises a variety of discrete sources and of astrophysical and cosmological backgrounds. Here, we restrict attention only to coalescing BBH across their mass spectrum[3].

In exploring the GW Universe, we can define three windows: the *very – low* GW frequency window probed in pulsar timing array experiments, extending from nano-Hz up to 10^{-7} Hz. In this bandwidth, we expect to detect the uncorrelated GW signal from a population of inspiraling SMBH binaries of $10^{8-9}\,M_{\odot}$ centuries prior to coalescence, and possibly to detect some individual source. This stochastic background originates from SMBH binaries in interacting galaxies at redshifts $z \sim 1$. Detecting this radiation will provide insight into the role of mergers in shaping galaxies and on the still unexplored dynamics of SMBHs at parsec scales when they form a Keplerian binary (De Rosa *et al.* 2020).

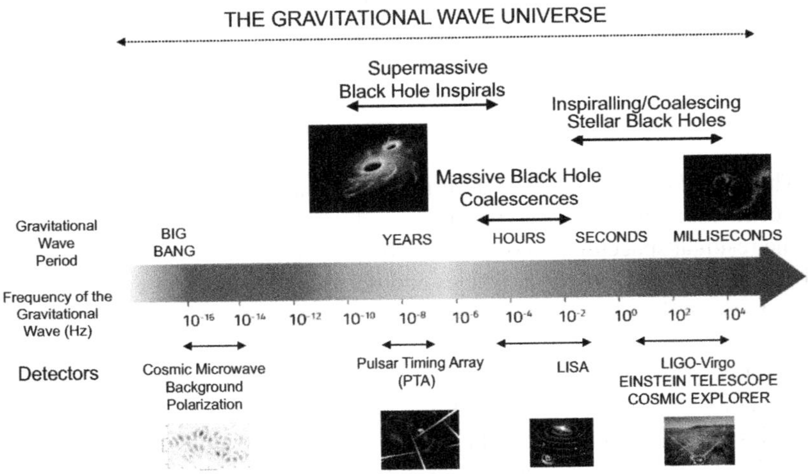

Figure 6.3. *The Black Hole Gravitational Wave Universe. The figure shows the broad-band frequency spectrum of GWs and associated periods. BBH mergers will be detected in the LISA domain, in current and third-generation ground-based interferometers, and in pulsar timing array (PTA) experiments. For the sake of completeness, the figure also shows the possibility of detecting indirectly a primordial GW background from inflation, through measurements of the polarization of the cosmic microwave background radiation*

3. In the text and figures, we included SMBH binaries that can be detected in forthcoming pulsar timing array experiments, but we refrain from describing this population, and refer the reader to Arzoumanian *et al.* (2016), Kelley *et al.* (2018) and Kelley *et al.* (2019).

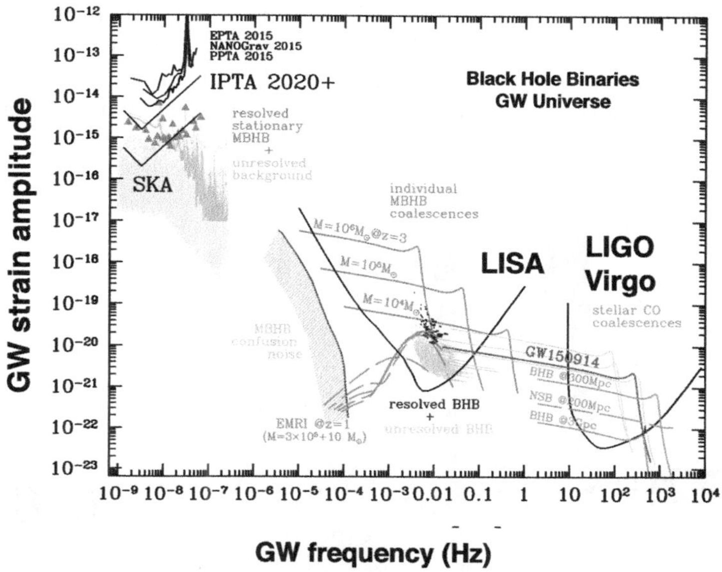

Figure 6.4. *A multi-band view of the characteristic strain of the GW signals from coalescing BBHs as a function of the GW frequency. Black lines indicate the strain sensitivity of the LIGO, LISA observatories, and the International Pulsar Timing Arrays (IPTA 2020+ and future SKA). The LIGO/Virgo interferometers detect the GW signal from coalescing sBHs in the frequency bandwidth between 10 Hz and 1,000 Hz; LISA will detect the coalescence signal of* $10^{4-7}\,M_{\odot}$ *binaries at frequencies between 40* μ*Hz and 0.1 Hz, and the nearly monochromatic GW signal from sBH binaries like GW150914, years/months prior to coalescence in the high-frequency bandwidth (Sesana 2016). GW signals are overlaid from* $z = 3$ *binaries with different masses and signals from GW150914-like binaries. At very low frequencies, PTA experiments will detect the stochastic background from a population of* $z \sim 1$ *SMBH inspiraling binaries of* $10^{8-9}\,M_{\odot}$*, and some individual source*

In the *low* GW frequency bandwidth explored by the ESA-led space mission LISA (Amaro-Seoane *et al.* 2017; Colpi *et al.* 2019), ranging between 40 μHz and 0.1 Hz, we expect to detect the GW signal from coalescing massive black hole binaries of $10^{4-7}\,M_{\odot}$, and track this new population out to very large redshifts. LISA has the sensitivity to also detect the nearly monochromatic signal from GW15014-like sBH coalescences years/months before they enter the LIGO/Virgo bandwidth.

Finally, the *high*-frequency bandwidth explored by ground-based interferometers is the privileged one to detect coalescing sBH binaries (Abbott *et al.* 2019), possibly also beyond the pair-instability gap (Mangiagli *et al.* 2019; Sathyaprakash *et al.* 2019). Figures 6.3 and 6.4 give an overview of the GW Universe. In particular, Figure 6.4 shows the characteristic strain amplitude evolution expected from coalescing BBHs as a function of the GW frequency for selected binaries in various mass intervals. During the inspiral phase, the chirp is characterized by a strain amplitude in the Fourier domain $\tilde{h}_{(+,\times)} \propto f_{\rm gw}^{-7/6}$, decreasing with the GW frequency due to their rapid inspiral, that is, their short-lived persistence at a given frequency bin as they approach coalescence. Near merger the amplitude rises before falling when the new black hole with its new horizon has formed.

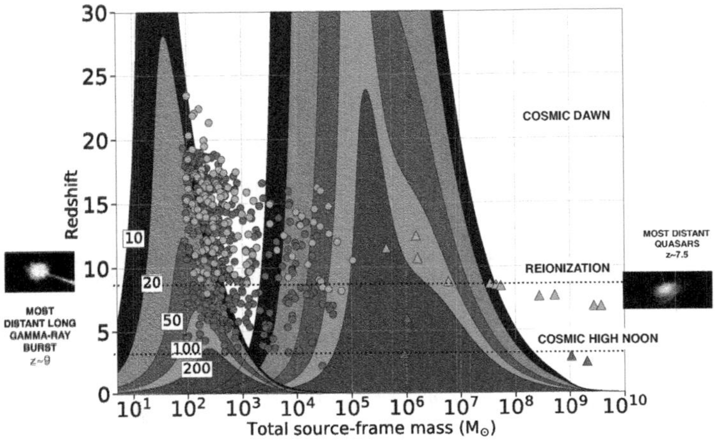

Figure 6.5. *The vastness of the LISA (Amaro-Seoane et al. 2017; Colpi et al. 2019) and Einstein Telescope (Punturo et al. 2010) cosmic exploration volume. The figure shows contours of constant signal-to-noise ratios (SNR-sky and polarization averaged) of equal mass, non-spinning black hole binary coalescences in the LISA and Einstein Telescope observatories, plotted in the total rest-frame mass of the black hole binary (x-axis) and redshift (y-axis) plane. Sources can be discovered at redshift beyond the cosmic reionization epoch with SNR between 20 and 50, and at lower redshifts, during the peak of the star formation and AGN activity (the "cosmic high noon"), at $z \sim 2$ with SNR as large as 1,000 for the LISA black holes and 200 for the IMBH in the Einstein Telescope window. Green dots highlight merging halos that host light (circle) and heavy (triangle) binary seeds at their center. Bright green (dark green) points are extracted from a simulation of a rare high-density region of 10^{12} $M_\odot$ forming a quasar at redshift $z \sim 6$ ($z \sim 2$) (Valiante et al. 2018). Detecting sources in the two bandwidths will provide invaluable insight into the formation and growth of black hole seeds, as discussed in Valiante et al. (2021)*

Third-generation ground-based interferometers together with LISA will let us explore the entire black hole GW Universe, having the sensitivity to detect coalescing sBH binaries, IMBH binaries and low-mass SMBHs across all cosmic ages, far beyond the redshift of cosmic reionization, if they form. Figure 6.5 shows the incredible reach of the LISA mission and of the Einstein Telescope. As shown in the figure, sources with a signal-to-noise ratio larger than ten can be seen out to $z \sim 20$. These are heavy sBH binaries of total mass in the hundreds and binaries of 10^{5-7} $M_{\odot}$, in the low-mass tail of the SMBH spectrum. The lower-redshift sources ($z < 5$) in the LISA band will be detected with signal-to-noise ratios as high as a few 1,000 to enable exquisite parameter estimation and precision test of gravity in the strong-field dynamical regime of general relativity. The cloud of green and dark green dots show how a population of black hole seeds, driven by the assembly of halos, evolves in the redshift versus mass plane. Dots pinpoint places where interacting dark matter halos host a binary black hole. They are extracted from a state-of-the-art semi-analytical model (Valiante *et al.* in preparation) that follows the growth of light (dots) and heavy (triangle) seeds when they experience both accretion phases and mergers. Green dots belong to a halo that was able to grow a quasar at redshift $z \sim 6$, while the dark green dots refer to a halo forming a quasar at $z \sim 2$.

6.5. Black hole dynamics in merging galaxies

In this section, we briefly review the outcome of current studies on the dynamics of massive black holes in merging galaxies highlighting the role of stars and gas in ruling their path towards coalescence (Colpi 2014).

When two galaxies, each hosting a central massive black hole, collide and merge, their black holes start a long journey that can bring them from separations of ~ 100 kpc down to sub-milliparsec scales, below which GWs drive the binary to coalescence. However, this voyage has bottlenecks and failures, as illustrated in Figure 6.6. Thus, a central question arises: *Do black holes merge as their host galaxies merge and what is the rate?* The answer to this question is astrophysically challenging and needs to be addressed in a cosmological context, matching the evolution of galaxy halos with that of black holes.

The growth of dark matter halos through mergers with other halos in our current cosmological concordance model gives us the scenery within which black hole dynamics develops and evolves. The halo merger rate depends on the mass ratio of the interacting halos and peaks at different redshifts for different halo masses: major mergers, that is, mergers of halos of comparable masses (from 1 : 1 down to 1 : 4) are rare, while minor mergers (mass ratio less than 1 : 10) are more common. However, the dynamical evolution is much faster in major mergers than in minor mergers. Thus, in the former fast sinking of the black holes may occur, while in the latter case, if the mass ratio of the merging halos is too small, the satellite is tidally disrupted early on in its dynamical evolution and its central black hole is left on a wide orbit, too far from the center of the larger galaxy to ever merge with the central companion. A population of wandering massive black holes is therefore predicted to exist in galaxies as well as a population of merging ones.

A black hole in a galaxy is a massive, tiny object as its gravitational sphere of influence close to $GM_{\rm BH}/\sigma_*^2$ (with σ_* being the velocity dispersion of the stars surrounding the black hole) is of only a few parsecs for a M_{BH} of $10^6\,M_\odot$ and a $\sigma_* \sim 100\,km\,s^{-1}$. Despite that, the study of its dynamics, driven by its interaction with stars and gas, depends on both local and global properties, as accretion flows and stellar drag/refilling are ruled by large-scale motion/instabilities and relaxation processes inside the host galaxy. We distinguish four phases for the dynamics of black holes in merging galaxies[4].

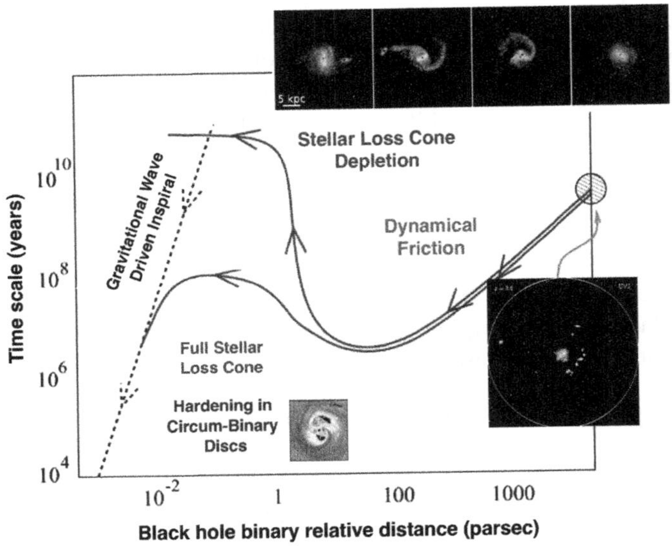

Figure 6.6. *A sketch of the characteristic timescales associated with the three phases of the dynamical evolution of massive BBHs inside the newly formed galaxy in the aftermath of a merger. The figure depicts the early phase of dynamical friction experienced by each black hole before forming a Keplerian binary. It further highlights the potential stalling of the binary on parsec scales in case of loss cone depletion or full loss cone replenishment. The dotted line gives the coalescence time driven by GW emission. The upper-right inset shows a high-resolution numerical simulation of two interacting galaxies with central black holes (Tremmel et al. 2017); the lower-right inset shows the initial condition of a numerical simulation of a galaxy group, which hosts a rapid merger of two massive black holes (Mayer et al. 2015); the central inset shows the snapshot of a GRMHD simulation of a circumbinary disk surrounding a black hole binary in the verge of merging (Bowen et al. 2017) (courtesy of M. Tremmel, L. Mayer and R. Bowen)*

4. A comprehensive overview of the dynamics of black holes in merging galaxies can be found in Colpi (2014) and De Rosa *et al.* (2020). In De Rosa *et al.* (2020), the reader can find a comprehensive review of both the observational status concerning the identification of dual and binary AGN and the current theoretical understanding of the SMBH dynamics from the most recent, state-of-the-art simulations. A rather complete set of references is present in this long review article. Here, we outline key concepts and the main findings.

(I) The *pairing* phase is identified as the early phase, which characterizes the sinking of the two black holes inside the merging galaxies. Dynamical friction against the common dark matter backgrounds causes the galactic disks/bulges to merge into a unique structure. In this phase, the black holes are dragged naturally in the central regions of the new, violently relaxed galaxy and their rate of pairing is the rate at which galaxies complete their merger. In minor mergers, inefficient pairing can occur when the light satellite galaxy is tidally disrupted by the gravitational field of the primary. In this case, the "naked" black hole wanders in the outskirts of the main galaxy. It may further happen that heavy black hole seeds with mass $M_{\rm BH} \sim 10^4\,{\rm M}_\odot$, born in dark matter halos at redshift as large as $z \sim 9$, display highly erratic dynamics (Pfister *et al.* 2019). Here, the dynamics is dominated by the stellar component, whose distribution is irregular and patchy. This induces stochasticity in the orbit and the black hole may be anywhere in the proto-galaxy. The possibility of a third encounter with another galaxy may enhance the chance of a merger; however, multiple seeds each inherited from a different merger may coexist in the high redshift universe. In low-redshift encounters, instead, the "effective mass" gained by each black hole due to gas-accretion and/or the formation of a massive star cluster resulting from major central gas inflows surrounding the black hole(s) can be instrumental for their effective pairing (Mayer *et al.* 2015).

(II) The *binary* phase begins when the black holes are at the center of the newly formed remnant galaxy and stellar dynamical friction acts individually on each black hole, causing progressive loss of orbital energy and angular momentum. When the mass in stars and/or gas enclosed in the hole's orbit falls below the mass of the two black holes, a *Keplerian* binary forms. Typically, this occurs when their separation is less or around a parsec, that is, at a distance $a_{\rm Kep} \sim G\nu M_{\rm BH}/\sigma_*^2$. The binary is in the midst of either a rotating, triaxial stellar bulge, or inside a gaseous massive nuclear disk (depending on the type of encounter, whether gas-rich/poor). A "stalled" black hole pair would arise when the final separation is too large for a Keplerian binary to form. Unless a close encounter with a third black hole occurs in a multiple merger, stalling can go on forever (Bonetti *et al.* 2019). It may further happen that in gas-rich galaxy collisions at redshift $z \sim 2-4$, the black holes move at the center of a clumpy, massive disk site of star-formation. Scattering off giant gas clouds and/or stellar clusters of comparable mass can cause a random walk of the black hole orbit, leading either to off-plane ejection and delayed binary formation, or accelerated orbital decay depending on many details of the dynamics of the encounter (Fiacconi *et al.* 2013; Roškar *et al.* 2015). This causes a broadening of the typical residence time across the pairing and binary formation phase, with times longer than hundreds of millions of years. During this phase, the black holes can accrete gas, thus changing the mass ratio between the two black holes and their dynamics (Capelo *et al.* 2015).

(III) The *hardening* phase starts soon after the binary has formed. In this phase, the persistence and crossing into the GW regime hinges on exchanges of energy and angular momentum between the binary and single stars moving on low-angular momentum orbits in the nucleus of the galaxy (Begelman *et al.* 1980). Recent results of direct N-body simulations are giving an optimistic view of what has been considered to be the main bottleneck of the binary evolution for almost 40 years: the "final parsec problem", that is, the stalling of the binary due to loss cone depletion in spherical galaxies characterized by relaxation times longer than the Hubble time. The delay in stellar refueling was considered a leading cause to stalling. However, simulations of black hole hardening in triaxial stellar backgrounds have recently indicated that a departure from spherical symmetry in the stellar nucleus of the galaxy (after the merger) keeps the rate of the interaction of stars with the binary at a high enough level so that the binary continues to shrink. A sufficiently high number of stars moving on centrophilic orbits appears to be present in non-spherical potentials to let the binary enter the GW-driven phase (Vasiliev 2016). This is illustrated in Figure 6.7, where upper limits on the binary lifetime as a function of $M_{\rm BH}$ are inferred at the crossing of the stellar hardening time with the coalescence time, using physically motivated recipes for the underlying stellar profile in order to connect the binary black hole mass with properties of the host galaxy (Biava *et al.* 2019). This investigation indicates that lifetimes can be highly uncertain and range between tens of Myrs up to the Hubble time.

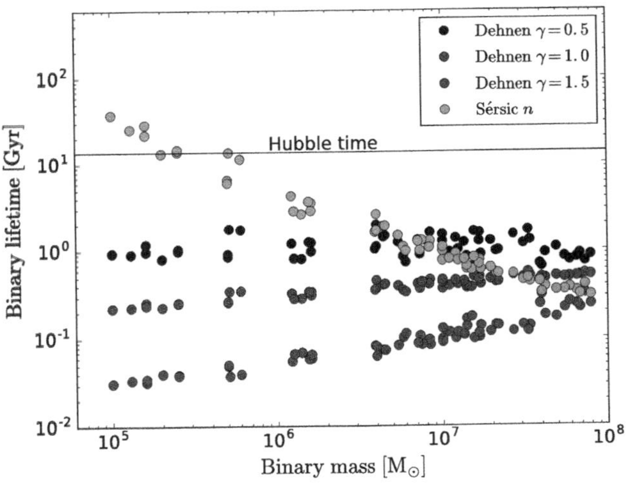

Figure 6.7. *Upper limits on the lifetime of binary black holes with total mass* $M_{\rm BH} = 10^{5-8}\,\mathrm{M}_\odot$ *in the hardening phase. Lifetimes are inferred exploring two different underlying stellar density profiles, known in the literature as Dehnen and Prugniel & Simien profiles and using local scaling relations between the mass of the black holes and several quantities of their hosts, as the velocity dispersion, the Sérsic index and the stellar mass. Lifetimes are computed when the hardening time by individual stellar scattering off the binary equals the GW timescale. The figure shows the great amount of uncertainties we have in estimating the binary lifetimes, which sensibly depend on the density of stars near the black hole gravitational sphere of influence (Biava et al. 2019)*

The binary can further/alternatively harden in its interaction with a circumbinary disk that may form during the merger. Gas-dynamical torques provide an alternative avenue to the hardening of the binary, which contracts at a pace related to the mass inflow rate onto the inner edge of the disk (Haiman *et al.* 2009; Tang *et al.* 2018). If the mass ratio $q << 1$, the less massive black hole behaves as a fluid element within the accretion disk of the primary, thus experiencing what is known as type I migration. If the mass ratio $q > 0.01$, then the formation of a central cavity slows down the evolution of the binary. The energy and angular momentum transfer between the binary and the circumbinary disk are mediated by streams leaking from the inner rim into the cavity. The rate of binary hardening depends on the thermodynamics of the streams crossing the gap, which are partially accreted by the two black holes to form mini-disks and partially flung back to the disk, extracting binary angular momentum (Roedig *et al.* 2012). Simulations and analytical studies show that accretion into the central cavity is efficient enough to bring the binary into the GW-driven domain on timescales less than a Hubble time (Haiman *et al.* 2009).

(IV) The *gravitational – wave* inspiral phase: this phase starts when the coalescence time ruled by the emission of GWs becomes shorter than the hardening timescale. This typically occurs at separations close to 10^{-3} pc and smaller. The two black holes almost detach from the exterior world and evolve as if in isolation up to coalescence. This phase can last millions to billions of years prior to merging. If gas enshrouds the binary, the binary carries the circumbinary disk down to coalescence (Bowen *et al.* 2017; Tang *et al.* 2018).

In the LISA band, the rate of merging black holes per unit total mass, mass ratio, and redshift is extremely difficult to calculate, as the origin and evolution of seeds are uncertain, as are the dynamics of black hole pairing over a wide range of scales, from 100 pc down to 100 μpc. Coalescences may occur in a rich variety of galactic halos, either gas-rich or gas-poor, in galaxies having different morphologies and stellar contents. Progress in this field is highly desirable through the union of the most advanced numerical simulations of galaxy assembly with physically motivated sub-grid recipes (Dayal *et al.* 2019) to account for unresolved physics on the smallest scales, or using semi-analytical models based on the extended Press–Schechter formalism (Klein *et al.* 2016; Valiante *et al.* 2018). Figure 6.8 gives the expected number of black hole binary coalescences per unit time and redshift (left) and their distribution in mass as measured in the observer frame (right). They are inferred adopting different seeding mechanisms and a recipe for the time delay distribution, as described in the legend. Rates can vary between a few to 10–20 per year. It is still difficult to quantify and correlate the merger rate in the Einstein Telescope bandwidth with that of LISA, but progress is expected in this field.

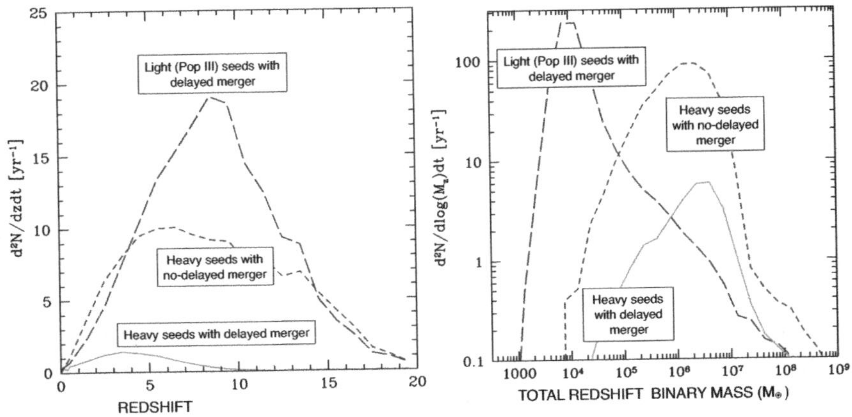

Figure 6.8. *The rate of black hole binary coalescences inferred from the semi-analytical model by Klein et al. (2016), which simulates a population of light seeds (Pop III), and of heavy seeds and their evolution through mergers and accretion episodes. According to the labels, the merger rates from both light and heavy seeds include delays between the merger of the halos and that of the black holes. For comparison, the merger rate has also been calculated with no delay. The left panel provides the merger rates per unit time and redshift. The right panel gives the mass distribution of the events as a function of the total mass of the binary in the observer frame*

6.6. Multi-messenger view of a massive black hole merger

Black hole coalescences in the LISA bandwidth may not occur in vacuum, and low-redshift $z \sim 2$ binaries of $10^{5-7}\,M_\odot$ surrounded by a circumbinary disk may outshine in the optical and X-rays during the inspiral and merger proper (Schnittman 2011; d'Ascoli *et al.* 2018; Tang *et al.* 2018), becoming key targets for EM follow-up, with an advance warning of hours (McGee *et al.* 2020). These mergers will be localized within 10 or even 0.4 deg^2, corresponding to the field of view of the Large Synoptic Survey Telescope and of the *Athena* Wide Field Instrument. The science with contemporaneous EM and GW observations promises to be spectacular. It has the potential to discover the yet unknown periodic emission from shocked gas surrounding the two massive black holes in the violently changing spacetime before merger, flashes, bursts and jetted emission at merger, and also post-merger afterglow signatures (Rossi *et al.* 2010; Kelly *et al.* 2017, 2019; d'Ascoli *et al.* 2018). Linking masses and spins determined with exquisite precision by the GW signal with EM emission will be paramount.

6.7. Concluding remarks

A new, golden era of exploration of the universe has started, following the detection of the first signal from the cosmos by the LIGO Scientific Collaboration and Virgo Collaboration (Abbott *et al.* 2016, 2019). Thanks to major advances in experimental physics, laser interferometry has proved to be successful and powerful in detecting the minuscule perturbations of spacetime produced by the most extreme sources: the coalescence of sBH binaries. The discovery of GW150914 represents the culmination of an experimental research able to achieve strain noise sensitivities at the level of 10^{-23} $Hz^{-1/2}$ around 200 Hz.

On June 8, 2016, ESA announced to the public the first results of the LISA Pathfinder in-flight experiment, designed to measure the level of residual acceleration noise on two test masses free-falling with respect to a local inertial frame. The results demonstrated that the two masses remain in "free fall" with a relative acceleration noise corresponding to $0.56 \pm 0.01 \times 10^{-15} g$ $Hz^{-1/2}$, with g being the gravitational acceleration on Earth (Armano *et al.* 2016). The results of the LISA-Pathfinder experiment have beautifully shown that current technology meets the requirement for a space-based gravitational wave observatory. This has led to the selection of LISA as the third large ESA mission (Amaro-Seoane *et al.* 2017). With an expected launch of LISA in the mid-thirties and the construction on ground of the third generations of interferometers, the GW Universe will be scrutinized across all accessible frequency bands from mHz and kHz, and possibly also at nano-Hz with the PTA experiments. The high-, low- and very low-frequency gravitational universes that encompass the multi-band GW Universe should not be perceived as disconnected. They are instead inter-winded. A new, golden era of exploration of the universe has started that promises a quantum leap in understanding the origin and evolution of the astrophysical black holes and of quasars and AGN in the broad context of the cosmological assembly of galaxies.

6.8. References

Abbott, B.P. *et al.* (2016). Observation of gravitational waves from a binary black hole merger. *Physical Review Letters*, 116, 061102.

Abbott, B.P. *et al.* (2019). Binary black hole population properties inferred from the first and second observing runs of advanced LIGO and advanced Virgo. *The Astrophysical Journal Letters*, 882, L24.

Abbott, R. *et al.* (2020). GW190521: A binary black hole merger with a total mass of 150 $M_{\odot}$. *Physical Review Letters*, 125(10), 101102. Doi: 10.1103/PhysRevLett.125.101102.

Abbott, R. *et al.* (2021). GWTC-2: Compact binary coalescences observed by LIGO and Virgo during the first half of the third observing run. *Physical Review X*, 11(2), 021053. Doi: 10.1103/PhysRevX.11.021053.

Amaro-Seoane, P. *et al.* (2017). Laser interferometer space antenna. arXiv e-prints arXiv:1702.00786.

Armano, M. *et al.* (2016). Sub-femto-g free fall for space-based gravitational wave observatories: LISA pathfinder results. *Physical Review Letters*, 116(23), 231101.

Arzoumanian, Z. *et al.* (2016). The NANOGrav nine-year data set: Limits on the isotropic stochastic gravitational wave background. *The Astrophysical Journal*, 821(1), 13.

d'Ascoli, S. *et al.* (2018). Electromagnetic emission from supermassive binary black holes approaching merger. *The Astrophysical Journal*, 865(2), 140.

Baldassare, V.F. *et al.* (2015). A 50,000 solar mass black hole in the nucleus of RGG 118. *The Astrophysical Journal*, 809, L14.

Bañados, E. *et al.* (2018). An 800-million-solar-mass black hole in a significantly neutral Universe at a redshift of 7.5. *Nature*, 553(7689), 473–476.

Barack, L. *et al.* (2019). Black holes, gravitational waves and fundamental physics: A roadmap. *Classical and Quantum Gravity*, 36(14), 143001.

Barausse, E. *et al.* (2012). On the mass radiated by coalescing black hole binaries. *The Astrophysical Journal*, 758(1), 63.

Begelman, M.C. *et al.* (1980). Massive black hole binaries in active galactic nuclei. *Nature*, 287(5780), 307–309.

Berti, E. *et al.* (2015). Testing general relativity with present and future astrophysical observations. *Classical and Quantum Gravity*, 32(24), 243001.

Biava, N. *et al.* (2019). The lifetime of binary black holes in Sérsic galaxy models. *Monthly Notices of the Royal Astronomical Society*, 487(4), 4985–4994. Doi: 10.1093/mnras/stz1614.

Blanchet, L. (2014). Gravitational radiation from post-Newtonian sources and inspiralling compact binaries. *Living Reviews in Relativity*, 17(1) [Online]. Available at: https://doi.org/10.12942/lrr-2014-2.

Bonetti, M. *et al.* (2019). Post-Newtonian evolution of massive black hole triplets in galactic nuclei IV – Implications for LISA. *Monthly Notices of the Royal Astronomical Society*, 486(3), 4044–4060.

Bowen, D.B. *et al.* (2017). Relativistic dynamics and mass exchange in binary black hole mini-disks. *The Astrophysical Journal*, 838(1), 42.

Buonanno, A. and Sathyaprakash, B.S. (2014). Sources of gravitational waves: Theory and observations. arXiv e-prints, arXiv:1410.7832.

Capelo, P.R. *et al.* (2015). Growth and activity of black holes in galaxy mergers with varying mass ratios. *Monthly Notices of the Royal Astronomical Society*, 447(3), 2123–2143.

Colpi, M. (2014). Massive binary black holes in galactic nuclei and their path to coalescence. *Space Science Reviews*, 183, 189–221.

Colpi, M. *et al.* (2019). The gravitational view of massive black hole mergers. *Bulletin of the American Astronomical Society*, 51(3), 432.

Dayal, P. *et al.* (2019). The hierarchical assembly of galaxies and black holes in the first billion years: Predictions for the era of gravitational wave astronomy. *Monthly Notices of the Royal Astronomical Society*, 486(2), 2336–2350.

De Rosa, A. *et al.* (2020). The quest for dual and binary supermassive black holes: A multi-messenger view. arXiv e-prints, arXiv:2001.06293.

Devecchi, B. *et al.* (2012). High-redshift formation and evolution of central massive objects II – The census of BH seeds. *Monthly Notices of the Royal Astronomical Society*, 421(2), 1465–1475.

Fiacconi, D. *et al.* (2013). Massive black hole pairs in clumpy, self-gravitating circumnuclear disks: Stochastic orbital decay. *The Astrophysical Journal*, 777(1), L14.

Genzel, R. *et al.* (2010). The Galactic Center massive black hole and nuclear star cluster. *Reviews of Modern Physics*, 82(4), 3121–3195.

Haiman, Z. *et al.* (2009). The population of viscosity- and gravitational wave-driven supermassive black hole binaries among luminous active galactic nuclei. *The Astrophysical Journal*, 700(2), 1952–1969.

Heger, A. *et al.* (2000). Presupernova evolution of rotating massive stars I. Numerical method and evolution of the internal stellar structure. *The Astrophysical Journal*, 528, 368–396.

Hirano, S. *et al.* (2015). Primordial star formation under the influence of far ultraviolet radiation: 1540 cosmological haloes and the stellar mass distribution. *Monthly Notices of the Royal Astronomical Society*, 448(1), 568–587.

Kelley, L.Z. *et al.* (2018). Single sources in the low-frequency gravitational wave sky: Properties and time to detection by pulsar timing arrays. *Monthly Notices of the Royal Astronomical Society*, 477(1), 964–976.

Kelley, L.Z. *et al.* (2019). Multi-messenger astrophysics with pulsar timing arrays. *Bulletin of the American Astronomical Society*, 51(3), 490.

Kelly, B.J. *et al.* (2017). Prompt electromagnetic transients from binary black hole mergers. *Physical Review D*, 96(12), 123003.

Kelly, B.J. *et al.* (2019). Immediate EM counterparts of binary black hole mergers in the LISA regime. *American Astronomical Society Meeting Abstracts*, 233, p. 141.07.

Klein, A. *et al.* (2016). Science with the space-based interferometer eLISA: Supermassive black hole binaries. *Physical Review D*, 93(2), 024003.

Kormendy, J. and Ho, L.C. (2013). Coevolution (or not) of supermassive black holes and host galaxies. *Annual Review of Astronomy and Astrophysics*, 51, 511–653.

Lousto, C.O. *et al.* (2010). Remnant masses, spins and recoils from the merger of generic black hole binaries. *Classical and Quantum Gravity*, 27(11), 114006.

Lupi, A. *et al.* (2014). Constraining the high-redshift formation of black hole seeds in nuclear star clusters with gas inflows. *Monthly Notices of the Royal Astronomical Society*, 442(4), 3616–3626.

Mangiagli, A. *et al.* (2019). Merger rate of stellar black hole binaries above the pair-instability mass gap. *The Astrophysical Journal*, 883(1), L27.

Marconi, A. *et al.* (2004). Local supermassive black holes, relics of active galactic nuclei and the X-ray background. *Monthly Notices of the Royal Astronomical Society*, 351, 169–185.

Mayer, L. *et al.* (2015). Direct formation of supermassive black holes in metal-enriched gas at the heart of high-redshift galaxy mergers. *The Astrophysical Journal*, 810(1), 51.

McGee, S. *et al.* (2020). Linking gravitational waves and X-ray phenomena with joint LISA and Athena observations. *Nature Astronomy*, 4, 26–31.

Merloni, A. (2016). Observing supermassive black holes across cosmic time: From phenomenology to physics. In *Lecture Notes in Physics*, Haardt, F., Gorini, V., Moschella, U., Treves, A., Colpi, M. (eds). Lecture Notes in Physics vol. 905, Springer Verlag, Berlin.

Nguyen, D.D. *et al.* (2019). Improved dynamical constraints on the masses of the central black holes in nearby low-mass early-type galactic nuclei and the first black hole determination for NGC 205. *The Astrophysical Journal*, 872(1), 104.

Özel, F. *et al.* (2010). The black hole mass distribution in the galaxy. *The Astrophysical Journal*, 725, 1918–1927.

Pfister, H. *et al.* (2019). The erratic dynamical life of black hole seeds in high-redshift galaxies. *Monthly Notices of the Royal Astronomical Society*, 486(1), 101–111.

Punturo, M. *et al.* (2010). The Einstein Telescope: A third-generation gravitational wave observatory. *Classical and Quantum Gravity*, 27(19), 194002.

Reitze, D. *et al.* (2019). Cosmic explorer: The U.S. contribution to gravitational-wave astronomy beyond LIGO. *Bulletin of the American Astronomical Society*, 51, 35.

Roedig, C. *et al.* (2012). Evolution of binary black holes in self gravitating discs. Dissecting the torques. *Astronomy & Astrophysics*, 545, A127.

Roškar, R. *et al.* (2015). Orbital decay of supermassive black hole binaries in clumpy multiphase merger remnants. *Monthly Notices of the Royal Astronomical Society*, 449(1), 494–505.

Rossi, E.M. *et al.* (2010). Black hole mergers: The first light. *Monthly Notices of the Royal Astronomical Society*, 401(3), 2021–2035.

Sathyaprakash, B.S. and Schutz, B.F. (2009). Physics, astrophysics and cosmology with gravitational waves. *Living Reviews in Relativity*, 12.

Sathyaprakash, B. *et al.* (2012). Scientific objectives of Einstein Telescope. *Classical and Quantum Gravity*, 29(12), 124013.

Sathyaprakash, B. *et al.* (2019). Multimessenger universe with gravitational waves from binary systems. *Bulletin of the American Astronomical Society*, 51(3), 276.

Schnittman, J.D. (2011). Electromagnetic counterparts to black hole mergers. *Classical and Quantum Gravity*, 28(9), 094021.

Sesana, A. (2016). Prospects for multiband gravitational-wave astronomy after GW150914. *Physical Review Letters*, 116(23), 231102.

Sesana, A. *et al.* (2014). Linking the spin evolution of massive black holes to galaxy kinematics. *The Astrophysical Journal*, 794, 104.

Spera, M. and Mapelli, M. (2017). Very massive stars, pair-instability supernovae and intermediate-mass black holes with the SEVN code. arXiv e-prints, arXiv:1706.06109.

Tamanini, N. *et al.* (2016). Science with the space-based interferometer eLISA III: Probing the expansion of the universe using gravitational wave standard sirens. *Journal of Cosmology and Astroparticle Physics*, 2016(4), 002.

Tang, Y. *et al.* (2018). The late inspiral of supermassive black hole binaries with circumbinary gas discs in the LISA band. *Monthly Notices of the Royal Astronomical Society*, 476(2), 2249–2257.

Tremmel, M. *et al.* (2017). The Romulus cosmological simulations: A physical approach to the formation, dynamics and accretion models of SMBHs. *Monthly Notices of the Royal Astronomical Society*, 470(1), 1121–1139.

Umeda, H. *et al.* (2016). The final fates of accreting supermassive stars. *The Astrophysical Journal*, 830(2), L34.

Valiante, R. *et al.* (2016). From the first stars to the first black holes. *Monthly Notices of the Royal Astronomical Society*, 457(3), 3356–3371.

Valiante, R. *et al.* (2017). On the formation of the first quasars. *Publications of the Astronomical Society of Australia*, 34, e031.

Valiante, R. *et al.* (2018). Chasing the observational signatures of seed black holes at redhifts greater than 7: Candidate statistics. *Monthly Notices of the Royal Astronomical Society*, 474(3), 3825–3834.

Valiante, R. *et al.* (2021). Unveiling early black hole growth with multifrequency gravitational wave observations. *MNRAS*, 500(3), 4095–4109. Doi: 10.1093/mnras/staa3395.

Vasiliev, E. (2016). Evolution of binary supermassive black holes and the final-parsec problem. In *Star Clusters and Black Holes in Galaxies Across Cosmic Time*, Meiron, Y., Li, S., Liu, F.K., Spurzem, R. (eds). *IAU Symposium*, 312, 92–100.

Volonteri, M. and Begelman, M.C. (2010). Quasi-stars and the cosmic evolution of massive black holes. *Monthly Notices of the Royal Astronomical Society*, 409, 1022–1032.

Volonteri, M. *et al.* (2021). The origins of massive black holes. *Nature Reviews Physics*, 3(11), 732–743. Doi: 10.1038/s42254-021-00364-9.

Whalen, D.J. *et al.* (2013). Finding the first cosmic explosions I. Pair-instability supernovae. *The Astrophysical Journal*, 777, 110.

Wu, X.-B. *et al.* (2015). An ultraluminous quasar with a twelve-billion-solar-mass black hole at redshift 6.30. *Nature*, 518, 512–515.

List of Authors

Suzy COLLIN-ZAHN
LUTH
Paris Observatory
Meudon
France

Monica COLPI
University of Milano-Bicocca
Milan
Italy

Francoise COMBES
LERMA
Paris Observatory
France

Jean-Pierre LASOTA
Institut d'Astrophysique de Paris
France
and
Nicolaus Copernicus
Astronomical Center
Warsaw
Poland

Mark MORRIS
Department of Physics and Astronomy
UCLA
Los Angeles
USA

Hélène SOL
LUTH
Paris Observatory
CNRS
France

Index

K, L

M, N

O, P

Q, R

S, T

U, V, X

rinted and bound by CPI Group (UK) Ltd, Croydon, CR0 4YY
13/09/2026
14961331-0001